Extinction in Our Times

EXTINCTION IN OUR TIMES

Global Amphibian Decline

James P. Collins
Martha L. Crump

OXFORD
UNIVERSITY PRESS

2009

OXFORD
UNIVERSITY PRESS

Oxford University Press, Inc., publishes works that further
Oxford University's objective of excellence
in research, scholarship, and education.

Oxford New York
Auckland Cape Town Dar es Salaam Hong Kong Karachi
Kuala Lumpur Madrid Melbourne Mexico City Nairobi
New Delhi Shanghai Taipei Toronto

With offices in
Argentina Austria Brazil Chile Czech Republic France Greece
Guatemala Hungary Italy Japan Poland Portugal Singapore
South Korea Switzerland Thailand Turkey Ukraine Vietnam

Published by Oxford University Press, Inc.
198 Madison Avenue, New York, NY 10016

www.oup.com

Collins, James P.
Extinction in our times : global amphibian decline / James P. Collins and Martha L. Crump.
p. cm.
Includes bibliographical references and index.
ISBN 978-0-19-531694-0
1. Amphibian declines. 2. Extinct amphibians.
I. Crump, Martha L. II. Title.
QL644.7.C65 2009
597.8'1788—dc22 2008035277

9 8 7 6 5 4 3 2 1

Printed in the United States of America
on acid-free paper

We dedicate this book to the many researchers, educators, and conservationists whose efforts focus on amphibian declines.

Contents

The fifth inhabitant of Mexico
(Dedicated to Jan Washburn)

Emmylou Harris' blues
and Lyle Lovett's lyrics ricochet
around my room like firecrackers at a funeral.

Except that it's 4 am and a red-eyed tree frog
is going to leap off the wall
into my fizzing grapefruit drink,
maybe swim around awhile, maybe lip synch Lyle Lovett.

After all, the frog is looking decidedly sexy
in its green body suit, and me,
well, I'm kind of lonely,
a pushover for any amphibian, especially one
speaking fluent Spanish
from the rain forests of Costa Rica.

In the background of Frida Kahlo's
Four Inhabitants of Mexico,
hidden in the adobe houses and church,
in the damp shade, in the crevices,
were Mexican frogs—green ones, poisonous ones,

ones that fit into the stunning speckled
mouth of an orchid, glass ones, ones with
red toes . . .

Hidden because they could see into the future
(Frida painted this picture in 1938),
the future
where they would become one of the mysterious
disappearing acts of the late twentieth century . . .

Amphibians are the Crown of the Evolutionary Tree!

If I ever get to heaven I'm not going in
unless the frogs are there first.

(From Jan Conn, *What Dante Did with Loss.* Montreal: Véhicule Press, 1994, pp. 91–92. Reprinted with permission.)

Foreword

Awareness of the amphibian decline crisis first coalesced as many scientists at the First World Congress of Herpetology in 1989 came to realize that they were not alone in observing amphibian declines and disappearances. The *Washington Post* ran a front page story on the phenomenon, and appropriately so. Something different and quite odd was going on. What was it, and what did it mean for people?

The amphibian decline phenomenon has been a bit of a scientific puzzle. In lucid prose accessible to anyone, this volume does a superlative job of laying it out in all its complexity. Clearly, there are multiple and interacting factors, but an amphibian chytrid fungus is a dominant player, wiping out populations almost as rapidly as influenza virus spreading through a subway car. That is notable, because ecologists and paleontologists rarely think of pathogens as a cause of extinction. In this respect alone, the decline of some modern amphibian species to extinction is worth our attention—and a lot of research—to understand why it is happening in amphibians and what other species might be at similar risk.

As early as 1996, William Laurance and two colleagues put forth the disease hypothesis with respect to declines and extinctions of Australian frog populations (the first recognized wave of major declines). He deserves great credit for his prescience but suffered a lot of dismissive responses. That is the way of science, of course: One person's hypothesis or result is tested by others. And the results don't always stand up. But we should perpetually applaud those who are willing to put forth a new idea, for without them, progress in science and science-based conservation would be very slow indeed.

Amphibian declines are happening quite rapidly, advancing at a measurable rate that exceeds that of birds and mammals. It is also clearly something that cannot be adequately understood without boots-on field work combined with laboratory experimentation and theory. The authors of this book, Jim Collins and Marty Crump, are role models of scientists who are at home in all those realms, drawing on those different aspects of science not only to pursue their own part of the research but also to understand the larger problem.

Fairly early on, the question arose: Would science as usual—represented by the individual investigator who is driven by curiosity—suffice to illuminate what the amphibian crisis is all about? Clearly, the answer is no, because there are multiple pieces to the puzzle. Team research is needed, and there is plenty of room for contributions from individual investigator-scientists.

As his Science Advisor, I once drew the attention of Bruce Babbitt, Secretary of the Interior in the Clinton administration, to amphibian declines. Always attentive to the role of science, he called an interagency meeting with the National Science Foundation, the Environmental Protection Agency, and other government agencies to listen to scientists such as George Rabb and David Wake lay out the situation and the state of knowledge. Carole Browner, then Administrator of the Environmental Protection Agency, was quick to understand that this problem was too big and too complex for it to be confined within the purview of a single agency.

Furthermore, determining the cause of the declines and disappearances did not require a major shift from basic to applied science. Personally, I find the distinction between the two to be greatly overdrawn. Indeed, my own research on habitat fragmentation in the Amazon is at once both very practical conservation biology and a test of the theory of island biogeography: The same data set applies to both. I have little time for the distinction between "pure" and applied research, as if the latter were impure. Rather, I am firmly in the camp of Donald Stokes (Chapter 8), who recognizes the great synergies generated between basic and applied science. It should not be a matter of "either/or." Instead, it is important to think in bigger and more inclusive terms.

Outside of (and before) government involvement, amphibian biologists came together to address amphibian declines. Indeed, the biologists essentially reorganized themselves and how they work. I find great hope in this, having often despaired about how many of my fellow organismal biologists seemed oblivious to the larger biodiversity crisis. Recently, the Amphibian Survival Alliance has come into existence to assist in taking the science and conservation plans forward to conservation action. Collins and Crump tell the story of how scientists and educators rallied around global amphibian declines, and they tell it in a way that highlights how this community's response can serve as a model for other communities that address other groups of organisms threatened with extinction.

As the authors note, the question has arisen in this situation as in others: When does a researcher stop being a scientist and begin being an advocate? When doing science, I work and think like a scientist. But I am also a citizen, and it is not only correct but, in my view, also is incumbent upon us as scientist-citizens to state what we think the implications of science are. If we don't, then who will?

It is heartening, as the scientific understanding of amphibian declines has advanced, that the field of conservation is now at a point where it can rescue amphibian species before they are doomed. The conservation ideal, of course, is to restore the populations in the wild. At this moment, with the ubiquitous presence of amphibian chytrid fungus, this is currently not an option.

Some question the point of keeping the most threatened species from extinction by maintaining them in captivity. Preventing extinction may not always be possible, of course, but, although what may be learned from a captive population is clearly less than one part of a functioning natural ecosystem, there is plenty to be learned nonetheless.

A related perspective, and congruent with Collins and Crump's position in this book, is this: If society universally keeps art objects and other artifacts of our own species in museums, it can be considered no less appropriate to have living museums of samples of biological diversity. Indeed, it would be very self-centered of humans not to, and I very much expect that those who criticize the captive populations would be the first to come running to see a living dinosaur were it to be in captivity somewhere.

Biodiversity in general represents a living library from which the life sciences are developed. Harvard University's Edward O. Wilson laments that at the very time that evolution is capable of understanding itself, it is (we are) busily destroying the evidence from which that knowledge can be built. In virtually every moment of our daily lives we benefit from some aspect of the life sciences derived from studies of species other than ourselves. Hopefully, we are smart enough to recognize those practical reasons as well as the larger ones of ethical relationships with the rest of life on Earth, plus the sheer wonder of it all. Collins and Crump summarize the complex interactions among science, ethics, and policy as they all relate to global amphibian decline. The presentation of this story as an integrated whole is a distinctive and needed addition to our way of thinking about these problems.

There is hardly a person who is not fascinated by or loves frogs. Let us hope that the authors' clear, prescient, and exceptional treatment of amphibian declines—the first documented major extinction event in modern times across an entire vertebrate class of animals—will awaken people to the crisis for life on Earth. If so, by prompting more public awareness and with the help of the scientists tackling this puzzle, the story told in this book about amphibian declines will have done a lot for life on Earth and for future generations of humanity.

Thomas E. Lovejoy is President of the H. John Heinz III Center for Science, Economics and the Environment. July 2008.

Dr. Lovejoy has been the World Bank's Chief Biodiversity Advisor and Executive Vice President of the World Wildlife Fund–U.S. He has served on science and environmental councils or committees under the Reagan, Bush, and Clinton administrations.

Preface

Extinction is a fact of life. Paleontologists estimate that more than 99 percent of the species that have ever lived on Earth are now gone, mainly as a result of five major episodes of mass extinction. The last ended about 60 million years ago. Some are calling Earth's most recent biodiversity losses a sixth extinction event with mainly human causes. The role of humans in the Pleistocene extinction of large mammals—a time that included the last major glaciations and the appearance of humans—is still debated. But, it is clear that over the last 300 years humans have been a direct and indirect cause of extinction.

Many of us imagine extinction as an infrequent, mostly ancient, and geographically distant process with little impact on our daily lives. Global losses of biodiversity in our own lifetimes are changing that mindset, however, as an increasing number of extinctions are documented.

Recent extinctions are caused mainly by predation and competition by introduced exotic species, overharvesting, land use change, and pollution. As climate change progresses through the 21st century, it will cause more local and global extinctions.

If extinction is the result of an obvious cause, for example overhunting of American passenger pigeons, we can understand why the species is gone even if we cannot predict the consequences for the ecological community. But if a species dies out for seemingly unknown reasons, the world looks and feels like a less safe place, and the species' loss creates a strong sense of uncertainty. If that species has gone extinct for no obvious reason, which other ones might be next?

By the late 1980s, herpetologists realized that something unusual was happening with amphibians. Populations of frogs, toads, and salamanders were declining or disappearing relatively quickly, and in some cases rather suddenly. Through informal conversations with colleagues, we learned that our own various observations were not unique. Lack of a clear explanation made the losses a mystery. Alarmed, we felt compelled to identify the causes of these losses and, if possible, to stop or reverse them.

Step one was to ask, have we seen this before? Scientists had certainly seen animal and plant populations lost to habitat destruction. We knew of cases of overcollecting that reduced populations and had recorded the decline of native species after exotic organisms entered a habitat. But before the late 20th century, no one had reported such frequent and rapid losses of amphibian populations in undisturbed natural areas. Golden toads (*Bufo periglenes*) disappeared within two years from a protected reserve in Monteverde, Costa Rica, with no apparent habitat changes. What mechanism could account for such a rapid decline to extinction?

Amphibian losses in protected areas were a source of disbelief, the cause a mystery. In the case of the golden toads from Monteverde, there were not even dead bodies. The animals simply seemed to vanish. In other instances, surprising numbers of carcasses were found—surprising because dead bodies do not last long in tropical and subtropical habitats. The large numbers lost of many species, in just a few years, and around the globe, suggested that—whatever the reasons—the disappearances had happened quickly enough to produce a sufficient number of corpses to overwhelm scavengers and decomposers. Hypothesized culprits—toxins, climate change, and infectious disease—could cross reserve barriers undetected to kill organisms. The suspicion that these causes might be at work was new and worrisome—and therefore deserving of significant research.

In the pages that follow, we review our deepening understanding of the causes of amphibian declines and extinctions over the past two decades. Increasing evidence suggests that we are witnessing a modern extinction event across an entire vertebrate class. The enigmatic amphibian declines in the midst of protected natural areas now have some rational explanations, or at least hypotheses with more tightly framed dimensions.

Overall, evidence suggests that human actions are a major cause of at least some amphibian declines—either directly or indirectly. Using amphibian declines as a lens, we focus our story on environmental change, conservation of biodiversity, and a range of related ecological, evolutionary, ethical, philosophical, and sociological issues.

Three human-related hypotheses to explain the causes of some amphibian declines are introduction of exotic species, commercial use, and land use change. These three hypotheses are long-standing, well-known causes of biodiversity losses that would place amphibians in the company of many other species. Research during the past decade has clarified the contributions of these three causes, providing a context for understanding amphibian declines and also extinctions relative to losses in other groups. These causes explain part of the "mystery," and they tell us that some amphibian losses are part of a much larger story of diminishing biodiversity.

But infectious disease also has had an unexpected role in the extinction of frogs and probably salamanders. This discovery affords new insights into how a pathogen can emerge and spread in ways that can drive some populations and species extinct. It is an important lesson that we must understand better than we do now. A major research group in Queensland, Australia, has characterized the impact of infectious disease on frogs as "the most spectacular loss of vertebrate biodiversity due to disease in recorded history." What would lead to such a startling conclusion? Is it warranted, or is it just hyperbole?

Modern amphibians provide a unique opportunity to study the biology of extinction. We have not a moment to lose. Amphibians are one of the first examples in which humanity has encountered extinction as a modern, global phenomenon. Amphibian declines and disappearances are important to all of us, because population fluctuations may reflect previously unknown problems in the environment. We know little concerning the consequences of species extinction on the capacity of ecosystems to deliver the goods and services on which all life depends, but we do know that amphibians play a key role in ecosystem health.

Our story focuses on change: environmental change, change in density of amphibian populations worldwide, change in scientists' perceptions and hypotheses regarding the causes of amphibian declines, and change in how we are responding to amphibian declines as part of a general loss of biodiversity. Scientific debate, testing of alternative hypotheses, and collaboration by scientists from diverse fields of expertise have been key to understanding amphibian declines. Our goal is to offer an objective analysis and a context for understanding the changes and their causes. To reach it, we summarize research done by many individual scientists, groups of investigators and their students, and international research teams.

We intend our analysis to be a road map and hope that future research can progress even faster as it builds upon a clearer vision of what we already know. Our wish is that our readers gain a better understanding of the forces shaping Earth's biodiversity. The story of global amphibian declines begins with a scientific mystery: What is causing the declines? It ends with another mystery: With the information we now have in hand, will we have the wisdom to respond thoughtfully, creatively, and effectively?

In a final note, because this book is for scientists as well as nonscientists, we want to comment on our use of the binomial names for species. For the most part we use the pre-Frost et al. binomials from the published literature, and to avoid confusion, do not use the names recently recommended by Darrel Frost and his colleagues at the American Museum of Natural History (Frost et al., "The Amphibian Tree of Life") in their proposed phylogeny of living amphibians. Instead, in the Index we list the names recommended by Frost and colleagues alongside the pre-Frost et al. published name.

James P. Collins, Tempe
Martha L. Crump, Flagstaff

Acknowledgments

Ever since we began writing this book in 2002, our editor, Peter Prescott, has been a constant source of encouragement. He believed not only that we could do it but that we would do it. Had it not been for his gentle prodding, we might still be trying to catch up on publications focusing on declining amphibian populations. If we had told the story a half-decade ago, we would have had fewer pieces of the puzzle. If we were to wait another half-decade, we would be missing fewer pieces and the picture might be clearer. Peter convinced us that the time was now to write the story.

Over the course of these many years of manuscript preparation, we have had the support of our home institutions: Arizona State University and Northern Arizona University. The National Science Foundation and the Arizona Game and Fish Department Heritage Fund Program provided further support.

We thank many colleagues who have helped us along the way. Ross Alford, Ana Carnaval, Cynthia Carey, Peter Feinsinger, Tim Halliday, Karen Lips, and Danna Schock answered our questions and provided information. Douglas Erwin drew our attention to the hyperdisease argument. Michael Lannoo kindly provided a prepublication copy of his book, *Malformed Frogs*. We thank the ASU Visualization Laboratory and its director, Charles Kazilek. Sabine Deviche prepared a number of the images.

Ross Alford, Forrest Brem, Roberto Brenes, Cynthia Carey, Dante Fenolio, Karen Lips, Joseph Mendelson, Martin Ouellet, Stephen Richards, and Danna Schock generously allowed us to reproduce their photographs.

For reading early drafts of sections and entire chapters, we thank Ross Alford, Cynthia Carey, Carol Collins, Cheryl Dybas, Peter Feinsinger, Lee Kats, Karen Lips, Alan Pounds, Rick Relyea, and Vance Vredenburg. Three reviewers read the entire manuscript for Oxford University Press and offered helpful feedback—Peter Daszak, Karen Lips, and David Skelly. Any omissions, misinterpretations, or errors remain ours to shoulder.

We thank the scientists and nonscientists who have dedicated portions of their lives to carrying out research concerning the causes of amphibian declines, educating the general public on the significance of declines, and initiating conservation efforts to mitigate declines. Finally, we thank our families for their patience, help, and understanding while we worked on this manuscript over the years.

Extinction in Our Times

1

Declining Amphibian Populations and the Biodiversity Crisis

Conservation biologists, ecologists, and evolutionary biologists focus on the world's biodiversity. Biodiversity is a buzzword, but what does it really mean? Introduced a couple of decades ago, the term has since taken on numerous definitions, some more biological than political, some more political than biological. To most nonbiologists and many biologists, biodiversity means the number of kinds of living organisms, or species richness. In this sense, an area with 800 species of butterflies has higher biodiversity than an area with 100 species.

The concept is much more than simply number of species, however. Reed Noss and Allen Cooperrider, leading conservation biologists and independent consultants, offered the following working definition of biodiversity: "Biodiversity is the variety of life and its processes. It includes the variety of living organisms, the genetic differences among them, the communities and ecosystems in which they occur, and the ecological and evolutionary processes that keep them functioning, yet ever changing and adapting."[1] Thus, loss of biodiversity is more than the loss of species. It is also the loss of genetic and ecosystem diversity, the alteration of structure, and the disruption of ecological and evolutionary processes.

How do we address all the components of biodiversity conservation? In practice, we usually do not. For better or worse, it is much easier, faster, and cheaper to focus on just one scale and consider just one component of biodiversity. Most often, we select the scale of species, and the component we measure is assemblage composition—identity, abundance, and variety of organisms in a given area. And so we have come full circle. In practice, conservation biologists most often end up measuring the number and identity of species and calling that biodiversity. We will focus on amphibian biodiversity in this restricted sense.

Geography of Amphibian Biodiversity

Estimates of the number of species of organisms living on Earth today vary from 10 million to 100 million.[2] Scientists have formally named only about 1.7 million species, however.[3] Insects make up about 72 percent of the named species of animals, vertebrates only about 5 percent, and amphibians—numbering approximately 6347 species—about 0.5 percent.[4] The total number of amphibian species increases as scientists continue to discover and describe new species.

Amphibians consist of three orders (Figure 1.1). The 5602 species of frogs and toads belong to the order Anura, from the Greek words *an* and *oura,* meaning "without tail." Most anurans have strong back legs, well suited for jumping. The 571 species of salamanders belong to the order Urodela, from the Greek words *uro* and *delos,* meaning "tail evident." Unlike adult anurans, all adult (and larval) salamanders have tails. The 174 species of caecilians belong to the third order, Gymnophiona. Caecilians are unfamiliar to most people because these amphibians live only in tropical regions, where they are usually hidden from view in water or underground. The order name comes from the Greek words *gymnos* and *ophis,* meaning "naked serpent." Caecilians resemble snakes or large earthworms. They are long, skinny animals with no legs. Some of these amphibians have small scales beneath the surface of their mucus-covered skin. Because very little is known concerning population biology of caecilians or even their distribution, we will focus on anurans and salamanders.

Amphibians live on every continent except Antarctica, but their diversity varies among regions. Certain areas of the world are centers of biological diversity. In 1988, Norman Myers, at Oxford University, introduced the concept of conservation "hot spots."[5] He defined a hot spot as an area that both (*a*) houses a large number of endemic species (species restricted to a specific region or locality) and (*b*) is under extreme threat of habitat loss. William Duellman, at the University of Kansas, identified 43 areas worldwide as having exceptionally high amphibian species richness, high endemism, or both (Figure 1.2; Table 1.1).[6] Nineteen of these are in the Western Hemisphere, where 52 percent of the anuran species, 71 percent of the salamander species, and 53 percent of the caecilian species occur. The Neotropical region alone is home to 54 percent of the world's species of

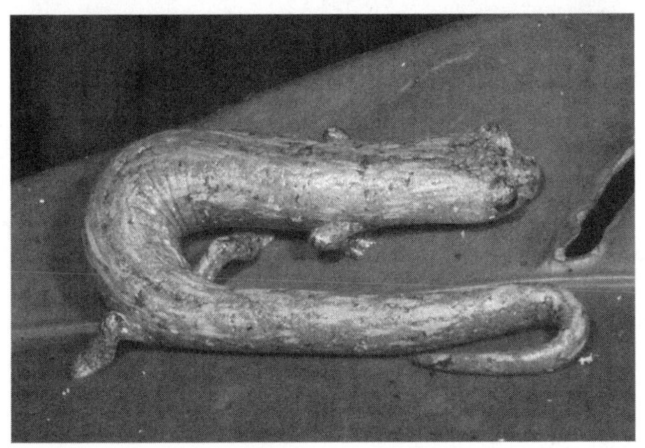

Figure 1.1 Representatives of the three amphibian orders. *Top*: Anura (*Hyla boans*). Photograph by Martha Crump. *Middle*: Urodela (*Bolitoglossa altamazonica*). Photograph by Dante Fenolio. *Bottom*: Gymnophiona (*Geotrypetes seraphini*). Photograph by Dante Fenolio.

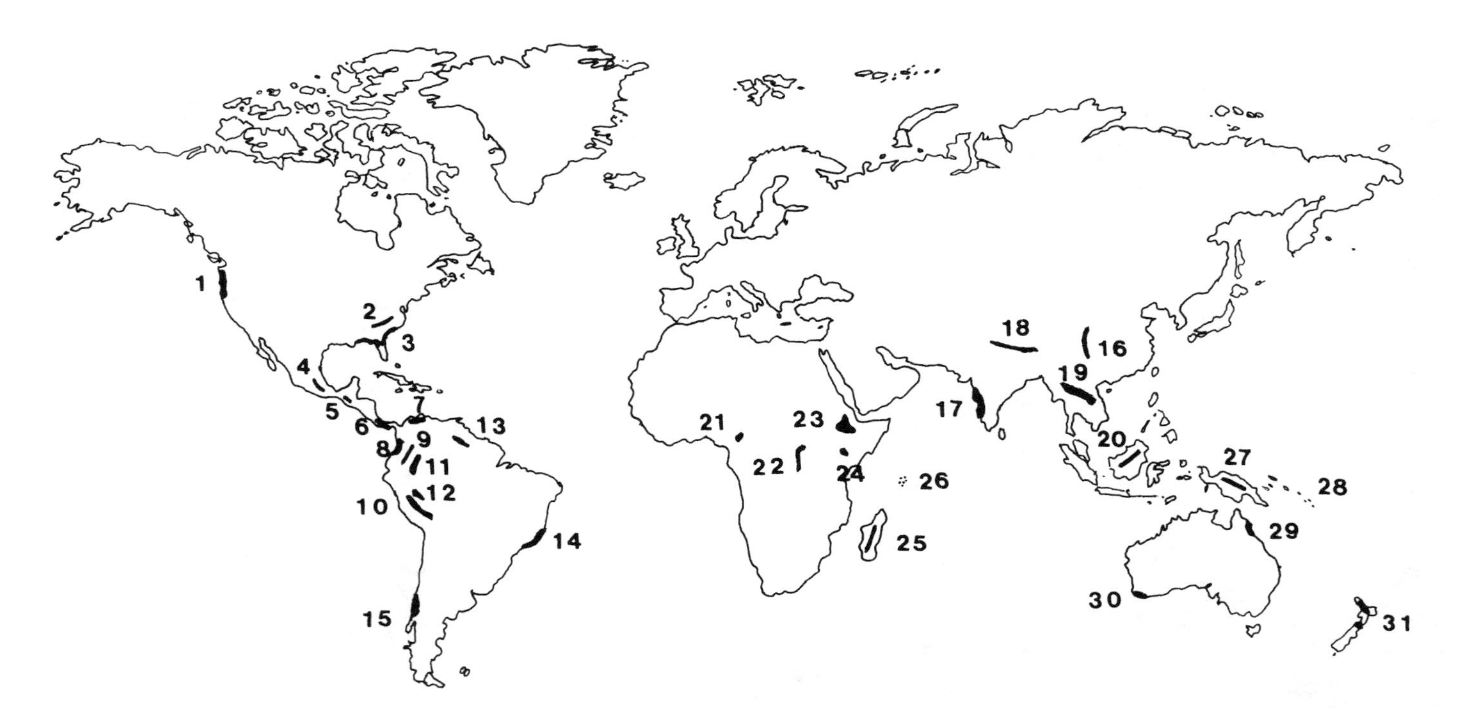

Figure 1.2 Areas of high amphibian species richness, endemism, or both. (See Table 1.1 for identification of numbers.) Redrawn from Duellman 1999.

TABLE 1.1 Regions of the World with Amphibian Species Richness of at Least 50 Species and/or Endemism of at Least 75 Percent[a]

Key	Region	No. of Species	No. of Endemics	% of Endemics
	North America			
1	Pacific-Cascades-Sierra Nevada ranges, USA	52	43	83
2	Southern Appalachian Mountains & associated plateaus, USA	101	37	37
3	Southern Coastal Plain, USA	68	27	40
	Middle America			
4	Southern Sierra Madre Oriental, Mexico	118	74	63
5	Highlands of western nuclear Central America	126	70	56
6	Highlands of Costa Rica & western Panama	133	68	51
	South America			
7	Sierra Nevada de Santa Marta, Colombia	19	19	100
8	Cordillera Occidental, Colombia & Ecuador	200	156	78
9	Cordillera Oriental, southern Colombia & Ecuador	159	147	92
10	Tropical southern Andes Mountains, Bolivia & Peru	132	101	77
11	Upper Amazon Basin, Ecuador & northern Peru	198	119	60
12	Upper Amazon Basin, southern Peru	102	22	18
13	Guiana Highlands, Venezuela & Guianas	76	71	93
14	Atlantic Coastal Forest, Brazil	334	310	93
15	Austral Temperate Forest, Argentina & Chile	32	32	100
	Eurasia			
16	Hengduanshan Mountains, SW China	111	52	47
17	Western Ghats, India	114	84	74
18	Southern Himalayan slopes, India	68	26	38
19	Highlands of Burma, Thailand, & Vietnam	129	50	39
20	Highland of Borneo, Indonesia, & Malaysia	137	90	66
	Africa			
21	Cameroonian Highlands, Cameroon	62	39	63
22	Central Highlands, eastern rim of Zaire Basin	69	50	72
23	Ethiopian Highlands, Ethiopia	32	25	78
24	Eastern Highlands, Kenya	57	32	56
25	Central Highlands, Madagascar	66	61	92
26	Seychelles Islands	12	11	92
	Australo-Papuan Region			
27	Central Highlands, New Guinea	205	193	94
28	Solomon Islands	26	24	92
29	Northeastern Queensland, Australia	56	29	52
30	Southwestern Australia	25	19	76
31	North Island & islands in Cook Strait, NZ	4	4	100

Source: Duellman 1999, Table 1-6.

[a]See also Figure 1–2.

amphibians. The remaining 24 species-rich areas occur in Eurasia, Africa, and the Australo-Papuan region. Amphibian endemism is generally higher in tropical than in temperate regions, and in montane as opposed to lowland areas. Many of these 43 areas overlap with regions that are diverse in other organisms as well, such as western Ecuador; Atlantic Coastal Forest, Brazil; Madagascar; Cape Region, South Africa; and Western Ghats, India. Many areas that have the greatest numbers of amphibian species and/or high levels of endemism are experiencing extensive habitat loss—that is, they are "conservation hot spots" in both senses.

The "Biodiversity Crisis": Extinction Is Forever

When conservation biologists speak of the "biodiversity crisis," they are referring to the rapid and accelerating loss of species and habitats. Ecologists and conservation biologists warn that we are in the early phase of the sixth major extinction episode in Earth's history. We are losing species at an alarming rate.[7] Edward O. Wilson at Harvard University speculated based on the fossil record that the "background" extinction rate (that which "normally" occurs) is about 1 species per 1 million species per year.[8] He suggested that deforestation in rain forest habitats has increased the extinction rate to between 1000 and 10,000 times this background level. The current extinction episode might eventually surpass some of the great mass extinctions of the past, including that which erased the dinosaurs 65 to 70 million years ago. The present extinction episode differs from past ones in that human activities are the cause of most current extinctions.

Estimates vary, but some of the world's most respected conservation biologists fear that by the year 2100 two thirds of all plant and animal species could be headed for extinction.[9] Tropical regions represent a major concern, because an estimated 80 percent of the world's species live in the tropics.[10] The most conservative estimate Wilson offers for loss of biological diversity in tropical forests around the world is 27,000 species each year, which amounts to 74 each day, or 3 each hour.[11]

Extinction estimates vary widely and are hard to evaluate because it is very difficult to know for sure that a species has gone extinct. Another problem is that anywhere from 85 to 95 percent of the Earth's species have not yet been described. Furthermore, different scientists use different numbers for the total estimated number of species on Earth and for the rate and extent of habitat loss.

The International Union for the Conservation of Nature and Natural Resources (IUCN), through its Species Survival Commission, assesses the conservation status of species on a global scale. The Union maintains the IUCN Red List of Threatened Species, a catalog of taxa that face a high risk of global extinction. Scientists evaluate species and designate the most threatened as "Critically Endangered," "Endangered," or "Vulnerable." Species that cannot be evaluated because of insufficient information are labeled "Data Deficient." Although the Red List is one of our best sources of information on the conservation status of species, the "official" figures are far from ideal. For many groups, only certain species have been evaluated. The list also includes species that have been evaluated over only a fraction of their ranges.[12]

For these reasons, the 2008 Red List of 16,928 species of animals and plants facing extinction is conservative. Scientists who are familiar with how these data are collected suggest that many more species face extinction, especially among plants and invertebrates. All species of birds and mammals were evaluated for the 2008 list. According to the data, 12 percent of described birds and 21 percent of described mammals were considered to be threatened with extinction. Nearly 99 percent of all described amphibian species were evaluated; of those, 30 percent were considered threatened.

Since the year 1500, according to the 2008 Red List, 754 species of animals are known to have gone extinct or are extinct in the wild, including the following vertebrates: 78 mammals, 138 birds, 22 reptiles, 39 amphibians, and 103 bony fishes. The numbers are woefully incomplete, especially for invertebrates. Why is there such a discrepancy between Wilson's estimated loss of biodiversity and the much lower figures provided by the IUCN Red List? Again, it is very difficult to document that a species is extinct.

A study entitled, "Sri Lanka: An Amphibian Hot Spot," illustrates the difficulty of documenting extinct species.[13] The 2002 report made newspaper headlines. Scientists had discovered more than 100 new species of Old World tree frogs (Rhacophoridae)—many times more than the previous number of known species—on the small island nation of Sri Lanka in the Indian Ocean, about 32 km off the southeast coast of India. What did not make the title or newspaper headlines is that one author, in an effort to identify his specimens, had traveled to museums that housed preserved frogs collected from Sri Lanka more than a century ago. There he found preserved specimens of as many as 100 more species that were never observed during the survey that unveiled the 100 new rhacophorids.[14] The scientists speculated that these old specimens most likely represent species now extinct, which is hardly surprising since the island has lost more than 95 percent of its original rain forest. These "ghost" species were never recorded as extinct simply because they had never been described in the first place—just preserved in glass jars.

Clearly, humans directly or indirectly cause many or most species' declines and extinctions. In 1950, an estimated 2.5 billion people lived on Earth. By 2008, the figure had risen to more than 6.7 billion people. Our numbers may reach 9 billion by 2042. The more people, the more demands we put on the environment and the more we threaten other species. The result is extinction of the least resistant or least adaptable species.

Land use change represents the single largest human-caused threat to biodiversity.[15] Not only do we destroy every type of habitat, we also fragment and modify habitats through logging, burning, overgrazing, and other activities, thereby diminishing its quality for many organisms. Fragmentation results in smaller, more isolated populations that have reduced opportunity for dispersal and increased risk of extinction.

A second leading cause of species' declines and extinctions is the effect of species we introduce, deliberately or accidentally, to areas where they do not normally occur.[16] In some cases, we purposely introduce animals to new areas for biological control. Other introductions are accidental, such as when exotic pets escape. Animals and plants expand their ranges when we modify habitats. Introduced species can eliminate native species through predation, competition, hybridization, or disease.

A third threat leading to species declines and extinctions is overexploitation by humans for food, medicines, pets, or products such as skin or fur.[17] Contamination of air, water, and land with organic chemicals and other toxins represents a fourth threat to biodiversity.[18] A fifth threat, and one of great long-term importance, is global climate change, including global warming.[19] Emerging infectious diseases is a sixth factor resulting in loss of biodiversity.[20]

Habitat loss. Introduced species. Overexploitation. Contamination. Global warming. Disease. Not only can each of these factors decrease biodiversity, but also the factors can act synergistically. As we will discuss later, these threats to biodiversity in general also apply to amphibians in particular. We will take an especially close look at an emerging infectious disease called chytridiomycosis, caused by a chytrid fungus that infects amphibians.

Why Should We Care about Loss of Biodiversity?

Conservation biologists, philosophers, environmental ethicists, and others offer several key reasons to conserve biodiversity.[21] One argument is that organisms have direct economic value for humans.[22] We use plants and animals for medicines, food, clothes, building materials, recreation, and other luxuries and necessities. But what if an organism that is of no use to us for food or hides is screened for useful medicinal compounds and found to have none? Do we sanction its extermination? Why must a plant or animal be of direct economic benefit to humans to have worth? Economic value alone is not the only reason to preserve biodiversity.

Another reason often given (but not universally accepted) to conserve biodiversity is that organisms, as components of ecosystems, provide services, and their interactions with other organisms contribute to the overall healthy functioning of ecosystems. These services include maintenance of the atmosphere's gaseous composition, production and maintenance of soils, disposal of wastes, recycling of nutrients, control of pest species by predators and parasites, and pollination.[23] On a practical level, biologists want to know just how much the loss of a few species will reduce the quality of services within a specific ecosystem. Two schools of thought prevail.

One idea, called the "rivet hypothesis," is that every species contributes to ecosystem integrity.[24] The analogy is that biological species are like rivets in an aircraft. Only a limited number of rivets can be removed from an aircraft before it falls apart. Similarly, as species are lost, at some point ecosystem function becomes damaged.

The alternative view, called the "redundant species hypothesis," suggests that high species richness is not necessary to ecosystem function. The argument is that as long as the biomass of primary producers, consumers, decomposers, and other trophic levels is maintained, ecosystems can function perfectly well with fewer species. Up to a point, species loss can be tolerated because of the considerable redundancy built into ecosystems. In this view, ecosystem function is driven by the interactions of relatively few species—not by every living organism within an ecosystem.[25] Some ecologists suggest that even if all the organisms now considered to be threatened with extinction did indeed go extinct, other plants and animals would take over their roles and ecosystems would continue to function with scarcely a hitch.[26]

Going from theory and empirical data for specific ecosystems to generalizations about all ecosystems, however, is a huge leap. Scientists are a long way from agreeing on which species (if any particular ones) are crucial to maintain productive ecosystems.[27]

Aesthetics represents a third reason to conserve biodiversity. Humans, no matter what culture, unconsciously seek connections with other forms of life—a phenomenon Edward O. Wilson calls "biophilia."[28] We appreciate the beauty of nature in the uniqueness of each species. Other organisms add pleasure to our lives. We use flowers to express our deepest emotions of love and respect. A deep inhalation of the delightful fragrance of roses makes our spirits soar. Animals have long served as our faithful companions, making us happier and healthier. Ever since humans were humans, we have worshipped animals and admired their strength, energy, and beauty. We need biodiversity because the richness of nature adds meaning to our lives.

A fourth reason to conserve biodiversity invokes ethics. Most people understand this argument, a particularly powerful and convincing one, because it is central to the value system of most religions and philosophies.[29] Basically, the belief is that each species has intrinsic value. Each has a right to exist, and we are obliged to respect and protect all of nature—with the possible exception of the smallpox virus, the polio virus, and their ilk. The World Charter for Nature, adopted by the United Nations General Assembly in 1982 and signed by well over 100 nations, stated this belief eloquently: "Every form of life is unique, warranting respect regardless of its worth to man, and, to accord other organisms such recognition, man must be guided by a moral code of action."[30]

From this framework of biodiversity and its loss, we move on to the focus of our story—declining amphibian populations.

Why Should We Care if Amphibians Decline?

People differ in their concern over amphibian declines. At one extreme, Peter Daszak, who at the time was at the Institute of Ecology, University of Georgia, and his colleagues proposed that global declines of amphibian populations are "perhaps one of the most pressing and enigmatic environmental problems of the late 20th century."[31] At the other extreme, an editor emeritus of a Tennessee newspaper wrote: "We read an article recently that indicated that frogs are becoming an endangered species. The information in the article indicated that scientists (interested in frogs) were unable to explain what has happened. Personally, we have not missed the frogs as we have little contact with them."[32]

If we could chat with that editor emeritus, how would we convince him that conservation of amphibian diversity is important? Why should we care if amphibians disappear? Reasons include the same four types of arguments discussed for biodiversity in general: economics, ecosystem function, aesthetics, and ethics.

We should care if amphibians disappear because we use them for our own benefit.[33] Every year we use huge quantities of frogs (mostly Ranidae and Pipidae) in medical research and for teaching purposes. Isolation, identification, and characterization of novel chemical compounds that occur in the granular glands of anuran skin have led to the development of new drugs for human use. In 2001, The Food and Agriculture Organization of the United Nations estimated the average quantity of frog legs sold for human consumption to be 4716 metric tons annually from 1987 to 1997.[34] We buy millions of amphibians each year for pets: poison dart frogs, pacman frogs, White's tree frogs, and fire-bellied toads to name but a few.

Amphibians play a key role in energy flow and nutrient cycling, in both aquatic and terrestrial environments. They are efficient at converting food into growth and reproduction, they are both predators and prey, and they are sometimes extremely abundant. Amphibians serve as "conveyor belts" by transferring energy from invertebrates to predators higher up on the food chain.[35] This transfer of energy is efficient because amphibians expend relatively little energy to maintain themselves. About 50 percent of the energy an amphibian gets from food is converted into new tissue.[36] That, in turn, is transferred to the next level in the food chain when a predator eats the amphibian (Figure 1.3). In contrast, birds and mammals have low

Figure 1.3 *Oxybelis* (vine snake) eating *Hyla raniceps*. Photograph by Dante Fenolio.

conversion efficiency. They use up to 98 percent of their ingested energy just to maintain their body temperatures. Therefore, as little as 2 percent of ingested energy is converted to new animal tissue and passed on up the food chain.

Amphibians provide the world a valuable service through their eating habits. Tadpoles eat tremendous quantities of algae. In doing so, they alter the dynamics of aquatic ecosystems and reduce the rate of natural eutrophication (over-enrichment of water with nutrients, resulting in excessive algal growth and oxygen depletion). Most adult amphibians eat insects and other arthropod prey. A population of 1000 cricket frogs (*Acris crepitans,* small tree frogs about 3 to 3.8 cm long) consumes an estimated 4.8 million small insects and other arthropods annually.[37]

Thus, both because of what they eat and because they serve as food for fish, reptiles, birds, mammals, and other animals, amphibians play a central role in the food web. Cascading effects will be felt throughout both aquatic and terrestrial food webs if large numbers of amphibians disappear.

A world without amphibians would be an aesthetically less interesting place. Frogs serve as good luck charms for people all over the world, because of their association with rain. Frogs symbolize fertility—some species produce more than 20,000 eggs at a time. They represent resurrection, because they seemingly appear out of nowhere after heavy rains. Frogs are "magic." How else could they transform from aquatic, algae-eating swimmers into terrestrial, carnivorous jumpers? From tadpole form to frog form?

Amphibians, especially frogs and toads, provide inspiration for our artistic endeavors. Many cultures have folk tales about a kissed toad turning into a handsome prince. Who could forget Mr. Toad in Kenneth Grahame's *The Wind in the Willows,* or Beatrix Potter's Jeremy Fisher? How many millions of children have read and loved Arnold Lobel's *Frog and Toad Are Friends* and its sequels? Amphibians provide inspiration for our music, from the English folksong "Froggy Went a-Courtin'" to the 1906 ragtime hit "Frog Legs Rag." Frog and toad motifs decorate our clothes and appear in just about every kind of art. Archeologists worldwide unearth ceramic vases and vessels with anuran designs, and frog and toad images are woven into tapestries and carved into wood and stone (Figure 1.4).

Finally, from an ethical standpoint, we are obliged to respect and protect amphibians, just as we should respect and protect all other organisms. It is easier for people to respect organisms considered to be "good" than the "bad" ones such as malaria-carrying mosquitoes. As just indicated, most cultures value amphibians.

Figure 1.4 Zuni frog fetish. Photograph by Martha Crump.

The Scope of Amphibian Declines

Silence of the Frogs. The Case of the Disappearing Frogs. Amphibians Out on a Limb. Frog Population Leaps Downward. Where Have All the Frogs and Toads Gone? These and other catchy titles of newspaper and magazine articles have grabbed the public's attention for 20 years. Articles written for the general public often present an alarmist viewpoint, so it is important to ask: What really is happening?

Due to widespread media coverage beginning in the late 1980s, the popular perception is that amphibian population declines are a new phenomenon. Investigators, however, have reported declines throughout at least the past century. More than 100 years ago Mary Dickerson, the first Curator of Amphibians and Reptiles at the American Museum of Natural History, wrote that bullfrogs (*Rana catesbeiana*) were less common than many other frogs in the eastern United States, in part because of heavy hunting pressure on them for human food.[38] Hunting bullfrogs was profitable—a dozen at the market sold for US$1 to $4. She warned that such heavy hunting pressure on bullfrogs might eventually cause their extinction. In 1955, James Oliver, Curator of Reptiles at the New York Zoological Society, likewise noted that bullfrog populations were declining in places where they were hunted for their legs, and for this reason many states had instituted laws to regulate bullfrog hunting.[39] He also noted that collection of amphibians from the wild for pets had diminished populations. Decades ago, authors of field guides warned that amphibian populations were declining in the United States due to habitat destruction.[40]

More than 35 years ago, Erich Gibbs, Director of Ultrascience, Inc, in Skokie, Illinois, and his colleagues warned that "The live frog is almost dead."[41] They referred to the situation as a "frog crisis" and estimated that leopard frogs (*Rana pipiens*) and bullfrogs (*Rana catesbeiana*) had declined some 50 percent from about 1960 to 1970. Suspected causes for these declines included habitat destruction; environmental contamination from insecticides, fertilizers, and biological and chemical wastes; and overcollecting. At the time the article was published, U.S. commercial suppliers shipped approximately 9 million leopard frogs and bullfrogs (about 327 metric tons) each year for research and educational purposes. That figure does not include the millions more that undoubtedly died during capture and handling.

Although naturalists and scientists have long documented declines of amphibian populations, one difference is that the past few decades have witnessed more frequent reports worldwide. Furthermore, many declines have been labeled mysterious, because they occurred in protected areas where there was no obvious cause such as habitat destruction. During the 1970s and 1980s, many amphibian biologists watched their study populations of frogs, toads, and salamanders decline or disappear—some mysteriously, some for obvious reasons. When herpetologists from around the world met in Canterbury, England, for the First World Congress of Herpetology in 1989, we compared notes and realized our stories were not unique. Something was happening to amphibians worldwide. Consider the following sampling of amphibian population declines. Some declines are from amphibian "hot spot" areas; others are from areas with much lower amphibian diversity. The numbers and capital letters correspond to the numbers on Figure 1.5.

1. Yosemite toads (*Bufo canorus*), found only in the high elevations of the Sierra Nevada in California, experienced up to a nine-fold decrease in population sizes between 1974 and 1982.[42] The California red-legged frog (*Rana aurora draytonii;* now *Rana draytonii*), once abundant, had disappeared from more than 70 percent of its pre-1975 historic range in California by 2001.[43]
2. Eleven populations of boreal toads (*Bufo boreas*) declined and eventually disappeared from the West Elk Mountains of Colorado between 1974 and 1982.[44] Between 1973 and 1982, nine populations of northern leopard frogs (*Rana pipiens*) from the Red Feather Lake region of Colorado disappeared.[45] During a 1986–1988 survey of northern leopard frogs from 33 historically known localities in Colorado, populations remained at only four sites.[46] Tiger salamanders (*Ambystoma tigrinum*) no longer occur in many high mountain lakes in Colorado.[47]

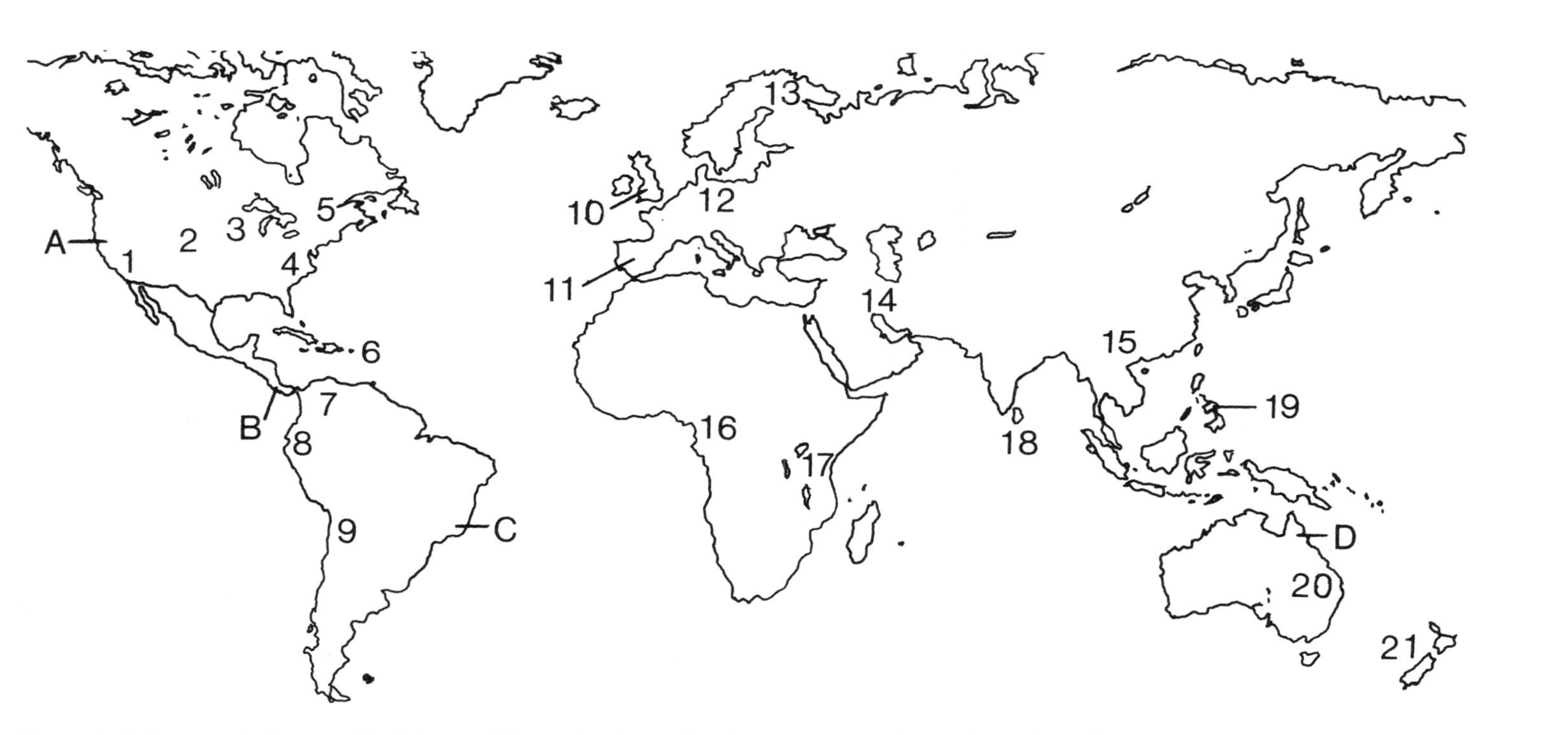

Figure 1.5 Representative worldwide amphibian declines. Numbers correspond to selected single species declines; capital letters correspond to assemblage declines. See text for identification of numbers and letters.

Tiger salamander populations also declined in ponds in the Elk Mountains of Colorado between 1982 and 1988.[48]

3. Blanchard's cricket frogs (*Acris crepitans blanchardi*), once abundant throughout much of the Midwest, are the amphibian species of most concern in the midwestern United States.[49] Many populations in Minnesota, Wisconsin, Iowa, Illinois, Indiana, and Michigan have declined or vanished. In a 1994 survey of 65 Wisconsin sites that historically had cricket frogs, only 2 still had notable choruses of calling males.[50]
4. Population densities of the green salamander (*Aneides aeneus*) declined synchronously across the Blue Ridge Escarpment in the southern Appalachian Mountains of North Carolina.[51] Monitoring studies at seven historical sites throughout the 1990s revealed that population densities had declined 98 percent since 1970. Populations of 44 species of *Plethodon* salamanders from hundreds of sites in eastern North America remained fairly stable from 1951 to 1985. However, surveys of 205 populations of 38 *Plethodon* species carried out during the 1990s at the same localities as surveys done before that decade revealed an average of only 41.6 percent as many salamanders per person per visit, compared to the earlier reports. One hundred and eighty populations appeared to have declined, while 25 populations increased in abundance. No salamanders were found at 32 of the localities showing declines.[52]
5. Whereas western chorus frogs (*Pseudacris triseriata*) were once abundant in southeastern Québec, Canada, they are now considered to be the rarest frog in the province. In a survey conducted throughout the species' known range in Québec in the springs of 1992 and 1993, observers heard the frogs at only 10 of the 279 sites they surveyed in the eastern region near Montréal, sites that previously had supported populations. In the Ottawa River lowlands, observers heard the frogs at 62 of 446 sites surveyed.[53]
6. Eight endemic species of *Eleutherodactylus* (commonly called dink frogs or robber frogs) have experienced drastic population declines in Puerto Rico, especially at high elevations.[54] Three species may be extinct: Karl's robber frog (*Eleutherodacylus karlschmidti*) has not been seen since 1974; the golden coqui (*Eleutherodactylus jasperi*) has not been seen since 1981; and the Villalba robber frog (*Eleutherodactylus eneidae*) has not been seen since 1990.
7. Harlequin frogs (genus *Atelopus*) range from Costa Rica south to Bolivia and eastward through the Amazon Basin into the Guyanas. Thirty of 53 species for which sufficient data are available to decipher population trends have disappeared entirely from their known

localities. Many other species have declined. Countries experiencing the sharpest declines in *Atelopus* populations include Costa Rica, Venezuela, and Ecuador.[55]

8. Populations of at least 24 species of anurans from the Andes of Ecuador have declined or gone extinct since the late 1980s. Examples include harlequin frogs (*Atelopus*), glass frogs (Centrolenidae), and dink frogs/robber frogs (*Eleutherodactylus*).[56]
9. Christian's marsupial frog (*Gastrotheca christiani*), endemic to the subtropical montane forests of northwestern Argentina, was known from five sites. The only site that has been monitored is situated just outside the border of Calilegua National Park. That population disappeared in 1996 and has not recovered.[57]
10. Natterjack toads (*Bufo calamita*) occur at the western edge of their geographic range in Britain. During the 20th century, their numbers declined sharply. An estimated 75 to 80 percent of natterjack populations that existed in 1900 had disappeared by the early 1970s. Currently, the toads persist in at least two sites in the United Kingdom: a heathland site in southern England and a coastal dune site in a very industrialized part of northwest England.[58]
11. From 1997 to 1999, common midwife toads (*Alytes obstetricans*) disappeared from 5 of 35 ponds where they had reproduced a few years earlier in the Peñalara Natural Park near Madrid, Spain.[59]
12. The European tree frog (*Hyla arborea*), widespread across much of central and southern Europe, is declining in many areas.[60] The species has declined by at least 50 percent in Lower Saxony, Germany, during recent years. An estimated 90 to 97 percent of European tree frog populations have disappeared from Denmark since 1945.
13. Populations of great crested newts (*Triturus cristatus*) have declined from many areas throughout their range, beginning in the 1970s. Declines have been reported from Britain, Norway, Finland, the Netherlands, Germany, and Switzerland.[61]
14. Populations of three species of Kurdistan newts (*Neurergus*) from the Zagros Mountains of western Iran have disappeared.[62]
15. Monitoring studies of the Gansu toad (*Bufo minshanicus*), Inkiapo frog (*Rana chensinensis*), and Songpan slow frog (*Nanorana pleskei*) from the Zoige Wetlands, Sichuan Province, China, over 5 years have revealed declines in the populations of all 3 species.[63]
16. The goliath frog (*Conraua goliath*), the world's largest frog at 30 cm in length, occurs only from southern Cameroon to equatorial Guinea, Africa. Populations of this species have declined drastically in recent years.[64]

17. Kihansi spray toads (*Nectophrynoides asperginis*), formerly found only within a 2-hectare wetland spray meadow along the Kihansi River Gorge in Tanzania, are now extinct in the wild.[65]
18. Once abundant in the mountainous tea-growing area of Sri Lanka, Gunther's whipping frog (*Polypedates eques*) and the Asian green frog (*Rana greeni*) are now almost absent.[66]
19. One reported decline on Negros Island in the Philippines is the Negros cave frog (*Platymantis spelaeus*), a large cave-dwelling frog endemic to the island.[67] From 1981 to 2000, the combined population size from two caves is estimated to have declined 90 percent. Forest frogs from the island are also declining: two species of wrinkled ground frogs (*Platymantis negrosensis* and *Platymantis dorsalis*).[68]
20. The gastric-brooding frog (*Rheobatrachus silus*), described from Australia in 1973, originally was so common in its stream habitat that one could see 100 frogs in a single night. These frogs disappeared from the rain forest in southeastern Queensland in 1981.[69] Not one of these unique frogs (females brood their young in their stomachs) has been found since.
21. The Coromandel New Zealand frog (*Leiopelma archeyi*), endemic to New Zealand, has declined drastically in the central Coromandel Range and at Tokatea Saddle, 30 km north.[70]

Many reports of amphibian declines have involved single species, often because an investigator was studying only that one species. We know, however, that in some cases many species within amphibian assemblages have declined or disappeared (see Figure 1.5). For example:

A. Comparison of recent survey results from the Sierra Nevada Mountains of California with data gathered in the early 1900s reveals a "collapse" in the frog fauna.[71] Populations of at least five of the seven species of frogs and toads in the Yosemite area have suffered serious declines. The foothill yellow-legged frog (*Rana boylii*) has disappeared from the area entirely. What was then considered the mountain yellow-legged frog (*Rana muscosa*), formerly the most abundant amphibian there, now occurs only in a few small populations. (This species has since been split into two species: *Rana muscosa* and *Rana sierrae,* each with more than 90 percent of their historic populations now gone.[72]) Boreal toads (*Bufo boreas*), described as "exceedingly abundant" in the early 1900s, now occur in small numbers at only one of the six sites where they were originally common.
B. Surveys conducted from 1990 to 1994 in the cloud forest of Monteverde, Costa Rica, reveal that 20 species of frogs and toads

(40 percent of the anuran fauna) had disappeared since 1987.[73] Among those were the golden toad (*Bufo periglenes*), three species of *Eleutherodactylus*, harlequin frogs (*Atelopus varius*), eight species of tree frogs, and several species of glass frogs. Numerous other species had declined in abundance.

C. Populations of frogs have disappeared or declined from sites in the Brazilian Atlantic Coastal Forest. Since 1979, several species of frogs have disappeared from the forests around Boracéia, in the southern part of the Atlantic Mountains.[74] Populations of several other species have drastically declined. Another five species of leptodactylids are assumed to be extinct around Boracéia. The Brazilian egg-carrying tree frog (*Fritziana ohausi*) declined and is possibly extinct in the region. Before 1981, 13 species of frogs occurred along forest streams near the town of Santa Teresa in the coastal mountains of Espirito Santo, north of Boracéia. By 1987–1988, eight of these species were missing or reduced in numbers.[75]

D. Frogs have declined from montane rain forests of northeastern Australia.[76] Since the late 1970s, populations of at least 14 species that breed in rain forest streams have sharply declined, some by more than 90 percent.[77] Some species that were endemic to montane areas have disappeared entirely (e.g., the armored mistfrog, *Litoria lorica,* the northern tinkerfrog, *Taudactylus rheophilus,* and the sharp-snouted day frog, *Taudactylus acutirostris*). Populations of other more widespread species have vanished from highland sites but persist in normal densities at lowland sites.

The preceding examples of declines and disappearances illustrate the wide taxonomic and geographic extent of loss of amphibian diversity. Some of these species and assemblages will be discussed in greater detail later. They will provide a framework for understanding the evolution of our thinking about the declining amphibian phenomenon and will showcase presumed causes of specific declines.

It is important to emphasize, however, that not all amphibians are declining. Some, such as Cuban tree frogs (*Osteopilus septentrionalis*) and marine or cane toads (*Bufo marinus*), are expanding their ranges. We will come back to this point in chapter 4. Many populations of frogs, toads, and salamanders are remaining stable or even increasing in numbers. Some adaptable and opportunistic amphibians thrive and breed in human-modified habitats such as roadside ditches and around our homes, where they find moisture, shelter, and food. Unfortunately, because we lack long-term population data for most amphibian species, we do not know how the majority of populations are faring.

This last point is especially true for caecilians as a group. In a review paper concerning conservation biology of caecilians, David Gower and Mark Wilkinson at The Natural History Museum, London, pointed out that no long-term field studies have ever been carried out on caecilians.[78] Furthermore, we have virtually no data on population size for any caecilian species anywhere in the world. Few field methods have been tested for sampling caecilians, baseline distribution data for most species are poor or nonexistent, and caecilian taxonomy is inadequate.

Gower and Wilkinson reported that although unequivocal declines have not been documented, some claims of caecilian declines have been made. In most cases the main cause of perceived declines is thought to be habitat destruction, although these reports are anecdotal and lack clear evidence. The authors concluded that the conservation status of most caecilian species must be considered "data deficient." Baseline studies currently are being conducted; quantitative field studies are badly needed to assess whether or not populations of these amphibians are declining.

Marvalee Wake, at the University of California, Berkeley, shared some of her anecdotal examples of caecilian declines associated with land use change in Central America with us.[79] For example, most of her successful, albeit low density, caecilian sites in Costa Rica have been converted to monoculture of oil palm and bamboo at low elevations and coffee at higher elevations, eliminating the mixed forest and ground litter conditions preferred by caecilians. When she last visited the sites more than a decade ago she found no caecilians. During the late 1970s and 1980s, one of Wake's field sites in northern Guatemala—a coffee finca—housed a high-density population of the terrestrial, burrowing species *Dermophis mexicanus*.[80] She and colleagues returned in 2006; in the time it would have taken to find more than 100 individuals 30 years ago, they found 4. The area was partially flooded, preventing caecilians from constructing burrows. The finca is now owned by Starbucks, which is using open growing of coffee rather than shaded. The change has reduced the amount of decaying organic matter for earthworms, food for these caecilians.

As mentioned above, not all amphibians are affected negatively by land use change. This generalization may include caecilians. For some terrestrial species, conversion of forest to agricultural areas that maintain moisture and shade and restrict soil erosion may provide appropriate living conditions.[81] And certainly several species of *Ichthyophis* thrive in rice paddies in southeast Asia.[82] Again, to understand the effect of land use change on caecilians—as well as other factors affecting population sizes—we need to carry out long-term field studies.

Tropical salamanders are another group for which we need more long-term population data in order to access species' status. Many species of high-elevation tropical plethodontid salamanders have narrow distributional ranges. David Wake, at the University of California, Berkeley, and Vance Vredenburg, at San Francisco State University, reported that certain plethodontids from northern Oaxaca, Mexico, occur at elevations above 2000 m in cloud forests that are being shifted upward by global warming. The authors speculated, "Global warming threatens to force them [the salamanders] off the mountain and into extinction."[83]

Are Amphibians Declining at Unexpectedly High Rates?

Are the amphibian declines described above—and others—truly unusual? After all, many amphibian populations normally fluctuate widely in response to climatic factors.[84] The few long-term studies available reveal that populations often decrease over several years, then shoot back up. This extreme fluctuation in population size is especially evident in aquatic-breeding amphibians.[85] Often it is difficult to distinguish between "normal" year-to-year population fluctuations and long-term downward spirals that may lead to extinction—a difficulty that greatly complicates any decision about where to maximize conservation measures.[86]

So, how can we determine whether a decline is something to worry about? In 1994, Andrew Blaustein at Oregon State University asked the rhetorical question: How many years of census data do we need before we try to restore or protect a declining amphibian population? He pointed out that a conservative approach of withholding intervention until we conclusively demonstrate that extinction rates are unusually high might lead to loss of populations or entire species. But if we jump the gun, we might mistakenly conclude that a global decline is occurring when in fact populations are simply fluctuating normally. Such action could waste money and credibility. Blaustein suggested that "one must balance the risk of lost credibility, which might seriously compromise future conservation efforts in this and other arenas, against the cost of failing to respond to a potentially serious environmental crisis."[87]

We need to know when population fluctuations pass beyond the "expected" range—that is, beyond what we expect based on natural oscillations combined with the impact of habitat destruction and degradation.[88] One way to test whether observed patterns of amphibian population declines differ from those expected from random fluctuations is through the use of statistical analyses called "null models."[89] A null model uses previously collected data—for example on an amphibian population before the era of major declines—to ask the following question: If there is no long-range, insidious decline going on, how probable is it that we would observe the year-to-year population declines we see right now?

Alan Pounds, Resident Scientist at the Tropical Science Center's Monteverde Cloud Forest Reserve in Costa Rica, and his colleagues constructed null models to interpret amphibian disappearances in the area of Monteverde, Costa Rica. There, from 1990 to 1994, censuses revealed that 20 species of frogs and toads (40 percent of the anuran fauna) had disappeared.[90] The authors constructed their null models based on data from long-term studies on other amphibians. They first estimated the likelihood that any one species would disappear (defined as absence of adults during 1 or more years). Then they calculated the probability that a given number of species would disappear at the same time. Their comparison of the observed number of disappearances to those predicted by null models led them to conclude that the observed disappearances at Monteverde were highly unlikely to have occurred if populations were just fluctuating "normally." The authors noted that if they had based their null model on the patterns of population fluctuations of temperate amphibians from the Savannah River Ecology Laboratory in South Carolina, one of the longest studies of amphibian population dynamics,[91] the chance of that many species' disappearing from Monteverde under normal circumstances would have been less than 1 in 100 billion. The collapse of this diverse amphibian assemblage was above and beyond the expected number of disappearing species based on natural population fluctuations.

The study by Pounds and colleagues illustrates how null models can serve as powerful tools to distinguish "natural" fluctuations from those outside the usual range, at least for an amphibian assemblage as a whole. Studies on other assemblages have also shown statistically that amphibian populations are declining at a faster rate than they were a few decades ago.[92]

Most herpetologists now agree that amphibian population declines and disappearances are part of the overall "biodiversity crisis." A 2004 Global Amphibian Assessment suggested that 1856 species of amphibians (32.5 percent) are globally threatened with extinction.[93] So, are amphibians declining faster than other groups of animals? As mentioned earlier, the 2008 Red List reported that amphibians as a group may have a greater frequency of threatened species (30 percent) compared to birds (12 percent) or mammals (21 percent).[94] But other groups of animals may be worse off. Whitfield Gibbons, at the University of Georgia's Savannah River Ecology Laboratory, and his colleagues argued that reptiles are disappearing faster than amphibians.[95] A 2006 Global Reptile Assessment carried out by scientists working in the Mediterranean Basin Biodiversity Hot Spot (Mediterranean coasts of North Africa, Southern Europe, and the Middle East) submitted nearly 50 percent of the species in the area for inclusion in the IUCN Red List of species threatened with extinction. They found that 46 of the area's 355 reptile species (13 percent) were not just threatened but critically endangered due to habitat destruction and other human activities such as introduced species and overcollecting. Insects and other invertebrates might be even more affected by environmental change and human activities. For example, 71 percent of native butterfly species have declined in Britain over the past 20 years.[96]

Malcolm McCallum, from Texas A & M University, recently compared the current rate of amphibian extinction to the historical amphibian extinction rate based on the fossil record.[97] His calculations suggest that the current extinction rate could be as much as 211 times the long-term historical rate. Using the 2004 Global Amphibian Assessment estimate of 33 percent of amphibian species as threatened with extinction, McCallum predicted that for the next 50 years the extinction rate would be 25,000 to 45,000 times the expected background rate based on the historical record! He concluded on a somber note: "These data indicate that the general trend of amphibian extinction suggests catastrophic future losses and uncertain opportunity for recovery." Using molecular methods, Kim Roelants at Vrije Universiteit Brussel and colleagues came to a similar conclusion: that "... recent amphibian extinctions ... are far too frequent to represent background extinction."[98]

Are Amphibians Truly "Canaries in a Coal Mine"?

Not long after the 1989 First World Congress of Herpetology, some biologists speculated that amphibians might serve as good indicators of environmental stress.[99] It was suggested to journalists that amphibians might provide us with early warning signals of serious environmental deterioration, just like "canaries in a coal mine." (Coal miners used to carry canaries into the tunnels. Because a canary is more sensitive than a human to carbon monoxide, if the bird collapsed on the floor of its cage the miner knew he had to get out quickly or succumb to carbon monoxide poisoning.) One journalist, Kathryn Phillips, hit the nail on the head: "By linking the frog to this classic story, scientists not only captured journalists' and their readers' attention, but managed to elevate the lowly frog's status from simple hairless creature to benevolent and important herald. Intentionally or not, it was great marketing."[100] But is good marketing based on good science?

A plausible case can be made that amphibians on average might be more sensitive than certain other animal groups to some environmental stresses. First, many amphibians spend part of their lives in water and part on land, leading to double jeopardy. Survival and reproduction of these species might be affected if either the aquatic or the terrestrial environment becomes more stressful, for example through aquatic contamination, increased drought, or elevated levels of ultraviolet radiation. Second, amphibian eggs are not protected by a hard or leathery shell. The gelatinous substance that covers the amphibian egg leaves the developing embryo fairly exposed to the environment. Third, amphibians breathe partly through their skin. Sufficiently permeable for oxygen to enter and carbon dioxide to exit, the skin also allows entry to other gases and/or toxic chemicals and allows water to exit, making amphibians more susceptible than some other animals to poisoning and desiccation. For these three reasons, amphibians have heightened sensitivity to certain environmental stresses when compared to some other animals. Nonetheless, they are hardy animals. After all, amphibians have been on Earth for some 350 million years.

The root of the "canary in a coal mine" metaphor—that amphibians might provide us with early warning signals of serious environmental deterioration—is unfounded. Not all declines can be linked directly to environmental stresses that would also affect other animals, including ourselves. As long ago as 1994, Joseph Pechmann and Henry Wilbur, ecologists at Duke University at the time, argued that evidence simply does not exist to support the claim that amphibians make good canaries.[101] This statement is still true today. The decline or extinction of an amphibian population indicates that something is—or was—affecting that population negatively, but the "something" might be different for each amphibian species or at least for each particular site experiencing declines.

The canary metaphor implies that amphibians could serve as "ecological indicators." An ecological indicator is a measurable feature of the environment that is closely linked to (*a*) the "health" of the biota as a whole and/or (*b*) the "ecological integrity" of the landscape as a whole. Ecological integrity refers not only to the health of the biota but also to the health of the entire complement of ecological interactions and processes, such as pollination, decomposition, and nutrient cycling.[102] An ecological indicator serves as a shortcut or surrogate measure of population trends or habitat quality. Instead of trying to evaluate the health of the landscape by monitoring every species of plant and animal and every possible ecological process, we should be able to tell, by measuring one or a few indicators, whether in general all is well or whether there is a problem. An ecological indicator should consistently provide us with reliable information on what is happening across the biota and/or the ecosystem as a whole, and it should do so more easily and directly than would other means.

Frequently, a particular group of animals or plants is used or proposed as an ecological indicator and labeled an "indicator group." Declines in diversity of species in the indicator group (and sometimes changes in their identities as well) indicate changes in the health of the rest of the biota and of the ecosystem, changes that might not be obvious at first glance.[103] Examples include aquatic insects as indicators of stream water quality and lichens on tree bark as indicators of air quality in urban landscapes. For a group of species to qualify as an ecological indicator, it should meet at least 10 criteria,[104] including the following 5:

- The group's characteristics (for example, diversity or abundance) should change quickly with environmental changes that might also affect the remaining biota and the ecosystem as a whole (that is, the group is sensitive to those factors that also would negatively affect other animals and plants and the entire ecosystem).
- Background variation in species diversity and population density of the indicators should not be so great as to mask significant variation resulting from environmental deterioration on the one hand or to trigger false alarms on the other.
- The natural history, geographic distribution, and taxonomy of the group should be fairly well known.
- The group should be equally active and/or accessible at all times when sampling might be carried out.
- The group should be easy and cost-effective to monitor, providing objective data abundant enough for rigorous statistical analysis.

Do amphibians meet these—and other—criteria? No. (*a*) There is no evidence for a direct relationship between factors responsible for amphibian declines singly or as a group and the critical factors that might negatively affect the rest of the biota and the ecosystem as a whole. (*b*) Background variation in species diversity and population density of amphibians is highly variable. (*c*) Many species of amphibians are difficult to identify, their natural history is unknown, and/or their taxonomy and geographic distributions are not well known, especially in the tropics, where most amphibians occur. (*d*) Amphibian activity is notoriously variable with respect to season, microhabitat, temperature, and rainfall. (*e*) With some exceptions, amphibians are neither easy nor cost-effective to monitor.

In 2005, Trevor Beebee, from the University of Sussex, and Richard Griffiths, from the University of Kent, made the following observation: "Some of the early amphibian decline literature therefore represents a paradox—amphibians were flagged as good biological indicators despite the fact that the authors concerned had little idea at the time what they were indicating!"[105] Beebee and Griffiths joined the growing rank of investigators who have argued that there is no evidence to suggest that amphibians should be more effective than other animals as indicators of biodiversity or habitat quality.

Marked changes in an ecological indicator, or indicator group, tell us that something is happening with the biota as a whole and/or with the ecological integrity of the landscape as a whole.[106] If amphibians do not meet the criteria for an ecological indicator group, what do marked changes in their diversity and abundance tell us? The changes tell us that something is happening with the amphibians themselves, whether the remaining biota is affected similarly or not. Instead of being an indicator group, amphibians are a target group: a group that is the target of our concern. Amphibians are "telling" us that they themselves are in trouble. Unlike the proverbial canary, an individual amphibian or an entire species may die because of a factor that has little or no effect on humans or on many other animals. Furthermore, as John Terborgh, then at Princeton University, pointed out, once frogs begin to die off, it is a late warning that something is wrong with them—not an early warning.[107]

Unfortunately, the reasoning that led to the original canary metaphor continues to fuel alarmist language and unfounded (and often outrageous) statements. For example, the Web site of an international conservation organization recently stated the following:

> "Amphibians are considered one of nature's most finely tuned indicators of general environmental conditions. Their highly permeable skin easily absorbs particles and pollutants and is sensitive to changes in both terrestrial and aquatic habitats. Amphibians are usually the first species to disappear from a habitat."[108]

The "canary in the coal mine" metaphor is not only incorrect, it is unnecessary. Many amphibians are in trouble, and environmental change appears to be the general cause, although the specific causes vary from species to species and from site to site. That alone should be enough to generate concern, especially because amphibians are fascinating and often beautiful creatures with a long association with human folklore and a much, much longer history on this planet than our own.

The fact that amphibians are not good indicator species does not diminish their conservation importance. They serve as a rare example of a vertebrate group for which we have the capacity and momentum to address and understand biodiversity loss on a global scale. Scientists studying declines and extinctions of other organisms will surely benefit from the lessons we are learning while addressing amphibian population declines—from seeking explanations for the underlying causes to implementing mitigation efforts.

Once scientists recognized the global amphibian decline problem, we responded by carrying out research, educating the public and other biologists, and advocating and instituting conservation measures. We will describe some of these efforts in the following chapter.

2

Rallying Around the Issue of Amphibian Declines

Although amphibian populations declined or disappeared before and during the 1960s, 1970s, and 1980s, it was not until the First World Congress of Herpetology, held in Canterbury, England, in September 1989, that investigators began to suspect the extent of the losses. Almost 1400 delegates from 60 countries attended this congress.[1] Participants presented a few papers addressing declines, but informal conversations in staircases, halls, meeting rooms, and pubs crystallized awareness of the bigger problem. Herpetologist after herpetologist shared stories of losing his or her study populations. Many were shocked to discover their stories were not unique. Amphibian populations fluctuate widely, so many of us had assumed that it was just a matter of time until weather conditions improved and "our" animals would bounce back. No need to panic; no need to publish about declines or disappearances. Just wait it out.

That is also what one of us (MLC) had assumed. Marty, then at the University of Florida, had recently lost her study populations of golden toads (*Bufo periglenes*) and harlequin frogs (*Atelopus varius*) from Monteverde, Costa Rica.[2] (We will return to these populations in chapter 6.) During the Canterbury congress, she and David Wake, from the University of California, Berkeley, discussed the disappearance of golden toads and harlequin frogs and of Wake's own trouble in the fall of 1988 finding frogs and salamanders in Monteverde. As we and other participants talked among ourselves about the loss of our study populations, we recognized that amphibian declines and disappearances were global and that protected as well as disturbed habitats were affected.

A few weeks after the congress, Wake attended a Board on Life Sciences meeting of the National Academy of Sciences/National Research Council in Washington, D.C. He told fellow board member Harold Morowitz, of George Mason University, the stories of declining amphibians that he had heard in Canterbury. Together, they decided to organize a meeting of scientists worldwide to address the nature and extent of amphibian declines and to determine whether concern about amphibian population declines justified a coordinated response by the scientific community.[3] Wake and Morowitz secured funding, and the U.S. National Research Council Workshop was held on February 19–20, 1990, at the National Academy's Beckman Center on the campus of the University of California, Irvine.

We focus on two questions in this chapter: (1) What is being done to document and understand amphibian declines? (2) What is being done to mitigate further declines? To answer these questions, we track how the scientific and lay communities have addressed the problem through research, education, and conservation. We take a look at programs that have either a North American or an international focus. In later chapters, we will see that a lot was happening in Australia and elsewhere as well.

The First Workshop

The Irvine workshop was the first organized effort to examine amphibian declines. About 40 scientists, including herpetologists, ecologists, wildlife conservation biologists, a geneticist, a climatologist, an ecosystems specialist, and a veterinary pathologist, attended the meeting. The workshop's focus was (*a*) to evaluate evidence that amphibians were declining worldwide; (*b*) to determine the geographic, taxonomic, and habitat distribution of declines; (*c*) to identify possible causes for declines; (*d*) to consider the implications of amphibian declines; and (*e*) to recommend action for addressing these declines. An underlying goal was to increase awareness among biologists that amphibians were disappearing.

Workshop participants concluded that, although much of the evidence was anecdotal, the multitude of reported declines and the widespread geographic nature of the declines suggested that something real was happening to amphibians. Beyond that, several major findings emerged.[4] Clearly, not all species and not all localities were affected. The group found no "smoking gun," no compelling evidence for a single factor responsible for all declines and extinctions. Habitat destruction was a major cause, but population declines were also happening in areas without obvious human disturbance. Other suspected anthropogenic causes for declines included chemical contamination and introduced species.

Workshop participants offered several long-term recommendations to address amphibian declines:

- Initiate long-term studies of amphibian populations, focusing on the significance of amphibians in ecosystems and their unique potential as "bioindicators in ecosystem studies," population biology, and identification of possible chemical and physical causes for declines in amphibian populations.
- Compile information on historical occurrence of amphibians, population density, and survivorship using data from museum records and the literature. Where data can be used as a basis for comparative studies, localities should be re-surveyed.
- Consider establishing an organization to focus on maintaining amphibian biodiversity.
- Design and develop educational programs to inform the public about the status of amphibian populations.

Workshop members discussed how they might generate public support to address the problem of amphibian declines and how they might obtain funds for research, education, and conservation programs.[5] It was at the Irvine workshop that the potentially strong selling point discussed in chapter 1 emerged: Amphibians might be good biological indicators of environmental health because their permeable skin could absorb toxic chemicals, many species are exposed to "double jeopardy" from environmental stresses because they have both aquatic and terrestrial stages in their life cycles, and their jelly-coated eggs do not afford much protection from the environment.[6] If they could be promoted as sensitive bioindicators of environmental threats that sooner or later would also affect less sensitive species, including humans, surely the plight of amphibians could seize the public's attention. Although the "canary in the coal mine" metaphor has not been supported by data, the bioindicator suggestion generated considerable media attention that has increased public awareness and generated financial support for the study of declining amphibian populations.

The Irvine meeting served as a catalyst. In the almost 20 years since the workshop, various individuals and groups have developed initiatives to address each of the four recommendations. People have responded in different ways, depending on their expertise, interests, and philosophies. Some have emphasized research to document population declines and discover their causes. Others have focused on public education to increase appreciation for amphibians and their habitats and to increase awareness of amphibian population declines. Still others have worked to protect habitats and declining species, sometimes through captive breeding and reintroduction programs. Some of the organizations, institutions, and programs established to address amphibian population declines are highlighted in Table 2.1.

Research

After the Irvine meeting, members of the scientific community worked to keep the issue of declining amphibian populations in the public eye. We also knew we could document declines and investigate their causes more efficiently if we worked through a network for sharing information. Because no such network existed at that time, one needed to be established.

George Rabb, at the time Director of the Chicago Zoological Society's Brookfield Zoo and an active member of the conservation community, envisioned an international working group to address amphibian population declines—a specialist group to be part of the International Union for the Conservation of Nature and Natural Resources (IUCN).[7] In December 1990, less than a year after the Irvine meeting, the Species Survival Commission of IUCN established the Declining Amphibian Populations Task Force (DAPTF).[8] Its mission statement was to determine the causes of the declines and to promote ways to halt or reverse them (see Table 2.1).

The task force saw itself as a "nerve center" with a global network of investigators feeding it information about amphibian declines. Two problems concerning data collection needed to be addressed, however. First, much of the information available on the population status of amphibians was anecdotal. Such reports were not going to convince anyone—skeptical scientists included—that a problem existed. Second, investigators needed standardized field methods and techniques to ensure that data would be comparable and reliable across sites. In response to this need, 47 scientists formed the DAPTF Working Group on Protocols and wrote a manual entitled *Measuring and Monitoring Biological Diversity: Standard Methods for Amphibians,* which was published by the Smithsonian Institution Press in 1994.[9] The manual was one of the first projects completed by the Task Force.

DAPTF established a database to receive and store information from field investigators. This database, the Declining Amphibian Database (DAD), contains fields for information on site (habitat) characteristics, natural history, census techniques, detailed population data, life stage affected, and date of first evidence of decline for each species entered.[10]

Field investigators belonging to about 100 DAPTF regional and subregional working groups have gathered and entered data into DAD. Several

TABLE 2.1 Organizations, Institutions, and Programs Designed to Focus on Amphibian Population Declines

Organization, Institution, or Program	Year Founded	Mission Statement/Purpose	Focus
Declining Amphibian Populations Task Force (DAPTF)*	1990	"To determine the nature, extent, and causes of declines in amphibians throughout the world, and to promote means by which the declines can be halted or reversed."	International
North American Amphibians Monitoring Program (NAAMP)	1994	1. Educate public about wetlands and frogs 2. Track population trends of anurans	North America
A Thousand Friends of Frogs	1995	1. Educate public about frogs and environment 2. Monitor populations and report malformations	Minnesota, USA
National Amphibian Conservation Center (NACC) (Detroit Zoological Institute)	1997	1. Educate 2. Maintain and breed endangered species 3. Serve as resource for research 4. Serve as model for future facilities	International
Partners in Amphibian and Reptile Conservation (PARC)	1999	"Develop a comprehensive strategy for the long-term conservation of U.S. amphibians, reptiles, and their habitats"	USA
Amphibian Research and Monitoring Initiative (ARMI)	2000	1. Set up national amphibian monitoring program on Federal lands 2. Develop sampling techniques and statistical analyses appropriate for determining status of amphibian populations 3. Identify possible causes for declines and malformations	USA
Global Amphibian Assessment (GAA)*	2000	Assess status of the world's amphibian species	International
Global Amphibian Specialist Group (GASG)*	2002	"To develop and implement a global conservation strategy for amphibians"	International
Amphibian Conservation Summit (ACS)	2005	Develop a plan to respond to amphibian declines—the Amphibian Conservation Action Plan	International
Amphibian Specialist Group (ASG)*	2006	"To conserve biodiversity by stimulating, developing, and executing practical programs to study, save, restore, and manage amphibians and their habitats around the world."	International
Amphibian Ark (AArk)	2006	"Working in partnerships to ensure the global survival of amphibians—focusing on those that cannot be safeguarded in nature"	International

*As of June 30, 2006, the DAPTF, GASG, and GAA make up the Amphibian Specialist Group.

regional working groups have summarized their findings in multiauthored books. DAPCAN, the Canada regional working group, published *Declines in Canadian Amphibian Populations: Designing a National Monitoring Strategy* in 1992. In 1997, *Amphibians in Decline: Canadian Studies of a Global Problem* was published. The Midwest Working Group (U.S.) published *Status and Conservation of Midwestern Amphibians* (1998).[11]

Not all DAPTF working groups focused on surveying amphibian populations. Other working groups included Disease and Pathology, Chemical Contaminants, Climate and Atmosphere, Education, and Captive Breeding.

In March 1992, the first issue of DAPTF's free, bimonthly newsletter, *Froglog*, was launched. Summaries of field studies, announcements of meetings and workshops, and citations of recently published papers and books concerning amphibian declines are among the topics published in *Froglog*.

Another focal activity of the DAPTF was its annual Seed Grant Program. Grants of $500 to $2000 were designed to jump-start projects addressing amphibian declines. The hope was that, after completing preliminary studies, investigators would be better able to obtain support for larger, longer-term studies. By the end of the DAPTF program in 2006, US $321,634 had been awarded in seed grants to 167 individual projects in 49 countries.[12] Many seed grantees have secured additional funding and have expanded their research. More than 100 studies supported by DAPTF seed grants have been published.

In addition to annual seed grant awards, the DAPTF maintained a Rapid Response Fund to provide support for investigators facing situations that require urgent action, such as discovery of mass mortality events and disease outbreaks. For example, Rapid Response funds have been awarded to investigators studying disease in amphibians from the Andes of Ecuador and from South Africa.

Notice that we have referred to the DAPTF in the past tense. On 30 June 2006, the DAPTF was subsumed into the Amphibian Specialist Group (ASG). We will return to the ASG in chapter 9.

A second series of workshops, held in 1998, influenced the direction of our thinking about amphibian declines and contributed significantly to understanding of the problem. Funded by the National Science Foundation, the workshops were initiated by one of us (JPC), Andrew Storfer, and Elizabeth Davidson, all from Arizona State University. By that time, suspected causes of amphibian declines included environmental change and emerging infectious disease.[13]

In May 1998, the workshop entitled "Amphibian Population Dynamics: Is the Threat of Extinction Increasing for Amphibians?" took place at the National Science Foundation in Arlington, Virginia. Participants included herpetologists, ecologists, physiologists, and investigators with expertise in ecotoxicology, infectious diseases, climate change, and science policy. Bruce Babbitt, then a member of President Clinton's cabinet as U.S. Secretary of the Interior, also attended. The group suggested that in many cases declines might be caused by synergistic interactions among several factors, especially involving pathogens and environmental change. For this reason, the group recommended that interdisciplinary, collaborative research programs be initiated to understand how multiple variables affect amphibian populations.[14] We will come back to this workshop and two that followed it in chapter 8.

After the May 1998 NSF-funded workshop, Babbitt convened an agency-level briefing on amphibian declines. This meeting led to the formation later that year of the federal interagency Taskforce on Amphibian Declines and Deformities (TADD) to coordinate the research activities of diverse agencies.[15] The rationale was that amphibian losses were not a problem for one agency to solve. Rather, teamwork was needed to develop answers and provide solutions. This taskforce was the first official recognition by the U.S. government that global problems might be responsible for amphibian declines. Most of the original funding for TADD was earmarked for surveying amphibian populations rather than investigating causes for amphibian morphological abnormalities. TADD "disappeared" after the new presidential administration took over.

In 2000, the activities formerly associated with TADD were taken up by the Amphibian Research and Monitoring Initiative (ARMI). The U.S. Geological Survey (USGS) coordinates this national program, which integrates the collective efforts of numerous U. S. federal agencies, including the National Park Service, the U.S. Fish and Wildlife Service, and the Bureau of Land Management.[16] The goals of ARMI are to monitor populations to detect changes in distribution and abundance of species on federal land and to study the causes of amphibian declines. ARMI supports research on various environmental conditions known to affect amphibians, such as pollutants, introduced species, and disease. It also provides information concerning amphibian conservation to the general public, resource managers, and policy makers.

Thanks to ARMI initiatives, more herpetologists are now being hired into USGS positions. A unique strength of the program is that ARMI encourages participation by state, university, and nongovernmental organizations. Herpetologists, hydrologists, chemists, and geographers work with local partners to monitor amphibians in national wildlife refuges, national parks, and national forests. The Patuxent Wildlife Research Center maintains a centralized database for data collected through ARMI monitoring programs.[17]

Information about amphibian population declines has been scattered geographically, is often anecdotal, and is limited in taxonomic scope. Scientists taking a global perspective realized that it was necessary both to do extensive survey and monitoring work and to encourage experts to share their knowledge. We needed to assess the status of all amphibians, not just the charismatic ones or those that happened to be in herpetologists' study sites. In response to these needs, the Global Amphibian Assessment (GAA) began in 2000 as a joint project of the Species Survival Commission of the IUCN, the Center for Applied Biodiversity Science of Conservation International, and NatureServe. By assessing the status of the world's amphibians, the GAA hoped to identify the most threatened amphibian species and the most critical habitats in need of preservation. It aimed to identify the relative importance of various threats to the survival of amphibians and to propose conservation measures aimed at reversing declines and extinctions. Ultimately, the long-term assessment was designed to provide a baseline for measuring the success of conservation efforts.

The GAA database contains fields for systematics, general information, distribution maps, national distribution, habitat, major threats, IUCN Red List assessment, and a bibliography for each species. To date, almost 600 scientists from 60 countries have contributed to the project. The first comprehensive data were released in 2004. Since then, the data have been revised and upgraded as new information has become available. Data gathered as part of the GAA are available to anyone with World Wide Web access.[18]

Basic research carried out by hundreds of investigators—some working alone, others as parts of large research teams—provides a critical link in our understanding of the extent and causes of amphibian declines. Much of the research carried out by U.S. investigators has been funded by the National Science Foundation. We will return to several research programs in chapter 6.

Education

Scientists by themselves cannot save amphibians—or any group of animals—from extinction. Everyone must be part of the solution. In order to gain public support, herpetologists have identified education as a high-priority activity.

One way to educate the public is through publicity—Web sites, television and radio broadcasts, zoo and natural history museum displays, and magazine and newspaper articles. A more profound way to encourage public appreciation for wetlands and for the animals that live in them is to familiarize people with ponds, marshes, and lakes. One means of doing this is by involving local volunteers in long-term amphibian monitoring projects.

In 1994, members of the DAPTF Midwest Working Group met with other biologists at the Indiana Dunes National Lakeshore to discuss amphibian population monitoring needs throughout North America.[19] The biologists wanted to establish a program to monitor the relative abundance and distribution of amphibians in such a way that the information would be statistically defensible and comparable across regions. After additional meetings, the North American Amphibians Monitoring Program (NAAMP) was established. Regional partners, including nonprofit organizations and state natural resource agencies, train volunteers to identify frogs by their calls. Volunteers then monitor local sites and record presence of frog species. Personnel from the USGS administer and coordinate the database and Web site. Participants are appreciating frogs and wetlands as never before, and scientists are accumulating data faster and tracking population trends more efficiently with the public's help. By 2008, the NAAMP calling survey was active in 23 states.

The general public also has participated in a monitoring program that reported frog morphological abnormalities. The project began with observations made by school children. In August 1995, middle school students on a field trip in south-central Minnesota discovered northern leopard frogs with hind limb abnormalities—some with missing legs, others with extra legs.[20] Half of the frogs the students caught that day were abnormal, often grotesquely so.

The teacher, Cindy Reinitz, later met with Judy Helgen from the Minnesota Pollution Control Agency and Tracy Fredin, Director of the Center for Global Environmental Education at Hamline University, to discuss opportunities for scientists, educators, and students to collaborate.[21] They reasoned that if these diverse groups worked together they might discover the causes of the abnormalities. From this meeting, the organization known as "A Thousand Friends of Frogs" was formed. Later that year, the Minnesota State Legislature provided funds to begin a survey protocol and a form for reporting abnormal frogs. The following year, the group added an amphibian monitoring project to the continuing program of reporting malformations.

Members of A Thousand Friends of Frogs have developed resources for educators to use in teaching about amphibian conservation and environmental awareness. In 1999, an online course, called Helping Your Local Amphibians (HYLA), taught educators about amphibian biology and amphibian declines.[22] (*Hyla* is a widespread genus of tree frogs in the family Hylidae.) The group's Web site encourages citizens to become involved in environmental issues, including monitoring frog populations, and it offers information for teachers and students regarding resources and scientific research.[23] A Thousand Friends of Frogs demonstrates the power of student and community involvement and the significant role of nonscientists in educating the general public about declining amphibians.

Conservation

The goal of conservation is to protect animals, plants, and their habitats. Whitfield Gibbons, from the Savannah River Ecology Laboratory, suggested that, "rather than sit, watch, and be discouraged by the global problem of human overpopulation, anyone who desires a world in which native amphibians occupy their natural habitats needs to make a counteractive effort of some sort. I believe the answer lies in PARC."[24]

In June 1998, Partners in Amphibian and Reptile Conservation (PARC) held its organizational meeting in Atlanta, Georgia. Anyone who could potentially affect amphibians' and reptiles' habitats or their well-being was welcome and encouraged to participate in PARC. Disparate groups—the academic community, government agencies, conservation groups, and private industry—have their own agendas and viewpoints. PARC's belief is that only by considering all viewpoints can we solve the problem of declining amphibians and reptiles. Only then can we carry out effective conservation.[25]

PARC includes partnerships among people from state and federal agencies, conservation organizations, nature centers, zoos and museums, the pet trade industry, universities and research laboratories, environmental consulting firms, forest industries, and the energy industry. The organization is divided into five regional working groups (midwest, southeast, southwest, northeast, and northwest) and six technical working groups (research, education and outreach, international, inventory and monitoring, management, and policy, regulation, and trade).

PARC has proven that people with diverse viewpoints can work together effectively. The partnership has formulated guidelines and position statements regarding the sustainable use of amphibians and reptiles; regulations for their collection, manipulation, possession, and sale; and the use and fate of animals used in teaching. Group members also have produced educational materials, developed habitat management guidelines, and written training modules along with inventory and monitoring manuals.

All three U.S. national herpetological societies (The Herpetologists' League, Society for the Study of Amphibians and Reptiles, and American Society of Ichthyologists and Herpetologists) have conservation committees. These committees interact with one another and focus on a variety of issues. For example, participants bring conservation concerns that affect amphibians to the attention of state and federal agencies and herpetologists worldwide, provide expert advice and educational information about matters ranging from habitat protection to commercial and private use of amphibians, and work to influence policy decisions concerning proposed national and international legislation involving amphibians. In short, these conservation committees are advocates for amphibians worldwide.

Zoos play an important role in keeping endangered birds and mammals alive, but not many zoos maintain and breed amphibians. Frogs, salamanders, and caecilians usually do not attract big crowds of visitors the way giant pandas do, and we do not know much about the nutritional and habitat requirements of most species. Maintaining and breeding captive populations can be expensive and labor-intensive, and such efforts require long-term institutional commitments. There are several notable exceptions, however—a few zoos that feature captive breeding of amphibians. Some of the most popular attractions are the colorful, diurnal poison dart frogs from the neotropics.

The National Amphibian Conservation Center (NACC) at the Detroit Zoological Institute, constructed in 1997, was the first major conservation facility dedicated entirely to exhibiting and conserving amphibians (Figure 2.1).[26] The center opened to the public in 2000 and by July 2008 was housing 41 species of amphibians. Almost half of the 12,000-square-foot facility consists of holding and breeding rooms, offices, and research space. A staff feeds the animals, cleans their aquaria, and manages the frogs' complicated life support systems, which include boilers, misters, chillers, and closed circulation for quarantine rooms. The center not only maintains and breeds endangered amphibians with the ultimate hope of reintroducing animals into the wild but also educates the public about amphibians and serves as a resource for scientists.

In 2006, the Conservation Breeding Specialist Group of IUCN, the Amphibian Specialist Group, and the World Association of Zoos and Aquariums established Amphibian Ark (AArk), with Kevin Zippel as Program Director.[27] Amphibian Ark is an international effort devoted to in situ and ex situ amphibian conservation. The goal is to maintain selected species in captivity until they can be safely reintroduced to the wild. Target species are those for which adequate protection in the wild is not possible at present. AArk hopes to house 500 individuals from each of 500 amphibian species. This ambitious project will require extensive funding. AArk designated 2008 as "Year of the Frog," hoping to raise public awareness of amphibian declines and to generate support for AArk's global conservation initiatives.

Figure 2.1 National Amphibian Conservation Center at the Detroit Zoological Institute. Photograph by Danna Schock.

Captive breeding programs offer an alternative to watching species go extinct in the wild, but they are not a panacea. For captive breeding to be effective, we must also preserve the animals' habitat so that someday the species can be reintroduced into the wild. Furthermore, in many cases it is not clear that captive breeding and reintroduction programs will be successful.[28] In some instances, projects have failed. In other cases, it is too early to determine success or failure.

One example is the Wyoming toad, *Bufo baxteri.* Historically, this anuran occurred in Albany County, Wyoming, around the town of Laramie, in an area of only about 2330 km^2. During the mid-1970s, the toads disappeared from most of their small range. In 1983, the species appeared to be close to extinction, and the following year it was listed as endangered under the Endangered Species Act.[29] By 1994, the toads had disappeared entirely from the wild.[30] Before the toads disappeared, however, personnel from the Wyoming Game and Fish Sybille Wildlife Research Unit in Sybille, Wyoming, established a captive breeding program. Other captive breeding programs were subsequently established elsewhere. Since 1995, more than 7000 tadpoles and toadlets have been released within their historical

Figure 2.2 Kihansi spray toad, *Nectophrynoides asperginis*. Photograph by Dante Fenolio.

range, under a U.S. Fish and Wildlife Service recovery plan. Most reintroductions have been to Mortenson Lake, the site of the last known wild population. Although biologists have heard toads calling at the lake, no one has observed reproduction since 1999. Census numbers are low, and investigators have found sick and dead toads. Some toads were reintroduced at two locations on the Hutton Lake National Wildlife Refuge, but they disappeared without establishing breeding populations. The Wyoming toad is the most endangered anuran in the United States due to its limited numbers, low genetic diversity, reliance on captive breeding programs, and persistence of amphibian chytrid fungus in their habitat.[31] Thanks to captive breeding, Wyoming toads are still with us, but recovery in the wild seems uncertain. (In chapter 5 we will return to the chytrid fungus associated with amphibian declines.)

The following story likewise reflects the uncertainty of captive breeding efforts for conservation purposes. The Bronx Zoo, the NACC at the Detroit Zoo, the Toledo Zoo, and several other zoos have focused on captive breeding of Kihansi spray toads (*Nectophrynoides asperginis*) (Figure 2.2). These anurans are small (10–18 mm body length), mustard-yellow, and ovoviviparous (larvae develop within the female's body and are born as toadlets). They were formerly found only within a 2-hectare wetland spray meadow along the Kihansi River Gorge in Tanzania.[32] A hydroelectric facility was built near the toads' habitat in 2000. After the dam cut off 90 percent of the water, there was not enough spray left to maintain the wetland. Engineers installed a huge sprinkler system to replace the waterfall, but only after the river had been diverted away from the gorge for nine months. The habitat had already dried by that point. In late November 2000, field staff members from the Wildlife Conservation Society collected 500 toads for breeding in U.S. zoos in hopes of providing a safety net against extinction. No one knew how to raise this species, though, and many toads died before zoo personnel perfected husbandry techniques. As of February 2006, there are about 280 Kihansi spray toads living at the Bronx and Toledo Zoos. But they cannot go home yet. In 2003, the dam's floodgates were opened briefly to flush sediments. Soon afterward, the wild toad population crashed from an estimated 18,000 individuals to zero. Kihansi spray toads are now extinct in the wild.[33] Tests showed that the sediments contained high levels of pesticides. Analyses of toad carcasses also revealed chytrid fungus. Will the toads born and raised 12,800 km away from their home in Tanzania spend the rest of their days confined to aquaria?

In 2002, the Species Survival Commission of IUCN created the Global Amphibian Specialist Group (GASG), its stated mission being to develop and implement a global conservation strategy for amphibians.[34] Why was another IUCN amphibian specialist group needed? The main focus of the DAPTF was research. The GASG would concentrate on conservation issues, for example by monitoring the conservation status of all amphibian species through IUCN Red Listing.[35] It would also develop a global action plan for protecting amphibians. The GASG trained conservation professionals, raised funds for support of global amphibian conservation, developed and implemented conservation strategies, and built a global network of amphibian experts. The GASG convened the Amphibian Conservation Summit in September 2005. Participants from around the world wrote an Amphibian Conservation Action Plan (discussed in chapter 9). On June 30, 2006, the GASG and the GAA merged, along with the DAPTF, into the Amphibian Specialist Group (ASG), an alliance that is also reviewed in chapter 9.[36]

Where are we now compared to September 1989, when the initial shock of global amphibian declines reverberated throughout the herpetological community? Investigators have focused on documenting amphibian declines and identifying their causes. Environmental educators, the media, teachers, and students have informed the general public about declines and increased appreciation for amphibians and their habitats. Conservation advocates have invested considerable time, energy, and resources to protect amphibians and their habitats, to breed endangered species in captivity and then reintroduce them into the wild, and to formulate conservation guidelines. The herpetological community, as we will learn further in chapters 8 and 9, is better organized, more prepared, and more knowledgeable now than it ever has been concerning amphibian declines. We do not mean to imply, however, that all of the organizations dedicated to addressing amphibian population declines—those mentioned here as well as others—have completed their goals. There is still much to do. In the following chapter, we will look at our current understanding of amphibian population declines.

3

Challenges, Correlates, and Hypotheses

Since the 1989 Canterbury congress, we have identified and surmounted some of the numerous challenges associated with the long-term research required to understand amphibian declines. In addressing the question of why some populations and species are declining while others are not, we have identified certain correlates that seem to increase risk of decline. We also have hypothesized six main causes for amphibian population declines. In this chapter, we focus on these three aspects: challenges, correlates, and hypotheses.

Challenges to Addressing Amphibian Declines

From the beginning, herpetologists knew that solving the mystery of amphibian population declines would be laborious and complex. Along the way, investigators have encountered intellectual, logistic, and financial challenges, many of which persist today.

Separating "Real" Declines from "Normal" Fluctuations

Herpetologists first had to agree that there was indeed a problem. As discussed in chapter 1, amphibian population sizes normally fluctuate from year to year. From 1989 through the early 1990s, skeptics argued that reported declines might be nothing more than "normal" downward fluctuations. Some argued that in most cases we could not separate human-caused population declines from background, or normal, population fluctuations.[1] This skepticism was productive because it forced us to scrutinize and evaluate the data and observations. To figure out how to pinpoint "real" declines—declines that exceeded the long-term average and were caused by change, whether related to humans, climate, or some other factor—we had to understand how much amphibian populations fluctuate normally. Clearly, long-term studies carried out at numerous sites worldwide would be critical to understanding what was happening.

We still have the challenge of distinguishing true declines from natural population fluctuations.[2] To meet this challenge, investigators sometimes compare their long-term monitoring data to an historical baseline of the species' occurrence and abundance—data gathered in the past. Sometimes, investigators compare current population declines to a "null model" (see chapter 1) by using natural fluctuations of amphibian populations as a baseline.

Public Support

By the mid-1990s most herpetologists agreed that there were too many unexplained disappearances occurring around the world to be attributed to simultaneous "normal" population fluctuations. To address our concerns productively, we needed to gain public support, to generate interest in people other than herpetologists. This need gave rise to various environmental education programs aimed at encouraging appreciation for wetlands and their resident amphibians. We discussed in chapter 2 some of these programs, including ones that directly involve the public in monitoring amphibians. Herpetologists still face the challenge of gaining support for the study of amphibian declines, but since 1989 we have gained much ground thanks to public education and media attention.

Funding

Getting financial support for scientific research on amphibians, whether field studies or laboratory experiments, has always been a challenge. Most funds for conservation work are awarded to researchers working on birds and mammals. As noted in chapter 1, however, amphibians are essential components of our ecosystems. They play an important role in energy flow and nutrient recycling, in both terrestrial and aquatic environments. Although this should be enough of a basis to secure funding, often it is not. Obtaining support for research on declining amphibian populations is especially difficult for researchers in developing countries, because public and governmental support does not always exist, and financial resources directed to conservation efforts are generally limited.

At the first workshop on declining amphibian populations in Irvine, California, in 1990, Morowitz is said to have advised herpetologists to "think big" in terms of funding.[3] He compared their needs to those of molecular biologists who successfully got the Human Genome Project enough recognition so that it became a specially funded project of the National Institutes of Health. Some amphibian biologists heeded his advice and applied for and received million-dollar-plus grants from the National Science Foundation to support their work on declining amphibian populations. The Amphibian Specialist Group demonstrates how herpetologists are now "thinking big" to expand the research, education, and conservation initiatives (mentioned in chapter 2; to be discussed further in chapter 9). At the Amphibian Conservation Summit held in September 2005, delegates estimated that over the next 5 years approximately US $409,000,000 would be needed for addressing the amphibian decline problem, from funding of long-term research to establishment of captive breeding programs. Appeals would be made to governments, private institutions, and individual donors.

Lack of Information

In many parts of the world, just trying to identify amphibians can be difficult, because field guides or taxonomic keys are not available. Often, a substantial fraction of the amphibian fauna has yet to be described formally. Reproductive characteristics and timing of breeding, activity periods, and geographic distribution are unknown for a large proportion of amphibians. We lack information on habitat requirements for many species that breed in aquatic habitats but spend the rest of the year in terrestrial sites. In many cases, the amphibians seemingly vanish between breeding seasons and also between metamorphosis and first breeding. How can we monitor amphibians if we cannot find them and cannot identify those we do find? In most areas of the world, we lack baseline data on the historical presence or abundance of species. Local populations may have disappeared, but this fact remains unreported because we do not know which amphibians formerly occurred in an area. Likewise, population declines may remain undocumented because we have no historical data on the species' abundance.

Communication

Investigators studying amphibian population declines need to collaborate with each other to share information and ideas, skills, and techniques to maximize efficiency and to avoid "reinventing the wheel." Because amphibian declines are part of the global decline of biodiversity, herpetologists can gain a broader perspective by working with scientists who study declines of other organisms, such as mammals, corals, or ferns. The more interdisciplinary collaborations we have, the faster we will unravel the possible causes for declines. Herpetologists working with climatologists, molecular biologists, epidemiologists, and scientists in other fields can address and answer a wider range of questions more effectively, but making the contacts, assimilating new information, and getting funding for large interdisciplinary studies are challenges.

Once an investigator gathers data, he or she needs to share the results with the scientific community. This can be difficult, especially in developing countries with limited national publication outlets. Results often stagnate in file drawers, or, if published, they appear in outlets with restricted distribution or in non–peer-reviewed journals.

What does a scientist do if he or she concludes that there is no evidence for a decline in a specific study population? Scientists have always found it difficult to publish results that do not support an expected pattern. But in order to determine whether amphibians as a group really are declining, we need to monitor as many populations as possible and document which ones are declining, remaining stable, and increasing. If we do not publish results documenting stable or increasing populations, we will paint a biased picture of worldwide changes. Furthermore, we need to identify characteristics and habitats associated with declines, in order to intervene and protect vulnerable populations and species.

Methods and Protocols

As noted in chapter 2, amphibian biologists recognized early on the need for standardized methods and protocols to ensure that data would be comparable and reliable across sites. To address this need, the DAPTF Working Group on Protocols wrote *Measuring and Monitoring Biological Diversity: Standard Methods for Amphibians*.[4] In addition to recommending techniques for monitoring amphibian populations, the manual provided recommendations for field design, sample sizes, estimating population sizes, and other statistical aspects of monitoring protocols.

Since 1994, when the manual was published, investigators have tested the recommended methods and protocols and offered improvements. Not everyone, however, has access to the updates or to the original manual. Even when the manual and more recent publications are available, not everyone can use the standard methods, because decisions about how to monitor often depend on available equipment, level of funding, how many field personnel are available, the natural history and habitat of the target species, and the physical terrain of the study sites. For these reasons, monitoring studies are still carried out using widely divergent field techniques and protocols, and it is still challenging to compare and interpret results across certain regions or taxa.

Logistics and Resources

Because population declines are happening worldwide, ideally we should monitor amphibians in all habitats on all continents. Carrying out fieldwork is hazardous to one's health—and life—in some parts of the world, however. In Afghanistan, Pakistan, and Iraq, an investigator currently encounters not only mine fields but also armed locals. In various African countries, civil war renders fieldwork virtually impossible. It is equally dangerous in parts of Colombia because of the operations of guerrilla and paramilitary groups. Therefore, it is not surprising that most monitoring is being done in certain areas of the world, such as North America, Central and South America (minus Colombia), Australia, parts of Africa, and Europe.

Researchers working where fieldwork can be carried out safely have their own logistical challenges. Investigators in developing countries often have limited access to field and laboratory equipment and supplies, technology, and computer software that many in developed countries take for granted. Transportation can be difficult because gasoline is unaffordable or vehicles in working condition are scarce or nonexistent. Availability of field personnel and time off from teaching or other duties to monitor amphibian populations also can present logistical problems.

Can We Identify Distinguishing Characteristics of Declining Species?

Within any given area experiencing declines or disappearances, only certain species are strongly affected. Why are some species disappearing, while others remain seemingly unaffected? The answer to this question is important, because, if we could identify characteristics of declining species, we might be able to use those characteristics to identify, and rescue, other vulnerable species. Sorting through what we know about several potentially relevant traits can help to tease out patterns.[5]

Geographic Distribution

If you're an amphibian, it is risky living in the mountains these days. Amphibian populations from tropical montane areas are declining at a significantly higher rate than expected. Most declines in the neotropics have occurred at high elevations: above 500 m in Central America and above 1000 m in the Andes of South America.[6] Often, populations of a neotropical species at high elevations are declining, but populations of the same species at lower elevations are not. Likewise, in eastern Australia, populations of a given species at high elevations seem to be more at risk of decline than populations at lower elevations. Something about high elevations seems to increase the risk of decline. For this reason, species restricted to high elevations are often particularly vulnerable.

Other geographic factors that influence the likelihood of a species' declining or going extinct are the number of populations in a particular region and the size of the area occupied by the species (its geographical range). Widespread species have numerous populations across a large area; if a population declines or goes extinct, many sources exist for immigrants to disperse and "rescue" the declining population. Species with narrow distributions have fewer populations, so if one population declines or goes extinct, there is less probability it will be "rescued" by immigrants from other populations.[7] The golden toad (*Bufo periglenes*) from Monteverde, Costa Rica, is an extreme example. Apparently a single population existed. When it disappeared, "rescue" was impossible. A comparison of species from four sites (including Monteverde) in Central America revealed that endemic species tend to decline, whereas widespread species tend to persist.[8] On a larger scale, Natalie Cooper, a masters student from Imperial College London, and her colleagues used a data set of 527 frog species compiled from the literature to test whether extinction risk could be predicted from geographic range size.[9] Not surprisingly, they found that the most threatened species tended to have small geographical ranges.

A population's location within the species' geographic range may affect the likelihood that it will decline or disappear. Populations at the extreme of the species' distributional range may literally be "living on the edge," under climatic conditions that are just barely within the species' physiological tolerances. These populations are likely to be especially vulnerable to even slight climate changes.[10]

Habitat

Are there any correlations between amphibian populations that have declined and the habitats where the animals live? Several assemblage-level studies allow us to examine patterns on a local level.

As discussed in chapter 1, Alan Pounds and his colleagues focused on amphibians that simultaneously disappeared from Monteverde, Costa Rica, beginning in 1987.[11] Their study began with populations of 50 species of frogs and toads in this cloud forest region (Figure 3.1). By 1990, 25 species

Figure 3.1 Cloud forest habitat, Monteverde, Costa Rica. Photograph by Martha Crump.

had disappeared. (During 1990–1994, five of these species recolonized the area.) Pounds' group found an association between the 25 missing species and habitat type. A total of 50 species of anurans was originally present: 18 were associated with streams; another 18 were associated with ponds, pools, or swamps; and 14 were completely independent of aquatic habitats. Of the 25 species that had disappeared, 11 were associated with streams (61 percent of that group). Another 11 were associated with still water (61 percent of that group.) Only 3 of the 14 completely terrestrial species (21 percent) had disappeared. Although these sample sizes are small, the risk of at least temporary local extinction was greater for species that depended on water.

Karen Lips, then a graduate student at the University of Miami, reported that numerous amphibian populations from the cloud forest of Las Tablas, Costa Rica, had drastically declined or disappeared from 1991 to

1996.[12] The populations most affected were of species that lay their eggs in streams and have aquatic larval development. Terrestrial species, such as *Eleutherodactylus* and some *Bolitoglossa* salamanders, that are independent of aquatic habitats were seemingly unaffected. Lips and her colleagues also surveyed amphibians from Fortuna, an upland site in Panama.[13] That study likewise revealed that species associated with riparian habitats had declined more than terrestrial species.

Population declines from another part of the world, eastern Australia, were also found to be associated with stream habitats. Authors of several studies pointed out that species that had declined or disappeared from the montane rainforests and wet tropics in Queensland were those that lived near or bred in running water.[14]

Simon Stuart from the IUCN Species Survival Commission and Conservation International, and his colleagues used the 2004 Global Amphibian Assessment data to examine patterns of declining amphibians on a global scale with regard to habitat.[15] They categorized 2650 species worldwide (46.1 percent of all known amphibian species at the time) as being associated with flowing water. Of the 435 species identified as "rapidly declining," 62.7 percent are associated with running water—a significant difference from the 46 percent expected by chance. The figures are even more dramatic for the subgroup of species for which no clear cause of decline has been identified: Among these 207 so-called "enigmatic-decline" species, 79.2 percent are associated with running water. Fewer species associated with still water bodies have declined compared to the frequency expected by chance alone. Although 2030 species (35.3 percent of all amphibian species known at the time) are associated with still water, only 24.2 percent of the "rapidly declining" species and 13.5 percent of the "enigmatic-decline" species are in this category. Therefore, on a global scale, amphibians associated with running water seem to be declining significantly more than expected (Figure 3.2).

Mode of Reproduction and Other Life History Characteristics

Amphibians display an amazing variety of modes of reproduction—a combination of their specific type of embryonic and larval development, site of egg deposition, egg and clutch characteristics, and type of parental care given, if any.[16] Might we detect a difference in vulnerability of species based on these reproductive characteristics?

Ronald Heyer at the National Museum of Natural History, Smithsonian Institution, and his colleagues speculated that severe frost was the most likely cause of local declines and extinctions of frogs at Boracéia in southeastern Brazil.[17] Although the sample sizes were small, it appears that those species exhibiting more terrestrial modes of reproduction were more

Figure 3.2 Río Lagarto, Costa Rica. Stream habitat of the population of harlequin frogs, *Atelopus varius,* that disappeared in 1988. Photograph by Martha Crump.

vulnerable to the climatic extremes. Of the five species that disappeared from the site, four have terrestrial modes of reproduction—either they are completely independent of water, or they lay terrestrial eggs that hatch into tadpoles that wriggle to water. The fifth species lays aquatic eggs, but the larvae have unusually long developmental times. In addition, populations of certain other species at Boracéia declined but did not disappear. Species in this latter group exhibit one of two modes of reproduction: in some, the female lays her eggs on leaves above water, and the tadpoles drop into the water below; in the others, the female lays terrestrial eggs that undergo direct development. In contrast to these local declines and extinctions, most of the species that lay their eggs in ponds and have aquatic tadpoles were still common during the census periods.

Peter Weygoldt, at the Institut für Biologie I (Zoologie) der Albert-Ludwigs-Universität in Freiburg, Germany, reported a similar association with mode of reproduction from a second site in southeastern Brazil. Weygoldt reported declines of frogs that breed in or near mountain streams around Santa Teresa.[18] Populations of 8 of 13 species declined or disappeared, most likely because of extremely dry winter weather. Four of the eight declining species lay few, large eggs and have at least a terrestrial egg stage. Weygoldt speculated that the other four declining species were particularly susceptible to drought because their aquatic larvae undergo long developmental periods. As Heyer found at Boracéia, the species whose populations did not decline were tree frogs with aquatic eggs and larvae. Again, although the sample sizes were small, the data suggest a pattern: species with terrestrial modes of reproduction seem to have been more vulnerable to unusually severe climatic conditions.

These two Brazilian studies reveal the opposite pattern to that found at certain sites in Central America and Australia (see "Habitat"). At the Costa Rica, Panama, and Australia sites, declining species were generally those that lay eggs in streams and have aquatic larvae. These contrasting observations point out the difficulty of trying to identify correlates of vulnerability based on selected assemblages and small sample sizes.

Do patterns exist on a larger scale? One of us (MLC) analyzed reproductive characteristics of declining anurans from the New World tropics.[19] Development pattern (larval or direct), clutch size, site of egg deposition, and site of larval development were included in the analysis. The numbers of declines associated with species representing each of three major modes of reproduction were very close to the expected values based on the overall rate of decline and the number of species represented in the New World tropics for each reproductive mode. Among mode 1 species—those with aquatic eggs and aquatic larvae (see Figure 3.3A), populations of 52 of 1029 species have declined. Among mode 2 species—those with nonaquatic eggs and aquatic larvae (see Figure 3.3B), populations of 21 of 413 species have declined. Among mode 3 species—those with neither eggs nor larvae aquatic (see Figure 3.3C), populations of 30 of 571 species have declined. The percentage declines for the three reproductive modes were 5.05, 5.08, and 5.25 percent, respectively. Based on this analysis and the opposite patterns reported in the studies previously discussed, we suggest that, between declining and nondeclining species, there is no obvious difference on a global scale in vulnerability associated with mode of reproduction.

Likewise, patterns do not exist for declines and certain other life history traits. For example, within the group of Australian rain forest frogs, evidence suggests that low clutch size (≤200 eggs) is associated with declining species.[20] Elsewhere in the world, however, ranids with large clutch sizes (many thousands of eggs) have declined. Furthermore, in

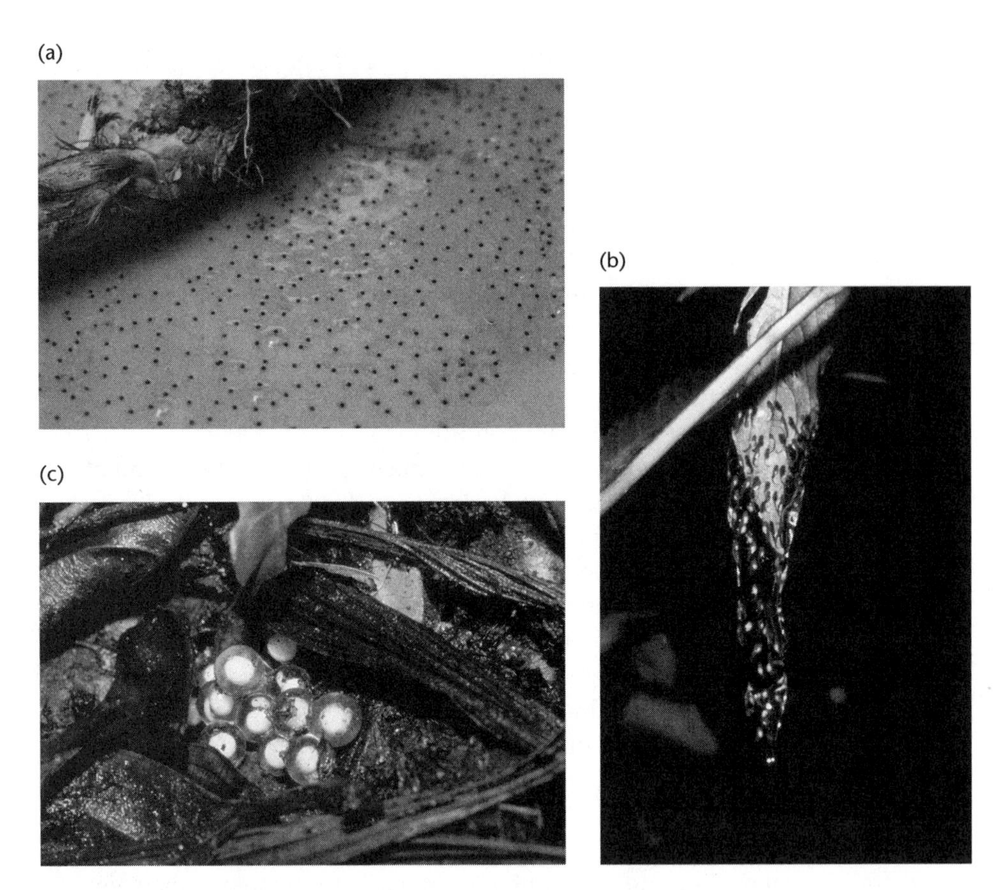

Figure 3.3 Egg clutches of each of the three main modes of reproduction. (a) *Dermatanotus* (microhylid), representing mode 1—aquatic eggs hatch into aquatic larvae. (b) *Hyla bokermanni* (hylid), representing mode 2—eggs are deposited out of water, larvae are aquatic. (c) *Eleutherodactylus* (leptodactylid), representing mode 3—neither eggs nor larvae are aquatic. Photographs by Martha Crump.

their comparison of endemic Australian frogs, Brad Murray and Grant Hose, both at the University of Technology Sydney, New South Wales, found that neither clutch size nor egg size was significantly related to decline.[21] Worldwide, populations of amphibians that produce large eggs as well as those that lay small eggs have declined. There is no obvious pattern between declining species and life span: populations of both short-lived and long-lived species have declined. Part of the problem with trying to tease out life history correlates is that certain traits tend to be associated with each other.

Taxonomy

Are certain families or genera of amphibians experiencing more population declines than expected based on the number of species in those taxa? Consider the 2006 IUCN Red List of Threatened Species for frogs and toads. The genus with the highest number of species listed as critically endangered (CR) is *Eleutherodactylus* (Brachycephalidae), with 76 species listed from a total of 488 (16 percent). Next is *Atelopus* (Bufonidae), with 62 species listed as CR and 3 extinct, out of a total of 78 (83 percent). Third is *Philautus* (Rhacophoridae), with 16 species listed as CR and 18 extinct from a total of 141 species (24 percent). Tied for fourth are *Craugastor* (Brachycephalidae), with 25 listed as CR and 2 extinct (23 percent of 115 species), and *Plectrohyla* (Hylidae), with 27 species listed as CR (66 percent of 41 species). In terms of percent decline, *Atelopus* and *Plectrohyla* are the most affected of the genera, experiencing numerous declines.

Table 3.1 summarizes the 2006 Red List data for anuran families experiencing population declines. Of the large families (containing >100 species), Brachycephalidae, Bufonidae, and Rhacophoridae are most affected. Clearly, certain taxa are in fact experiencing a higher proportion of declines than others. Remember, though, that our understanding of population declines is biased, because some species, genera, and families have been surveyed and monitored more extensively than others.

Working with the data available, Sarah Corey and Thomas Waite, both at Ohio State University, used a method called "phylogenetic autocorrelation analysis" to answer the question: Do related species share an intrinsic vulnerability to common kinds of threats? In other words, are closely related groups of species especially prone to extinction due to shared evolutionary history? They found that the superfamily Hyloidea (which includes the families Bufonidae, Brachycephalidae, and Hylidae among others) is most affected by autocorrelated threat. Furthermore, the phylogenetic clumping of risk is found most strongly in Central and South America and in Australia.[22]

Of the salamander families with 30 or more species, Ambystomatidae is the most affected, with one quarter of its species listed as critically endangered (Table 3.2). The two extinct salamanders, *Cynops wolterstorffi* and *Plethodon ainsworthi*, belong to Salamandridae and Plethodontidae, respectively. These large families have not experienced declines as severe as that of the Ambystomatidae.

Other Possible Distinguishing Characteristics

One might predict that species with thinner, more permeable skin would be more vulnerable to environmental changes. If so, their populations

TABLE 3.1 Percentage of Species Listed as Critically Endangered (CR) and Extinct from Each Anuran Family with Reported Declines

Family	No. Species CR	No. Species Extinct	No. Species CR + Extinct	No. Species in Family	% of Family
Alytidae	—	1	1	11	9.1
Amphignathodontidae	2	—	2	57	3.5
Aromobatidae	5	—	5	88	5.7
Athroleptidae	5	—	5	128	3.9
Batrachophrynidae	1	—	1	6	16.7
Brachycephalidae	115	2	117	793	14.8
Bufonidae	84	5	89	485	18.4
Centrolenidae	6	—	6	140	4.3
Ceratobatrachidae	1	—	1	82	1.2
Ceratophryidae	9	—	9	81	11.1
Cryptobatrachidae	1	—	1	21	4.8
Cycloramphidae	6	—	6	96	6.3
Dendrobatidae	15	—	15	161	9.3
Dicroglossidae	2	1	3	151	2.0
Heleophrynidae	2	—	2	6	33.3
Hylidae	66	1	67	814	8.2
Hyperoliidae	1	—	1	198	0.5
Leiopelmatidae	1	—	1	6	16.7
Leptodactylidae	4	—	4	91	4.4
Limnodynastidae	1	—	1	44	2.3
Mantellidae	7	—	7	162	4.3
Megophryidae	3	—	3	134	2.2
Micrixalidae	1	—	1	11	9.1
Microhylidae	6	—	6	413	1.5
Myobatrachidae	6	3	9	80	11.3
Petropedetidae	4	—	4	26	15.4
Pipidae	1	—	1	31	3.2
Pyxicephalidae	2	—	2	59	3.4
Ranidae	12	1	13	316	4.1
Rhacophoridae	18	18	36	272	13.2
TOTAL	387	32	419		

Data are from the 2006 IUCN Red List of Threatened Species.

might have declined more than the populations of species with thicker, less permeable skin. No obvious association exists, however. Populations of thin-skinned glass frogs (Centrolenidae) have declined, but so have many populations of thick-skinned toads (Bufonidae).

Body size does not predict vulnerability to decline. Populations of large, medium, and small species have declined. *Andrias*, the world's largest salamander at 1.5 m, has declined. But so have eight species of *Thorius*, the world's smallest salamanders, some measuring less than 40 mm total

TABLE 3.2 Percentage of Species Listed as Critically Endangered (CR) and Extinct from Each Salamander Family with Reported Declines

Family	No. Species CR	No. Species Extinct	No. Species CR + Extinct	No. Species in Family	% of Family
Ambystomatidae	9	—	9	35	25.7
Cryptobranchidae	1	—	1	3	33.3
Hynobiidae	5	—	5	49	10.2
Plethodontidae	36	1	37	375	9.9
Salamandridae	3	1	4	74	5.4
TOTAL	54	2	56		

Data are from the 2006 IUCN Red List of Threatened Species.

length. The world's largest frog, *Conraua goliath*, with a body length of 30 cm, is listed as critically endangered, but so are many *Eleutherodactylus*, poison dart frogs, and other anurans less than 30 mm long. Comparisons between populations of declining and nondeclining species from Australia have likewise revealed no difference in body size between the two groups.[23]

Populations of both diurnal and nocturnal species have declined. Analyses of declining populations of frogs from Costa Rica, Panama, and eastern Australia failed to reveal any difference in temporal period of activity between declining and nondeclining species.[24]

Most amphibians are opportunistic, generalized feeders: they eat just about any live prey they can fit into their mouths. Some species, however, are specialized, such as dendrobatids that eat primarily ants. Are specialized feeders more likely to exhibit declines? There have been no reports of association between declining species and breadth of diet or specific prey items eaten. Populations of both specialized and generalized feeders have declined.

Reviewing this list of characteristics that do not seem to be associated with declines, it might seem that the chance that a population will decline or even that an entire species will go extinct is random—unless it is a high-elevation, stream-breeding species with a small geographic range. In reality, amphibian species probably vary in their tolerance and response to environmental change. Declining species might be less able to tolerate habitat degradation, climate change, introduced pathogens, and other factors.[25]

Hypotheses to Explain the Causes for Amphibian Declines

There is no "smoking gun," no single cause to explain the worldwide amphibian population declines that have happened at an increasing rate over the past few decades. Researchers have hypothesized six main causes for observed declines. We will briefly introduce them here to set the stage for more detailed discussion in chapters 4 and 5. One of us (JPC) and Andrew Storfer, at Washington State University, sorted these six hypotheses into two main classes.[26]

Class I hypotheses are factors that have affected amphibian populations for at least the last century: (*a*) introduced species, (*b*) commercial over-exploitation, and (*c*) land use change. We have a fairly good understanding of how each of these factors directly affects some amphibians.

Class II hypotheses are factors that we think have negatively affected amphibian populations more recently: (*d*) environmental contaminants, (*e*) global climate change, and (*f*) emerging infectious diseases. We are only beginning to understand how Class II factors might cause amphibian population declines.

Introduced Species

For centuries, humans have directly and indirectly introduced animals into places where the species do not normally occur. These exotics have reduced or exterminated local populations of native amphibians.[27] Examples of such exotic species include bullfrogs, trout, and crayfish. These and other exotics can negatively affect native species in major ways. The introduced animals might be predators on one or more amphibian egg, larval, or post-metamorphic life history stages. Nonindigenous species might compete with larval or adult native amphibians for food, space, or other resources. Alien species could transfer pathogens or parasites to native species, or they could directly alter habitats. Finally, introduced amphibians sometimes hybridize with native amphibians, destroying genetic identity.

Over-exploitation

Overcollecting of some amphibian populations has caused declines.[28] One example is the extreme decline of many populations of leopard frogs, which are collected by biological supply houses for use in research and teaching. If collecting is to be done, it must be by sustainable harvesting—removing individuals from a population in such a way that the population will continue indefinitely. Unfortunately, we understand so little about the population biology of most amphibians that it is impossible to determine sustainable harvest levels. Even if a population is not extirpated, there may be too few animals left for them to find each other to breed. And even if some individuals breed, the depleted population almost surely will experience decreased genetic diversity.

Land Use Change

Land use change, including habitat modification, fragmentation, and destruction, is a major cause of reduced global biodiversity, including that of amphibians.[29] Such changes often modify habitats so they are no longer suitable for amphibians in terms of moisture and temperature requirements, food availability, refuges, or critical physical, chemical, and biotic characteristics. Changes can reduce the possibility of recolonization from other populations. Changes in land use can also kill amphibians directly, as the disturbances take place. Fragmentation can prevent access of animals to breeding sites, reduce recruitment, and lead to reduced genetic diversity within populations.

Environmental Contaminants

Another type of global change is the ever-increasing contamination of the environment.[30] Various pollutants, from elements such as mercury to human-produced compounds such as fertilizers and pesticides, harm amphibians. Some, such as acid rain, kill amphibian embryos and larvae. Contaminants also can have sublethal effects. They can stunt growth and development, cause anatomical deformities and behavioral abnormalities, or affect reproduction by disrupting hormone systems or causing gonadal abnormalities.

Global Climate Change

Some investigators have hypothesized that climate change is a cause of certain amphibian population declines.[31] Climate change might affect amphibians in numerous ways, both directly and indirectly. Direct effects might include temperatures so high that amphibians die outright or drought that animals cannot endure physiologically. Levels of ultraviolet B (UV-B) radiation might be lethal to some amphibians. Indirect effects could include reduced prey abundance and increased density of predators. Changes in the seasonal breeding patterns of species could result in heightened competition for calling and egg-laying sites, and ultimately in diminished recruitment. Competition among larvae could also lead to decreased recruitment. Climate change could disrupt periods of hibernation and estivation and the ability to find food. Changes in temperature might influence distribution and abundance and even lead to population extinctions. Warmer or cooler temperatures could encourage the emergence of new pathogens or increase susceptibility to new or old infectious diseases. Rather than temperature or moisture per se, stress resulting from environmental extremes could depress amphibians' immune systems and make them more susceptible to pathogens.

Emerging Infectious Diseases

Emerging infectious diseases include newly recognized diseases, diseases that have recently appeared in populations, and diseases that are rapidly increasing in incidence, virulence, or geographic range. Two major emerging diseases have been implicated as causes of population fluctuations and declines and even extinctions of amphibians: iridoviruses (genus *Ranavirus*) and a chytrid fungus (*Batrachochytrium dendrobatidis*).[32] These pathogens are often highly virulent, and they can kill their hosts directly.

Amphibian populations and species respond in diverse ways to introduced species, over-exploitation, land use change, environmental contaminants, climate change, and emerging infectious diseases. Although this variation adds complexity to the picture, it gives us the opportunity to make comparisons and carry out experiments. Some of these studies will be highlighted in later chapters.

Returning full circle to the first topic of this chapter, perhaps the greatest challenge of all is reversing amphibian declines given the suspected causes. As Tim Halliday at the Open University in the United Kingdom stated, the answer is not as simple as putting up a fence and declaring an area "protected."[33] Human squatters and poachers do not always respect boundaries or fences. Introduced species, climate change, environmental contamination, and infectious disease never do.[34] In the next two chapters, we will discuss in greater detail the six main threats to amphibians, as well as the complications that can occur when some of these factors act synergistically.

4

Introduced Species, Commerce, and Land Use Change

Humans influence Earth's environment in so many ways that it would be surprising if we had no role in amphibian population changes. It is easy to assume that the influence is negative, and there is evidence for that, but the story is more complex than just saying people are bad for frogs, salamanders, and caecilians. After all, while the ranges and population sizes of some amphibian species are decreasing, others are increasing. A change in land use that is unfavorable to one taxon may favor another and allow it to increase. Exotic species also can increase in population size and distribution in new habitats. Humans have a role in causing both kinds of change, and if we want a complete explanation for why there is a global concern about amphibians, we have to understand why some species are increasing even as others decrease.

Environmental changes cascade through ecosystems like a chain reaction with complex, nonlinear interactions. We are changing our environment in major ways and quickly—more quickly than the time scales on which many organisms, except those with short generation times, can respond evolutionarily. A slow evolutionary response time increases the chances of extinction, because organisms cannot adapt to the new conditions. We change habitats both locally and regionally, and, at least since the Industrial Revolution, we have changed our environment at a global scale.

Amphibians are not the only group of organisms experiencing the effects of humans, but amphibians are a rich model for studying environmental changes and how they affect wildlife populations. We must also realize that, even without human intervention, ecosystems and their species change all the time. It is against this background level of expected variation that we must assess recent alterations in amphibian populations and decide whether forces are at work that are accelerating change in ways that threaten the survival of amphibians.

In this chapter, we explore three forces—introduced species, commerce, and land use change. For millennia, these forces have altered the fates of wildlife populations. We will use examples from the last 150 years to illustrate how impacts of each on amphibian populations set the stage for the late 20th century global amphibian decline.

Introduced Species

Individual organisms move all the time, and the ranges of species also vary in response to such factors as climate change or the rise of mountain ranges that close or open access to new habitats. We will not focus here on species' movements resulting from these kinds of events, but rather the movement of animals by humans. Such movements interest ecologists and evolutionary biologists because they can initiate rapid, extensive changes in populations and even whole communities of organisms. Economists and policy makers are interested because of the sometimes negative effects of new weeds on agricultural crops, new pathogens on humans or domestic animals, and new species on natural environments (e.g., making communities more prone to fire).

A species is defined as native or indigenous to a region if humans had no role in how the organisms occupied their present range. Humans move species by accident, to improve wild stocks, and for purposes of aquaculture, sport, aesthetics, pets, or biological control. The words introduced, exotic, alien, transplanted, and nonindigenous all describe such species. The U.S. government recognizes "invasive species" as those that are "non-native (or alien) to the ecosystem under consideration and whose introduction causes or is likely to cause economic or environmental harm, or harm to human health."[1]

Some exotic species are moved great distances from their original range. For example, American bullfrogs (*Rana catesbeiana*) are indigenous to eastern Canada and the eastern United States, but humans have also introduced bullfrogs into the Caribbean, Europe, Mexico, South America, and Asia. Recent research suggests that an ancestor of the genus *Ptychadena* reached Madagascar from Africa probably by overseas rafting, where it evolved into several species; humans then brought the Mascarene ridged frog, *Ptychadena mascareniensis*, across hundreds to thousands of kilometers of the Indian Ocean to the Mascarene (Reunion and Mauritius) and Seychelles Islands.[2]

Equally important are species that are moved shorter distances but still outside their native range. People have introduced American bullfrogs into many parts of the western part of the country, far outside their native range in the east. Some species are moved outside their native community type. Trout species native to streams in the western United States are now exotic inhabitants of previously fishless ponds, lakes, or streams in the same regions or far away, even other countries. These movements are interesting because of the complex, often negative effects that the novel species have on native populations and communities. Amphibians illustrate such negative consequences.

New Arrivals in Foreign Lands

In 1990, Walter Meshaka was a graduate student at Florida International University when he became interested in Cuban tree frogs (*Osteopilus septentrionalis*) in Everglades National Park because the species should not be there.[3] Endemic to Cuba, Isle of Pines, Cayman Islands, and the Bahamas, the Cuban tree frog is an exotic species in Puerto Rico, several of the Lesser Antillean islands, Hawaii, and mainland Florida (Figure 4.1). People probably introduced the species in the lower Florida Keys in the 1800s, but it may have reached there without human aid. It is clear, however, that these tree frogs dispersed up the Florida Keys following Highway 1 construction,

Figure 4.1 Cuban tree frog (*Osteopilus septentrionalis*). Photograph by Dante Fenolio.

were in Dade County by the mid-1940s, and then entered the Everglades. South Florida is now home to this species, with isolated populations in the northern part of the state. Their spread illustrates how a species successfully colonizes a new region, how humans help this process, and some of the often negative effects that exotic amphibian species can have on natives.

Cuban tree frogs successfully occupied Florida largely because they have what biologists recognize as good colonizing traits: high fecundity, short generation time, tolerance for diverse physical conditions, similarity of new and old habitats, coexistence with humans, diverse diet, good competitive ability, and few predators. In addition, Cuban tree frogs are successful colonizers because they readily live on and around humans' buildings and disperse in natural and manmade cavities of trees used for landscaping.

Reaching a new locality is the first, and often the most difficult, step in colonization. We move amphibians accidentally, deliberately, or by a combination of both. Animals released deliberately and those that escape may establish feral populations. Eggs, juveniles, or adults moved with vegetation or in packaging may escape, which is how the greenhouse frog (*Eleutherodactylus planirostris*) probably entered Florida around 1875. Australians deliberately released marine or cane toads (*Bufo marinus*) into Queensland in 1935 as a biological control agent to eat beetles feeding on sugar cane. In reality, the toads rarely ate the beetles. The toads dispersed throughout Queensland, into New South Wales and are now observed as far away as the Northern Territory. American bullfrogs are moved as harvestable game animals, as pets, or for commercial frog farms, where they are grown for food.

But are these cases exceptional among amphibians? How often are frogs and salamanders moved outside their native range? Salamanders are not indigenous to Australia, New Guinea, or the West Indies, and these regions have no introduced salamander species (Table 4.1).[4] Salamanders have been introduced and have become established outside their native range within the United States, but no foreign species is established in the United States or Europe. Between 1 and 8 percent of the frog fauna for several large regions of the world are exotic species. By comparison, about 3 percent (23 of 810 species) of Nearctic birds are exotic species, and about 20 percent of U.S. freshwater fishes are in this category. No salamanders exceed their native ranges in Canada, but four U.S. species have expanded ranges as a result of movement by humans. One native Australian frog species, about 20 percent of Canadian frog species, and about 4 percent of continental U.S. species exceed their native ranges within each country also as a result of movement by people (Table 4.2).[5]

We now have three reference points for understanding how exotic species relate to amphibian declines. First, whereas some amphibian species

TABLE 4.1 Number of Nonindigenous Amphibian Species (Exotics) in Selected Regions

Region	Total Species	Frog Species		Salamander Species	
		Total	Introduced (% of total)	Total	Introduced
Australia	213	211	1 (<1%)	0	0
New Guinea	250	250	5 (2%)	0	0
Europe	77	44	2 (5%)	29	0
West Indies	174	174	8 (5%)	0	0
Nearctic	241	90	5 (6%)	151	0

Source: Duellman, W. E., ed. 1999. *Patterns of Distribution of Amphibians: A Global Perspective.* Baltimore: Johns Hopkins University Press.

TABLE 4.2 Number of Amphibian Species Native to a Country or Region That Have Been Moved Beyond Their Native Ranges

Region	Total Species	Frog Species		Salamander Species	
		Total	Range Changes	Total	Range Changes
Australia	211	211	1 (<1%)	0	0
Canada	45	24	5 (21%)	21	0
Continental USA	~217	~90	4 (4%)	127	4 (3%)
Europe	73	44	~4 (11%)	29	3 (10%)

Source: Duellman, *Patterns of Distribution of Amphibians.*

declined over the 20th century, a subset of species, mostly frogs, expanded their ranges, and some are still increasing. The marine toad, for example, is moving as many as 55 km/year into northern Australia.[6] For several large regions of the world, the fraction of the frog fauna composed of exotic species from other regions nearly equals the fraction of species that have been introduced outside their native range but within the region. Human actions have increased the ranges of about 24 amphibian species over the last century for several large regions with comprehensive records. Second, new areas occupied by introduced species are usually small (tens to hundreds of square kilometers). American bullfrogs and marine toads are exceptions to this generalization, occupying tens of thousands of square kilometers in several countries. Third, the amphibian pattern is not new, nor is it exceptional: Humans move many species, and amphibians have been moved outside their native ranges since at least the last half of the 19th century.

At a time when there is a great deal of worry about declining amphibian populations, it might be tempting to conclude that the increase in range of some species is good news. That would be a mistake.

Many studies have documented the negative effects of exotic species: They alter habitats, introduce parasites, change food webs, modify unique gene pools by interbreeding, and compete with and prey on native species. Is there evidence that non-native amphibians affect native amphibians in any of these ways? Do alien species other than amphibians affect native amphibians? Are nonindigenous species contributing to amphibian declines?

Impact of Nonindigenous Amphibian Species

Hawaii had no native amphibians before European contact. According to the U.S. Geological Survey's Nonindigeneous Aquatic Species database, the islands now have 16 exotic frog species,[7] with the earliest recorded introduction in 1857.[8] The arrival routes are varied. American bullfrogs were introduced there intentionally as a source of food, while the green and black dart-poison frog (*Dendrobates auratus*), wrinkled frog (*Rana rugosa*), and marine toad were introduced for biocontrol of insects. Cuban tree frogs were brought to Hawaii as pets, and escaped animals colonized native habitats. The greenhouse frog and the coquí frog (*Eleutherodactylus coquí*) were apparently unintended introductions arriving as stowaways in potted plants (Figure 4.2).[9]

Coquí frogs originally lived in commercial and residential areas. They now live in surrounding forests as well, raising questions about what effects

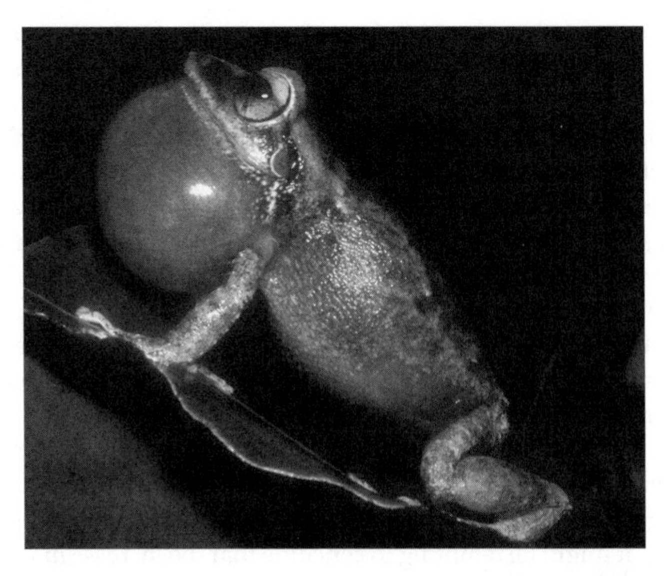

Figure 4.2 Coquí frog (*Eleutherodactylus coquí*). Photograph by Dante Fenolio.

the frogs might have on Hawaiian ecosystems. Karen Beard and her colleagues at Utah State University conducted experiments within the native range of coquíes, in the Luquillo Experimental Forest, Puerto Rico, that allow us to make some guesses about Hawaii.[10] They studied the effects of coquíes on aerial and litter invertebrates, plant growth, herbivory, and litter decomposition. Frogs reduced aerial invertebrates and leaf herbivory, but not litter invertebrates. Foliage production rates, measured as the number of new leaves and new leaf area produced, increased by 80 percent, and decomposition rates increased by 20 percent. The results suggest that *E. coqui* may affect ecosystem functions in Hawaii by decreasing some invertebrate species and increasing nutrient cycling rates. There is also a concern that native bird populations will decrease further, because coquíes prey on populations of insects eaten by native birds.[11] Tests are needed in Hawaiian ecosystems to confirm these suspicions.

The coquí frog is a striking study in contrasts. In Puerto Rico the name "coquí" is applied to 16 species of *Eleutherodactylus* endemic to the island. A Web site maintained by Proyecto Coquí describes the group as the "remarkable COQUI," then adds, "The coquíes are the ones [frogs] that add that special feeling to nighttime in Puerto Rico."[12] Hawaiians feel very differently about the new arrival. Hawaii had few calling insects and no calling frogs before Europeans arrived. An article on "Hawaii's hated frogs" noted that the calling frogs, *E. coqui,* might reach the "volume of moderate-to-heavy street traffic or the din in neighborhoods along aircraft takeoff and landing corridors."[13]

In Hawaii, where the introduced *E. coqui* has few predators, the frogs can reach five times the densities found in Puerto Rico. Unaccustomed to such nighttime noise and concerned that the frogs might overwhelm native ecosystems where species evolved without predaceous amphibians, the State of Hawaii, the U.S. Department of Agriculture, and the U.S. Fish and Wildlife Service have established programs to eradicate the coquí and greenhouse frogs.

Puerto Ricans value or appreciate the coquí frog very differently than do Hawaiians. Values, often expressed in economic or aesthetic terms, are part of the explanation for why humans move species, how new species are received, and why some amphibian species are declining while others are increasing. Meshaka's Cuban tree frog study can teach us some things relative to this last point.

Cuban tree frogs eat six frog species in the Everglades: green tree frog (*Hyla cinerea*), squirrel tree frog (*H. squirella*), leopard frog (*Rana sphenocephala*), southern toad (*Bufo terrestris*), and eastern narrow-mouth toad (*Gastrophryne carolinensis*); Cuban tree frogs are cannibalistic, so the sixth species is itself. Except for Cuban tree frogs, the other five species are native to Florida. Meshaka's studies revealed a common correlation: After Cuban tree frogs colonized an area in the Everglades, the two native tree frog species declined, and in some cases populations went extinct locally, apparently as a result of predation and not competition.[14]

Exotic species, especially predators, often negatively affect native species because the natives lack avoidance behaviors and become easy victims. Conservation biologists are starting to recognize this in general as ignorance of danger (naïveté) resulting from "first contact" with a novel predator. Meshaka reported just such naïve behavior by the two native *Hyla* in relation to Cuban tree frogs: The latter predictably ate the smaller, native tree frogs. Few species prey on Cuban tree frogs. It is the largest North American hylid, so its size is an advantage, both as a predator and for avoiding predators.

We now have two kinds of evidence that illustrate how exotic species negatively affect native frogs and salamanders. The first kind is based on correlative data like Meshaka's; namely, researchers observe that native species are rare or absent in habitats where an exotic species is common, in comparison to habitats without exotics. The conclusion that introduced species affect natives negatively is based on a correlation; no experiment is performed. Lee Kats and Ryan Ferrer at Pepperdine University reviewed the evidence for declines in native amphibian species after the appearance of exotic predators.[15] They found 10 examples based on correlations. Introductions of American bullfrogs correlated with declines in four U.S. frog species; introduction of fish such as trout and salmon correlated with declines in two Australian frog species, two U.S. frog species, and one U.S. salamander species; and declines in another U.S. salamander species followed introductions of crayfish and the guppy-like mosquitofish.

The second kind of evidence reviewed by Kats and Ferrer was from experiments that revealed the mechanisms by which exotic species cause declines of native populations. Experiments in the United States, Australia, and Europe demonstrated that when exotic, predaceous fish and crayfish were present, native frogs and salamanders reduced their activity, used different habitats, increased use of refuges, were shorter, weighed less and were smaller at metamorphosis, survived less well as larvae or metamorphosed animals, and showed more injuries. Exotic species also competed with natives for shared resources.

Observations and experiments identify the mechanisms that negatively affect natives and cause populations to decrease. They support the inference that exotic species cause native populations to decline, but another kind of experiment is needed to be certain. Imagine finding an area without Cuban tree frogs and subdividing it into several sections. One half of the sections are chosen randomly, and Cuban tree frogs are introduced into these "treated" sections. The other half of the sections have only native frogs; these "control areas" are a baseline that tell us how native populations are changing without the challenge of an exotic species. The number of adult frogs is monitored in treated and control sections. If exotic species negatively affect natives, we would predict that, on average, there will be fewer native frogs in the treated areas than in the controls. Ideally, behavioral observations would complement the experiment. Cuban tree frogs, for example, prey on native frogs. Observing such behavior during the experiment would provide a mechanism for explaining why native frogs decline.

Although this kind of experiment is desirable, for ethical reasons we do not want to introduce an exotic species into a new area. An appropriate experiment would have a similar control, but with the treatment being removal of an exotic species from one-half of the sections. We would then predict that, over time, the average density of native species would increase to the level of the average density of adults in control sections.

Only a few studies have combined baseline population data, experiments, and an understanding of how intruders negatively affect native species. It is also true that, regardless of the species introduced, exotic and native species may coexist, at least for a time, in part of the original range. Under these conditions, local biodiversity actually increases, but in many instances introduced species replace natives, leading to what ecologists call biotic homogenization. Frank Rahel at the University of Wyoming described this process as "the increased similarity of biotas over time caused by the replacement of native species with nonindigenous species, usually as a result of introductions by humans. Homogenization is the outcome of three interacting processes: introductions of nonnative species, extirpation of native species, and habitat alterations that facilitate these two processes."[16]

Introduced Trout Change a Mountain Ecosystem

Introduced fish are also well known for eliminating amphibians from parts of their native ranges. The decline of two frog species as a result of the introduction of exotic trout into fishless, high-elevation lakes in the U.S. Sierra Nevada is one of the best-documented examples.

Vance Vredenberg (now at San Francisco State University) and colleagues at the University of California in Berkeley recently separated the species, *Rana muscosa* or yellow-legged frog, into two species: Sierra Nevada yellow-legged frog (*Rana sierrae*) in the northern and central Sierra Nevada, and Sierra Madre yellow-legged frog (*R. muscosa*) in the southern Sierra Nevada and southern California. Using museum collections, they reviewed surveys taken since 1995 at 225 historical (1899–1994) localities and determined that 93 percent (146 localities) of *R. sierrae* populations and 95 percent (79 localities) of *R. muscosa* populations are extinct; in particular, *R. sierrae* is now extinct in Nevada.[17]

In the early 1900s, yellow-legged frogs were common and widespread above 1500 m in the Sierra Nevada Mountains of California and Nevada.[18] By 1910, scientists reported that frog populations had declined or disappeared from high-elevation lakes in this region. As we will see, the introduction of exotic predaceous fish affords a straightforward explanation for the historical population declines and, in some cases, local extinctions. But as we also will learn in chapters 5 and 7, recent studies in areas where populations are declining in the absence of fish indicate that multiple factors, including infectious disease and perhaps toxins, are acting alone or together to reduce population sizes and cause local extinctions.

California's Sierra Nevada is a mountain range of breathtaking beauty, with peaks that soar above 4000 m. The core of the range is a product of tectonic forces that date from the Jurassic Age about 150 million years ago, but it was Pleistocene Ice Age glaciers some 2 million years ago that carved the range's surface into a landscape with thousands of lakes and ponds. Historically, the aquatic habitats were fishless because of their high elevation and because of intermittent or absent surface water connections to lakes and streams at lower elevations that had fish. Several frog species (Sierra Nevada yellow-legged frog; Sierra Madre yellow-legged frog; Pacific tree frog, *Pseudacris regilla;* Yosemite toad, *Bufo canorus*) and two salamander species (Sierra newt, *Taricha torosa;* Ensatina salamander, *Ensatina eschscholtzii*) dispersed into these high elevation habitats as the glaciers retreated. In particular, the two species of yellow-legged frogs established thriving populations in the fishless wetlands. For our discussion, we will lump these two species together throughout the book and just refer to them as mountain yellow-legged frogs.

Starting in the late 1800s, these distinctive amphibian habitats were altered when California joined with federal agencies to establish a recreational fishery for non-native trout, primarily rainbow trout (*Oncorhynchus mykiss*) and brook trout (*Salvelinus fontinalis*), in the historically fishless lakes. Between 1870 and 1960, various methods, including the use of airplanes to drop fish into the most remote lakes, moved trout into these aquatic ecosystems that had been without fish, perhaps forever.

Early 20th century naturalists such as Joseph Grinnell at the University of California, Berkeley, suspected that fish caused declines by eating frogs, but no one systematically explored the question until late in the century. In the 1990s, David Bradford from the U.S. Environmental Protection Agency Laboratory in Las Vegas, Nevada, asked: Why do frogs occupy only some Sierra Nevada lakes and wetlands? Additional researchers, including

Roland Knapp from the University of California's Sierra Nevada Aquatic Research Laboratory and Vredenberg, joined the hunt for a cause. The following account summarizes the collective effort of these principal researchers and their colleagues.

Extensive field observations demonstrated a negative relationship between non-native trout and the occurrence of yellow-legged frogs and Pacific tree frogs. For one study in Yosemite National Park, all 2655 non-stream aquatic habitats were surveyed for presence or absence of fish and amphibians. Populations of toads (frogs in the family Bufonidae) and newts (salamanders in the family Salamandridae) seemed to be unaffected by the trout, perhaps because both are distasteful or toxic to many species, but more research is needed to be certain that trout are not influencing these species in more subtle ways.

After removal of trout from some lakes in Kings Canyon National Park, the frogs recolonized and populations grew, suggesting the negative effects of the introduced fish. The experimental removal of fish from lakes confirmed the correlation, and experimental diet studies of trout provided a mechanism: Fish eat yellow-legged frog tadpoles. As a result of this predation, juvenile and adult frog recruitment decreases, and population size declines, sometimes even to local extinction. Recall that this is just the sort of hypothetical experiment we proposed for assessing the impact of Cuban tree frogs on native populations. The impact of trout on amphibians is not unique to the Sierra Nevada; similar outcomes are reported from Costa Rica, Spain, and Australia.[19]

The effects of fish predation on these historically fishless communities extend throughout a lake's food web, because fish eat invertebrates as well as frogs, and to nearby lakes as well. The last point is important. Frog and salamander populations often include more than one lake or pond. Sometimes adults, but often juveniles, move among habitats, forming a regional network or "metapopulation" that links subpopulations in otherwise separate lakes, ponds, or wetlands. If trout eat all of the frogs in one or a few lakes, connections are broken, causing the negative effects in one place to cascade throughout the network, contributing to regional declines.

The evidence that fish caused yellow-legged frog declines affected policy makers, at least in Yosemite National Park: By 1991, fish stocking had stopped. Ironically, just as this policy was adopted, frogs began to disappear even in fishless habitats; in some cases, entire watersheds lost their frog populations. Declines like these in the midst of protected areas, where other human-influenced causes were not responsible, led to the sense of crisis that accompanied the late-1980s reports of amphibian declines. We will see in the next chapter that a pathogen specific to amphibians is moving through these Sierra Nevada frog populations and is a likely new, late

20th century cause of declines. The pathogen is yet another new species that these frog populations have probably never experienced before, one that will be a 21st century challenge to their survival.

Introduced species teach us several lessons. First, exotic species caused amphibian declines that occurred throughout the last century, with a scope that included several dozen species and received heightened attention by researchers and the public in the 1980s and 1990s. Populations were decreasing before the last decades of the 20th century, when global amphibian declines began to draw wide attention. Second, an exotic species affects each native amphibian species differently. Third, novel species may first affect communities within a small area, but their influence can spread widely, as the Cuban tree frogs, bullfrogs, cane toads, and trout illustrate.

Fourth, intentional introduction of exotic fishes in the Sierra Nevada was a policy decision that placed in conflict human values and the needs of native amphibian populations. Policies endorsing the introduction of exotic fish were based primarily on meeting economic and recreational goals, with less attention paid to ecosystem and landscape processes that are required to support amphibian biodiversity. Knapp and colleagues made the important point that their "results pose a difficult challenge for fisheries and wildlife managers interested in better balancing the conflicting goals of maintaining nonnative fisheries in wilderness areas while also minimizing the effects of these fisheries on natural processes."[20] Introduction of fish into high-elevation Sierra Nevada habitats in which they did not occur historically was a policy decision that exploited several species—the fish for commercial and recreational reasons, and the frogs that suffered the consequences.

Finally, populations remain vulnerable even when they are located in protected national parks and wilderness areas. As we have seen, Cuban tree frogs originally introduced into unprotected areas eventually crossed into the "protected" areas of Everglades National Park. In California, a decision to introduce fish into protected reserves disadvantaged frogs in high-elevation ecosystems.

Amphibians and Commerce

Frogs and salamanders are sold in large numbers as pets and for food, research, bait, leather, and teaching.[21] How do these markets affect native populations? And is commerce connected to population declines? Before answering these general questions, it is useful to review three areas in which the most extensive data are available: pets, food, and bait.

Amphibians and the Pet Trade

The pet trade moves millions of animals across the globe annually, and some invariably escape. Cuban tree frogs first entered Hawaii as pets. Animals that escape or are released may establish feral populations that compete with and prey on native species. A number of studies show that introduction of nonindigenous American bullfrogs predictably leads to this outcome. Pathogens are common in pet amphibians, and if these hosts escape, they could transmit diseases to native species.

In general, we should expect the impact of collecting to vary with the degree of exploitation and the resilience of the species. But assessment of this impact is easier said than done, and two cases illustrate why we need more research to improve our understanding of how collecting for the pet trade affects populations.

Gunther's triangle frogs (*Ceratobatrachus guentheri*), from the Solomon Islands, sell for as much as US $85 each.[22] In a 1999 report, Michael Tyler from the University of Adelaide, an authority on amphibians in Australia and the Papua-New Guinea region, observed that "Exploitation of the herpetofauna of the Solomon Islands for commercial trade is resulting in the exportation of thousands of *Ceratobatrachus;* the matter is causing considerable concern."[23] However, a 2004 record from the IUCN's Global Amphibian Assessment (GAA) concluded: "In the past this species was exported in large numbers to Europe and other regions for the pet trade. Despite this it remains widespread and exceptionally abundant throughout the Solomon Islands." These contrasting reports illustrate the need for more data documenting the status of wild *Ceratobatrachus* populations and how collecting affects long-term population size.[24]

Madagascar has a diverse amphibian fauna with a high degree of endemism. The Malagasy golden frog, *Mantella aurantiaca,* has a range of about 3000 km^2. The orange-, yellow-, or red-colored frogs are highly sought by collectors for the pet trade. John Jensen from the Georgia Department of Natural Resources and Carlos Camp from Piedmont College noted that all *Mantella* species are now listed under Appendix II of the Convention on International Trade in Endangered Species of Wild Fauna and Flora (CITES), because they may become threatened if their trade is not controlled, yet that trade is increasing. They also observed that the restricted range of the species is increasingly threatened by habitat destruction and cited two studies in which it was concluded that "the exploitation of this species for the pet trade may be a very significant threat."[25]

These conclusions make sense for a desirable species with a small range. But again, a 2004 record from the IUCN's GAA concluded that, for *Mantella aurantiaca,* "It is also possible that over-collecting for commercial and private purposes is a threat, but so far such harvesting has not had a visible effect on its populations."[26] Further, the same source states, "Limits on exportation of animals have been imposed, and the trade has been greatly reduced." When Simon Stuart, an organizer of the GAA, was asked about these reports, his response makes it clear how these cases are rarely straightforward. He noted that herpetologists in the region who know *Mantella* species feel that, in most cases, the trade is sustainable and probably acts as an incentive to preserve habitat. He went on, "The context in Madagascar, as in many other developing countries, is one of rural poverty leading to clearance of forest for subsistence cultivation, only to be followed by severe soil erosion and loss of farmland (as well as forest). Sustainable harvesting of *Mantella* frogs actually provides a financial incentive for conserving the forest (it being a lucrative income in the local context). Many of these species have tiny ranges all outside protected areas, so their future is in the hands of the rural people."[27]

Our understanding of the population biology of lower vertebrates is rarely developed to the point that we can predict sustainable harvest levels for exploited populations. Research is needed to understand the complex forces that determine how the pet trade influences local amphibian populations. It is also worth remembering that collecting from native habitats affects not only the population abundance of individual species but also community and ecosystem processes. So, even if harvesting of one species is judged sustainable, there are also wider effects to evaluate, because the individual animals removed are lost as members of the ecosystem, perhaps compromising the integrity of one or more of its components. Without such analyses, reports like those for *Ceratobatrachus guentheri* from the Solomon Islands and Malagasy golden frogs raise our suspicions that, under the right conditions, commerce and amphibian population declines might be linked. Supporting that claim requires a chain of evidence linking commercial use as a cause and population decline as an effect. The strongest case for such a chain comes from the marketing of frogs and salamanders as food or bait.

Collecting Frogs for Food

Owen Wister's western romance novel, *The Virginian,* is set in Wyoming between 1874 and 1890.[28] In one scene, a train heading to Billings, Montana, with the novel's protagonist—the Virginian—is delayed because of a washed-out bridge. Forced to forage for food, the Virginian walks to a nearby marsh and, to the astonishment of his fellow travelers, returns with a gunny sack full of frogs. "Well, my gracious!" said one passenger. "What fool eats a frog?" Soon, however, all were enjoying a feast of fried "frawgs legs," while the Virginian regaled them with stories of "Frogs' Legs a la

- Axolotls (*Ambystoma mexicanum*) are indigenous to high-elevation lakes around Mexico City, but Jeanne McKay at the University of Kent, Canterbury, U.K., reported that collecting for food, introduced exotic fish, and habitat degradation throughout the 20th century have greatly diminished native populations.[35]
- Large salamanders in the genus *Andrias* are eaten in Japan and China. In a 2007 report, Feng Xie and collaborators at the Chengdu Institute of Biology noted that overharvesting for food is a serious threat to the Chinese giant salamander (*Andrias davidianus*).[36] Overharvesting of this salamander is notable, because almost all other cases concern frogs.

How widespread is the collecting of animals from wild populations? In 2001, The Food and Agriculture Organization of the United Nations issued a report entitled, *The World Market for Frog Legs*.[37] The report was based on data collected until 1998, and it made the following four major points. The first one provides a context for understanding how frog harvesting and global amphibian decline might be related.

- Almost 95 percent of the world demand for frog legs and frog products is still supplied from wild stocks.
- The international trade in frog legs involves more than 30 countries and is valued at approximately US $48.7 million. From 1987 to 1998, the leading suppliers were Argentina, Bangladesh, Brazil, Cuba, India, Indonesia, Mexico, Romania, Taiwan Province of China, Thailand, Turkey, Uruguay, and the United States.
- Eleven species are the main focus of frog harvesting worldwide (Table 4.3).
- Almost the entire production of farmed frogs is sold in local or regional markets in the producing countries. Many of these are developing nations, and 5 of the 13 leading suppliers are Asian countries.

The presence of relatively fewer research scientists in economically developing nations means that we have almost no assessment of how these harvests affect local amphibian populations. The harvests are so large, however, that it is hard to imagine, in light of the red-legged frog example, that at least some local populations are not being depleted. Worldwide, from 1987 to 1997, an average of about 4716 metric tons of frogs were collected annually—some 200 times the tonnage at the peak year for harvesting of red-legged frogs—and the data on which this statistic is based do not include the major exporting nations of China and Vietnam.

TABLE 4.3 Frog Species Used Commonly as Food throughout the World

Species	Origin	Destinations	Uses
Rana catesbeiana	North America	USA, Mexico, Central America, South America, Asia	Human consumption, laboratory use
Rana pipiens	North America	USA, Mexico	Human consumption, laboratory use
Rana tigrina	Asia	Thailand, Taiwan, Indonesia	Human consumption
Rana rugulosa	Asia	China, Thailand	Human consumption
Rana hexadactyla	Asia	India	Human consumption
Rana temporaria	Europe	Europe	Human consumption, laboratory use
Rana esculenta	Europe	Europe	Human consumption, laboratory use
Rana grylio	North America	North America	Human consumption
Leptodactylus ocellatus	South America	Brazil	Human consumption
Leptodactylus labyrinthicus	South America	Brazil	Human consumption, laboratory use
Caudiverbera caudiverbera	South America	Chile	Human consumption

Assessing the Global Trade in Amphibians

Now we return to two questions raised at the start of our discussion of amphibians and commerce: How do these markets affect native populations? and Is commerce connected to population declines?

In an analysis of U.S. trade records from 1998 to 2002, Martin Schlaepfer, who was at Cornell University at the time, and his colleagues assessed the magnitude and content of the global commercial trade in amphibians and reptiles.[38] Their goal was to test whether collecting wild-caught individuals is sustainable. They argued that a first step is quantifying the number of animals imported and exported annually, to determine if there is a potential threat. The United States was their study model, because U.S. Fish and Wildlife Service maintains records of all legally imported and exported amphibians (although the annual records are destroyed after 5 years of storage), and the United States is probably the largest market with records. China may be larger, but similar records are unavailable from that country.

The salient findings of Schlaepfer's analysis for amphibians echo the United Nations report:

- During the 5-year period, 5.2 million kg and almost 15 million individuals were imported and declared as wild caught

- 96 percent of the trade was commercial, mainly for pets and food
- Few commonly traded species were monitored or regulated at the species level
- Most of the trade involved individuals from nine frog families and two salamander families[39]
- The United States represented about 15 percent of the world market made up of all other CITES party nations, but there was no global database or monitoring program for trade in non-CITES species between countries outside the United States
- The analysis did not include animals collected in the United States for the domestic or black markets.

Schlaepfer and colleagues concluded that the data "do not provide conclusive evidence of widespread, unsustainable collections. They do, however, reveal that the volume of animals taken from the wild (for the U. S. market alone, let alone globally) is large enough to potentially extirpate populations or species."[40] Clearly, more research is needed to understand when and at what harvest levels the taking of wild-caught animals is sustainable.

We know of no case in amphibians where a joint analysis of the population biology of the species, estimated sustainable yields, and expected economic costs and profits preceded establishment of a market. Exploitation of amphibian populations is usually an empirical experiment; that is, animals are harvested and sold, and collecting stops when the population size becomes so low that harvesting is no longer rewarding economically.[41] Schlaepfer's team made two recommendations to address this deficiency: (1) improve estimates of how many animals can in theory be harvested from wild populations, and (2) quantify with greater precision how many animals are removed. We agree. Such data are required to develop the mathematical models needed to predict when populations can support sustained collecting.

The ability to assess the effects of collecting will increase in importance as new and easier trade routes unite previously unconnected markets, linking species to ecosystems far beyond where they evolved. A recent report from Japan highlights this reality. In 2002, Bryan Stuart, a researcher at the Field Museum of Natural History in Chicago, described the Laos warty newt (*Paramesotriton laoensis*), a strikingly colored salamander species from Southeast Asia.[42] Following standard scientific methods, he described the new species and where he collected it. In March, 2006, Laos warty newts appeared for sale by Japanese pet traders for US $170 each. Apparently, a dealer noted the description of the new species in the technical scientific literature and used the report to locate animals for collection and sale. The tactic alarmed scientists because of how the research literature was used for commerce and because the species' range is unknown, so it is unclear whether it needs protection. The dealer justified his actions by observing that there were "lots of them [salamanders]" and, in fact, "local people eat them." As the world behaves more and more like one market, research is needed to assess the effects of globalization on biodiversity in terms of increased trade and the increased rate at which new species can be brought to commercial use.

It is likely that indigenous amphibians of the right size for human consumption have been hunted for food for centuries. Under the right conditions, harvesting could be sustainable, but the red-legged frog in California was hunted to unsustainable levels by the late 1800s. Land use changes and introduced bullfrogs combined with unsustainable harvesting to diminish populations even more.

We suspect that two broad classes of amphibian populations are being harvested. In the first class, animals are harvested from largely intact, native communities. We predict that in these cases commercially successful yields will rarely be sustainable, and native populations will decline to levels that threaten survival as small populations become vulnerable to other forces, including land use change, introduced species, diseases, and contaminants. The red-legged frog fits this scenario.

In the second group, animals are harvested from greatly modified native communities, such as tropical marshlands converted to rice paddies. Under these conditions, amphibian biodiversity is reduced because species are lost as complex tropical and subtropical habitats are converted to simpler agricultural communities, but, for the amphibian species that can coexist with agriculture, population numbers will stay high enough to support sustainable yields. American bullfrogs harvested from rice fields in Cuba seem to fit this model.

Collecting Amphibians for Bait

Amphibians are a minor part of the bait trade in comparison to artificial lures or other live bait such as fish or invertebrates. Still, amphibians are used as bait throughout the United States, where the commercial trade activities can be divided into midwestern/north-central, southeast, and western regions.[43]

Frogs are the preferred bait in the midwestern and north-central region, where there are harvest seasons, size restrictions, and collecting limits. There is a limited conservation concern for these species, but, as we will learn, collecting and moving these animals is likely also moving their pathogens. There is a smaller bait industry in the southeastern part of the country, where salamanders (genus *Desmognathus*) are sold as "spring lizards." Most states lack regulations for salamanders.

In the western United States, millions of tiger salamanders (*Ambystoma tigrinum*) of several subspecies are moved annually from the central plains to the intermountain west for sale as bait. As in the southeast, the western U.S. bait industry developed in response to fishing, mostly for sunfish, bass, and catfish indigenous to the eastern United States. The fish are introduced into ponds, lakes, rivers, streams, and large reservoirs. A 1970 study concluded that almost 2.5 million salamanders, valued at approximately US $500,000 (equivalent to US $2,766,489 in 2005 dollars[44]) were sold annually along the Colorado River bordering Arizona and California. Arizona tiger salamanders (*Ambystoma tigrinum nebulosum*) are collected in Arizona and sold within the state. Barred tiger salamanders (*Ambystoma tigrinum mavortium*) have been imported as bait into Arizona, Colorado, and California and deliberately or inadvertently released into native habitats and livestock ponds. Importing of salamanders into California and southeastern Arizona is presently prohibited.

The barred tiger salamander is now an established, nonindigenous subspecies that breeds with native races, altering the gene pools of California (*Ambystoma californiense*) and Sonora (*Ambystoma tigrinum stebbinsi*) tiger salamanders.[45] Both taxa are on the U.S. Endangered Species list. Fifty years ago, Bernard Martof, a University of Georgia herpetologist, warned of just this kind of alteration of gene pools. The introduced subspecies is causing the decline of native races, but not by killing animals or displacing them from habitats; rather, the genetic integrity of the native populations is altered by a mechanism that is subtle and not obvious to casual investigation, and which can be discovered only by sophisticated genetic analyses.[46]

Pathogens also move with bait animals, raising the possibility that native populations will be affected. Angela Picco, a doctoral student at Arizona State University now with the U.S. Fish and Wildlife Service, used amphibians as a model system to study how pathogens are moved in commerce.[47] Andrew Cunningham, Head of Wildlife Epidemiology at the Institute of Zoology in London, and his colleagues coined the term "pathogen pollution" to describe this anthropogenic movement of pathogens, often associated with commerce.[48] Picco studied Ambystoma tigrinum virus (ATV), an ideal candidate for pathogen pollution. These viruses cause die-offs of amphibian populations, and the potential for moving them outside their native range is high due to the trade in amphibians as fishing bait. Her results indicate that 85 percent of Arizona bait stores sold ranavirus-infected tiger salamander larvae during the 2005 fishing season.

James Jancovich from Arizona State University and Andrew Storfer from Washington State University led two teams conducting molecular analyses of ATV pathogens and tiger salamander populations throughout the western United States.[49] Except for a few populations, the salamanders and the virus showed clear patterns of co-evolution. The exceptional populations appear to be cases in which salamanders had been moved from their original population, probably as bait, and introduced into a new population. There is evidence, therefore, that ATV is moved via the tiger salamander bait trade, but the impact of this pathogen movement on possibly naïve hosts is unknown.

As yet, we have no example in which commerce alone has decreased amphibian population sizes to extinction—but the potential is there. Matthew Fisher and Trent Garner at Imperial College and the Institute of Zoology in London reported that 28 species of introduced frogs and salamanders carried the amphibian chytrid, which is a lethal pathogen that we will discuss in the next chapter.[50] They concluded that there is evidence that trade in amphibians is contributing to the spread of the pathogen. In general, concerns are raised by the large number of amphibians collected for food, the pathogens that move with exotic species, and the collecting for the pet trade of species distributed over small areas.

Commercial use of amphibians often affects small areas of the landscape. Humans, of course, also manipulate larger areas of the landscape in what is the greatest cause of amphibian declines—habitat alteration.

Land Use and Land Cover Changes

When Mark Twain published his short story, "The Celebrated Jumping Frog of Calaveras County" in the November, 18, 1865, edition of *The New York Saturday Press,* that California county was known for its gold mining camps in the foothills of the Sierra Nevada. Twain set the story in Angels Camp, and the California red-legged frog was at the story's center—"the celebrated frog." Fast-forward almost 150 years to a medium Twain never imagined, and on May, 1, 2006, the red-legged frog is a headline on the MSNBC News Web site: *"Habitat or homes? Frog jumps into debate. Species made famous by Mark Twain is being squeezed out."*[51] The point of the story is clear from its title and lead sentence: "The national debate over protecting fragile species comes to life here [Livermore, California], where upscale housing developments push ever deeper into the rumpled blanket of grassy hills at the eastern edge of the San Francisco Bay area." In 1996, the red-legged frog was listed as threatened under the U.S. Endangered Species Act, and about 1.6 million hectares were designated as critical habitat to ensure the species' survival by limiting development. Subsequent litigation reduced the area set aside as critical habitat to 182,000 hectares, allowing private landowners to develop areas that otherwise could serve as frog habitat. Red-legged frogs now occupy about 30 percent of their former range, and the California debate is one example of how land use and land cover changes affect amphibians and many other species.

Three points are useful to make before we discuss how land change affects amphibians. First, the natural world changes constantly and land can change as a result of marginal, indirect, or no human influence. The mechanisms can be rapid and violent, as in the eruption of the volcanic island of Krakatoa, or gradual, as when grassland slowly changes to forest after invasion by shrubs and then fast-growing trees, leading finally to the establishment of large, mature tree species typical of an old growth forest. Animal, fungal, and microbial species change with the plants. Ecologists call this gradual community change "succession," and it provides an important perspective, standard, or baseline for calibrating the rapid land changes that humans can initiate.

Second, when humans alter the way in which land is used, it is called simply land use change. Land cover change is alteration by humans of the physical or biotic nature of a site (e.g., conversion of a forest to grassland).[52] These processes can occur together or separately, and we will follow Peter Vitousek at Stanford University and label both "land use change."[53] Land use change can facilitate local and, eventually, regional extinction of populations and species by killing organisms, removing or transforming habitat, or preventing access of animals to breeding sites.

Third, except for marine environments and the Antarctic, most of Earth's freshwater and terrestrial biomes have amphibian species. Amphibians not only inhabit wetlands, but they also occupy deserts and high mountains. Some species live their entire lives in water, whereas others never venture near ponds, lakes, streams, or marshes. Instead, they have a life cycle in which eggs are laid and the embryos develop directly into miniature adults that then grow to maturity. Even in species that require water for development of larvae, the juvenile and adult metamorphosed stages often live away from water in terrestrial habitats. The point is that loss of aquatic habitats often negatively affects amphibian species, but most species also require associated terrestrial habitats. Land use change, therefore, includes aquatic and terrestrial habitats.

We will set the stage for understanding the relationship between land use change and population or species losses by first summarizing what would be expected if humans were not affecting things. Would changes still be expected? Indeed they would be, and it is important to recognize that amphibian populations can decline, even to local extinction, as a result of natural changes around them. Then we will review what we know about the transformation of landscapes by human actions and how that affects amphibian populations.

Monitoring Changes in Amphibian Population Sizes

Drawing firm conclusions about long-term population trends is not easy. Estimating a population's size is technically difficult, laborious, and complicated by the fact that many adult amphibian populations have "boom and bust" population cycles in which numbers may vary some 20 times over only a year or two.

As we discussed in chapter 1, a general problem in studying amphibians is separating the "noise" caused by annual variation in population size from the "signal" that is the long-term trend in population size (increasing, decreasing, or no change).[54] Declines in some years are expected even without negative pressures such as introduction of an exotic species, clear cutting of a forest, or spread of a toxin. After analyzing the data for hundreds of frog populations, Ross Alford and Stephen Richards at James Cook University reached the important conclusion that, on average, amphibian populations decrease in more years than they increase.[55] We might think, then, that "boom" years are key points for population persistence. David Green at McGill University also analyzed data for many populations.[56] He pointed out that average amphibian population size is not a function of relatively rare years of high recruitment offset by intervening years of gradual decline, so that the years with declines would outnumber years with increases without negative effect. It is more complicated than that, and here is why.

For many species of amphibians, breeding and recruitment are strongly influenced by water availability. In any year, the "decline" of a particular population may be nothing more than a year of reduced recruitment that is part of the expected ups and downs typical of many amphibian species in response to water availability. But the occasional "boom" year alone is not sufficient to offset years of reduced recruitment. Something important, but not well understood, happens in the mix of years of decreasing and increasing population size that is necessary to ensure persistence in the long term. Part of the explanation probably rests with the fact that we have a poor understanding, for most species, of how variation at one location relates to changes in the larger metapopulations in which many amphibians live. Specifically, population biologists speak of populations that are "sources" or "sinks." Some habitats, such as a pond or lake, are sources of individuals that colonize other ponds or lakes; the recipient populations are "sinks" in that fewer animals leave the habitat than enter it. In general, monitoring of the dynamics of local populations absent from the web of interconnecting habitats that constitute the metapopulation is not the best level of analysis for understanding processes at the regional or larger scales that control biodiversity.

These observations lead to a clear question: What is the best way to study populations that vary so significantly over years? Deciding whether a habitat is occupied by a species can be a challenging question, because not all individuals or even all species may be active when the location is sampled. The likelihood of detecting a species varies with the techniques used, the species, and the environmental variables that affect the species' activity. Long-term datasets increase our confidence in concluding that a species does or does not occupy a habitat, because observations over time increase the likelihood of sampling under just the right conditions needed for a species to be active and recorded. Long-term datasets that include a large number of habitats are the best way to detect meaningful trends for populations that vary greatly in number of individuals between years. Repeated sampling of populations, perhaps over decades depending on the amount of annual variation, may be needed to reach statistically meaningful conclusions.

A 40-year study of change in 14 species of frogs and salamanders at the E. S. George Reserve in Michigan exemplifies the long-term record needed for understanding variation in amphibian population sizes. Most importantly, the study documents the increases and decreases expected in systems not affected by humans through the introduction of exotic species, commercial use, or major land use changes.

The E. S. George Reserve in southeastern Michigan is a protected game reserve that has been maintained since 1930 by The University of Michigan. It comprises about 600 hectares of moraine and basin topography. Lowlands are marshes and bogs filled with a rich diversity of aquatic vegetation, including sedges, cattails, red maple, birch, and tamarack trees. Uplands are old fields and oak-hickory woodlands. There are some 50 permanent, semipermanent, and temporary aquatic habitats: marshes, swamps, bogs, and ponds. Between 1967 and 1974, one of us (JPC) and Henry Wilbur, who is now at the University of Virginia, surveyed annually all of the amphibians living in each habitat on the Reserve.[57] From 1988 to 1992, David Skelly, who is now at Yale University, and colleagues resurveyed all of the habitats and determined the net change in the number of breeding populations of 14 frog and salamander species in 37 ponds between the two surveys.[58]

Forty population colonizations and 34 population extinctions occurred between the two surveys. The number of amphibian populations stayed about the same, but the mix of species in each habitat changed. Why? Over time, especially small, shallow ponds and marshes filled with decaying vegetation and silt, making them shallower and less likely to hold water for long periods; put technically, the hydroperiod shortened. Likewise, open, sunny woodlands became shady over time as trees grew and filled the light gaps in the canopy. Ecological succession occurred as young, second-growth forests developed into older, mature forests. Skelly's group concluded that, as a result of succession leading to altered pond hydroperiod and canopy cover, the average amphibian species experienced about five colonizations and five extinctions between the two surveys. Said differently, over these 25 years in southeastern Michigan, a species of frog or salamander in an area protected from all but natural processes would appear as a "new" species in about five habitats and disappear from about the same number.

A long-term study like this one teaches us that amphibian populations decline even to local extinction as a result of natural habitat changes, often over relatively long time periods, such as decades. This is an important perspective to have as we review conditions in which humans play a major role in species declines through land use changes (clearing of forests, draining of wetlands) that can occur much faster—over days, weeks, or a year. We will start at the local and regional levels to get a sense of how land use change affects individuals and populations. Then we will assess how large forces extending over a greater scale affect populations and species.

Human Transformation of Landscapes and Its Effect on Amphibian Populations

In this section, we present examples of how land use change in various countries and regions has affected amphibians.

Starting in 1992, Stephen Hecnar and colleagues at Lakehead University began studying the variables controlling amphibian communities across 42,962 km^2 in southwestern Ontario, Canada.[59] Wetlands covered 69 percent of the southern and 23 percent of the northern part of the study area in presettlement times (the 1700s). Drainage of wetlands for agriculture began in the 1800s, and wetlands (often artificial agricultural ponds) came to cover 3 percent of the southern and 10 percent of the northern portion. Now, moving from south to north, there is increased forest cover, decreased human population density, and increased richness in amphibian species. Hecnar concluded: "The most important factor that has affected amphibian diversity in southwestern Ontario has occurred historically. The massive deforestation and wetland drainage of the 1800s destroyed most amphibian habitat. While the building of artificial ponds may have been beneficial (those not stocked with fish), the magnitude of habitat loss and change resulted in less diverse communities."[60]

The amount of regional woodlands, predation by fish, habitat isolation, and perhaps agricultural chemical use interact at local and regional scales to control amphibian diversity. Overall, however, "The primary process affecting local amphibian richness in southwestern Ontario appears to be historic habitat loss.... Continued deforestation in the more northern regions of southwestern Ontario will likely result in less diverse amphibian communities consisting mainly of green frogs [*Rana clamitans*], leopard frogs [*Rana pipiens*], and American toads [*Bufo americanus*]."[61] Historically, this region supported 13 species of frogs and salamanders. This fact leads to the more general concern that land use change that eliminates native woodlands and wetlands selects against endemic amphibian taxa and favors species that can adapt to human-dominated landscapes.

The negative effects of land use change on amphibian populations were at work in Ontario before the 1980s. Habitat conversion over more than a century had diminished amphibian biodiversity in southwestern Ontario across some 43,000 km^2. Habitat conversion is an issue in tropical areas as well: Between 1990 and 1997, an average of about 60,000 km^2 of humid tropical forest were converted to agricultural landscapes each year, and an average of about 23,000 km^2 were degraded.[62]

Gentile Ficetola and Fiorenza Bernardi at the University of Milano studied 84 wetlands in the Lombardy region surrounding Milan in northern Italy and showed how human-driven land use change negatively affected amphibians.[63] In the late 19th and early 20th centuries, there were nine amphibian species in the region, and all but one were abundant. At present, for the most common species, the edible frog (*Rana escuelenta*) and the Italian tree frog (*Hyla intermedia*), metapopulations remain intact because individuals move across the landscape using canals and hedgerows. Newts and toads are rare because they are sensitive to habitat alteration, which is common as a result of agricultural and industrial development. Because they do not move readily across the landscape, subpopulations of newts and toads are isolated. The authors concluded that "degradation of wetland ecological features (like pollution and fish release) and landscape alteration (like increased wetland isolation) cause the extinction of sensitive species from many patches."[64]

Tim Halliday at the Open University worked with one of us (JPC) and summarized data for the United Kingdom. Over the last 50 years, 20 percent of 350,000 actual or potential amphibian habitats were lost, at a rate of 1 and 3 percent annually (Table 4.4).[65]

Under the editorship of Michael Lannoo at the Indiana University School of Medicine, 215 scientists contributed to a 2005 summary of the conservation status of the 289 U.S. amphibian species, comprising 103 frog species and 186 salamander species, each with a species account that also described potential threats.[66] David Bradford analyzed most of these accounts and concluded that land use change—habitat loss, alteration, or fragmentation—was the most frequently cited cause of population declines and losses (about 77 percent of 91 frog species and about 91 percent of 176 salamander species analyzed).[67] The most-cited land changes

TABLE 4.4 Loss of Ponds (Actual or Potential Amphibian Habitats) at Selected Localities in the United Kingdom over the Last 50 Years

Location	% Lost	Annual Rate	Source
All UK	20% of ~350,000 (1958–1988)	0.7%	Swan & Oldham 1993
Sussex downland (rural)	18% of 33 (1977–1996)	0.9%	Beebee 1997
Huntingdonshire (rural)	99% (over 40 years)	2.5%	Nature Conservancy Council 1982
Milton Keynes (urban)	33% of 126 (1984–1994)	3.3%	Barnes & Halliday 1997
Geneva Basin *Triturus cristatus* populations (rural/urban area)	68% of 22 (1975–1997)	3.1%	Arntzen & Thorpe 1999

Source: Collins and Halliday, "Forecasting changes in amphibian biodiversity," 312.

were conversion of native landscapes to agriculture, urban development, and timber/silviculture. Exotic species were the second most frequently implicated factor for frog declines and the third most common for salamanders. Chemical contamination was the second most-cited factor for salamanders and the third for frogs. As we already discussed, the authors of the 2004 Global Amphibian Assessment reached a similar conclusion.[68] By far the greatest threat to all amphibians worldwide is habitat loss. Land use change is a major cause of reduced biodiversity globally, and amphibians are no exception.

Loss of native habitats, especially wetlands, has typified the 19th and 20th centuries, and it has every prospect of continuing. Amphibian populations have been declining for centuries as a result of land use changes, and the losses have gone largely undocumented.[69] It should be noted that land use change, especially for agriculture, exposes frogs and salamanders to toxins, a point we will discuss in the next chapter. Kenneth Dodd, who at the time was with the U.S. Geological Survey, and Lora Smith, also with the U. S. Geological Survey, made the important point that amphibian habitats must be thought of as all of the elements that affect their life history—both water and terrestrial sites; as they wrote, "The biotic integrity of amphibian habitats needs to be ensured."[70] Land use change is but one of the many ways in which humans change the Earth.

We are in the midst of significant global change, much of which is initiated and driven by humans. The patterns we reviewed are being seen worldwide, and things are likely to get worse before they get better. Michael Lannoo teamed with colleagues at the U.S. Geological Survey and the

University of Wisconsin-Milwaukee to analyze global land use change and amphibian distributions.[71] They showed that amphibian species diversity is highest in terrestrial and freshwater habitats in warm, moist regions such as Central America, northern and central South America, central Africa, and Southeast Asia. Before the 20th century, human population growth and spread was slower and occurred mainly in temperate zones, away from regions of high amphibian species richness. Throughout the 20th century, however, human population growth has accelerated together with its accompanying land use changes, mainly for agriculture, and especially in regions favored by amphibians. Overall, regions with the highest species richness of amphibians are undergoing the highest rates of landscape modification and will continue to do so for the immediate future. A clear prediction is that more amphibian declines are expected in the immediate future, as humans continue transforming Earth's ecosystems.

Habitat change, amphibian extinctions, and our incomplete knowledge of amphibian biodiversity were brought home by a 2002 study in Sri Lanka. Madhava Meegaskumbura at Boston University and Sri Lanka's Wildlife Heritage Trust led a research team in a project that we first discussed in chapter 1.[72] Surveys performed primarily in Sri Lanka's rain forest revealed some 140 frog species on the island—100 of which were previously unknown. In addition, a check of 19th century museum collections uncovered approximately 100 additional new species that were not recorded in the recent survey, presumably because they had become extinct. The authors concluded that this is not surprising, because the island has lost more than 95 percent of its rain forest habitat. Seventeen of Sri Lanka's frog species disappeared within the past decade—representing 50 percent of the 34 confirmed amphibian extinctions occurring worldwide in the past 5 centuries. Land use change is a major cause of extinctions in amphibians.

If exotic species, commercial use, and land use change were the only causes of amphibian declines, the issue would not have drawn the attention of scientists and the public as it did in the late 1980s and 1990s, because these factors are threatening global biodiversity of many taxa other than just frogs and salamanders. Something else was driving the interest in amphibian declines: the fact that species had disappeared in the midst of protected areas. By 2004, these losses came to be called the "enigmatic declines," and their unknown causes were the real driver of the acute interest in amphibian declines.[73] Untangling this mystery is the next step.

5

Contaminants, Global Change, and Emerging Infectious Diseases

In the late 1980s and 1990s, we were hearing reports of some amphibian species going extinct in only a few years; the search for the answer to the question, Why are they gone? was becoming paramount. But as we have seen already, the reality in science, as in any good detective novel, is complicated.

In the last chapter, we learned that historical explanations point to such causes as competition with exotic, introduced species or predation by the same; harvesting of wild animals for food or pets; and changes in land use. These mechanisms accounted for most of the damage to amphibian populations for much of the 20th century, and even today. These historical causes also typically leave clues—some as obvious as the mark of a chainsaw—from which an investigation can begin.

However, at least some of the amphibian declines and extinctions in the late 20th century left few, if any, clues. Animals went missing in national parks and nature reserves, where obvious causes such as habitat destruction did not apply; somehow, habitat protection, historically the best way to ensure a species' survival, was failing to safeguard some amphibians.

The realization that the standard historical explanations could not solve the mystery led researchers to consider explanations based on recent changes. Three suspects emerged: toxic chemicals in the environment; global change, particularly global warming or increased ultraviolet radiation; and emerging—in some sense, new—infectious disease. Each hypothesis had its champions, and given the complexity of ecological interactions, it was likely none would act alone. In addition, these suspected causes could have still undiscovered accomplices.[1]

Contaminants

As mentioned briefly in chapter 2, Minnesota schoolchildren from Le Sueur took a field trip to a farm pond in the summer of 1995. While collecting frogs, they found many that lacked legs or even had extra legs (Figure 5.1).[2] The deformed animals were surprising and unsettling. Were these animals signaling something about the water in which they lived and the environment surrounding the pond? And did these animals relate to global amphibian declines? The children's discovery evoked memories of Rachel Carson's *Silent Spring:* Were harmful chemicals being introduced into the environment?[3]

Environmental contamination is the introduction of manmade organic and inorganic chemicals into the environment, a field large enough to warrant many books. We will focus only on contaminants related to amphibians.

Why think about contaminants? Minnesota's frogs attracted a lot of media attention and a book by the journalist William Souder—*A Plague of*

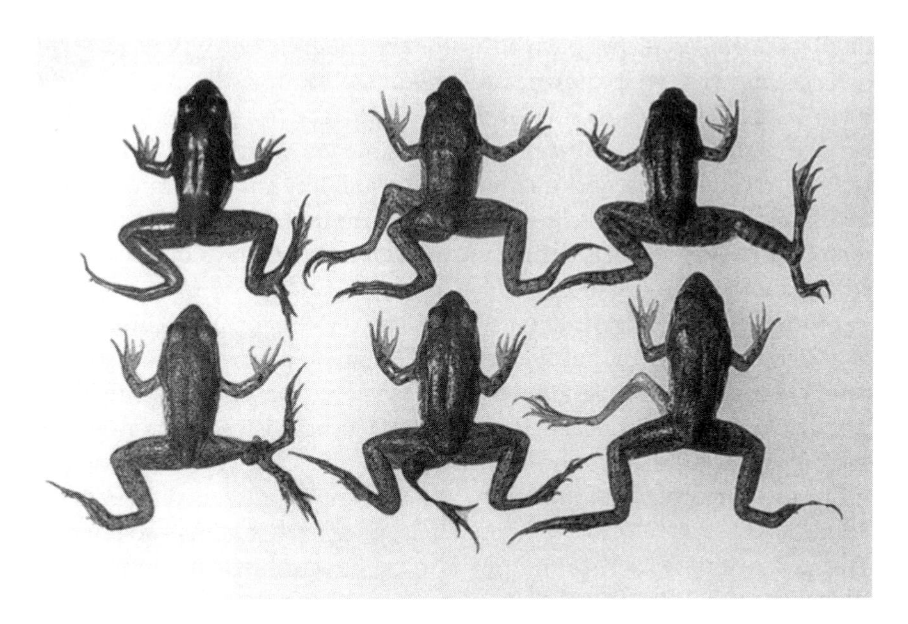

Figure 5.1 Deformed green frogs from Quebec. (After Souder, *A Plague of Frogs: The Horrifying True Story*; photograph by Martin Ouellet.)

Frogs. The Horrifying True Story.[4] The deformed frogs were a mystery because everything else about the pond seemed okay: There were no obvious disturbances in the area, no incriminating pipe discharging an unknown fluid, and no hazardous waste dump in sight. All of this suggested that something might be entering the area unnoticed, and one possibility was a contaminant—a toxin draining into the pond undetected or wafting in on air currents—that was interfering with normal frog development. But why should a contaminant come to mind?

The publication of *Silent Spring* in 1962 sensitized researchers and the general public to the influence of chemicals on wildlife. A classic example was the effect of the organic insecticide DDT (dichlorodiphenyltrichloroethane) on developing bird embryos, especially among predators high on the food chain. Top predators, such as bald eagles, accumulated DDT in quantities that led to thin eggshells and dead chicks. Extensive research connected DDT to declines in populations of eagles and other U.S. birds. In the closest thing to a large field experiment, U.S. raptor populations rebounded after the federal government restricted DDT usage.

Environmental toxins also affect humans. Japanese Minamata disease was recognized in the 1950s as a neurological syndrome caused by mercury poisoning. Deformities can result, especially in a congenital form of the disease that affects fetuses. The release of methyl mercury, an inorganic toxin, in the industrial wastewater of a Japanese chemical factory caused the disease. The mercury became concentrated in marine organisms in Minamata Bay, Japan, poisoning people who ate the animals.

These are two well-known cases involving an organic and inorganic toxin. It is easy to see how such examples might lead to a suspicion that toxins caused the Minnesota frog deformities. And knowing how DDT caused raptor populations to decline raised the possibility that deformed frogs, toxins, and amphibian declines might be connected. But there is a big difference between suspecting that a toxin causes deformities or population changes and demonstrating either result.

Studying the effects of chemicals in the environment and in organisms is difficult. When a chemical such as DDT enters a terrestrial or aquatic ecosystem, it retains its original chemical structure or separates into parts. The original form and its parts usually affect organisms differently, and separate analyses are required to isolate them. In some cases, the components are more powerful than the parental compounds.[5] Also, a chemical such as DDT is usually not delivered in its pure form, but instead another chemical with its own properties is used as a solvent or transporter. Research shows that the transporting chemical—in some cases, a surfactant that allows easier spreading—can be more hazardous than the pure chemical.[6]

When more than one chemical in combination enters an ecosystem, it is called a "chemical cocktail." Identifying the harmful chemical or chemicals within a cocktail is difficult, and in some cases there is a synergistic effect in which it is the mixture itself that is harmful and not one single chemical.[7] Chemicals originate from point sources, for example the chemical factory in Japan, or from nonpoint sources, of which the Central Valley of California is an example.

Thousands of metric tons of insecticides and herbicides, some 25 percent of the amount used in the United States, are applied annually in the Central Valley. These chemicals enter the atmosphere directly, as planes apply the chemicals to fields or when they adhere to soil particles and disperse as dust. Either way, the chemicals are transported on prevailing winds. At one scale, the valley is a discrete geological feature or a "point source"—imagine the valley viewed from outer space. But for the purposes of describing the transport of chemicals in Earth's atmosphere, scientists and policy makers treat such cases as nonpoint sources, meaning that the chemical is deposited kilometers downwind, and the "origin" is within a large volume of prevailing wind currents that include chemicals from the Central Valley and perhaps other sources. Distinguishing point sources and nonpoint sources is important, because the former are easier to manage than the latter. We will return to this example later.

Finally, once a chemical enters an organism's body, its effects vary. The same chemical acts differently depending on the species, the organism's age or developmental stage, environmental factors such as temperature, and the amount ingested. The latter factor may seem obvious: The more chemical ingested, the more lethal the toxin in a direct linear effect. In fact, for chemicals such as the widely used herbicide dioxin, low doses in some species produce beneficial physiological effects, whereas high doses are lethal in a classic nonlinear response that toxicologists call hormesis.[8]

Isolating a cause amid such complexity is difficult. Add the fact that each year many new chemicals are developed and introduced into the environment, and we have some idea of how hard it is to identify a chemical cause of wildlife deaths or deformities from among many candidates. Detection requires specialized analyses, and there is no single test to apply to a sample that will lead to the conclusion that a "toxin" is present. An investigator must often know or at least have a strong suspicion about what she or he is seeking. A test may require a particular tissue type (liver, brain, blood, muscle) that may be in short supply if only a few, small specimens are available. In addition, the amount of toxin applied varies, duration varies (e.g., a pulse or prolonged application), and persistence varies from minutes to days to years. It takes significant planning, skill, and often site-specific plus toxin-specific knowledge to identify contaminants in the field. The next section describes how an investigation often proceeds.

A Hierarchical Research Strategy

A general strategy used to study how chemicals affect amphibians, and other organisms in general, is what Michelle Boone and Christine Bridges at the University of Missouri called a hierarchical approach.[9] Researchers expose individual organisms to chemicals to understand basic physiological responses and then integrate these findings with studies of individuals exposed to chemicals under diverse environmental conditions. Ideally, a conclusion regarding a chemical's effect integrates all outcomes as a test for consistency of response. A researcher then wants to connect an understanding of how toxins act in an individual's body with changes in population sizes of amphibians in the field.

Step 1 is often assessing a chemical's effect on performance or development of organisms in the laboratory. Toxicology is the study of how chemical substances and their doses adversely affect traits such as survival, development, reproduction, likelihood of genetic mutations, or behavior. Researchers typically study one species under controlled laboratory conditions. Ideally, variation in the toxin's effects is assessed among individuals in a population and in multiple populations, among genetically different families within one population, and among species.

After exposing a sample of organisms to a test chemical, a typical measure of outcome is the LD50, or time until 50 percent of the test population dies. Thousands of such studies have been reported for vertebrates and invertebrates. The U.S. Environmental Protection Agency maintains a database of ecological toxicity records dating from 1926 that summarizes the effects on organisms of organic and inorganic toxic substances.[10] For amphibians, there are tests assessing chemical effects relative to temperature, pH, and ultraviolet (UV) light levels. For example, in nine species of frogs in the genus *Rana,* time to death after exposure to a lethal dose of the pesticide carbaryl varied from 5 hours in wood frogs to 35 hours in northern red-legged frogs. Pesticide tolerance varies a good deal among frog species, within species, and within populations of a single species.[11]

In a 2003 review, Donald Sparling, who was with the U.S. Geological Survey, summarized LD50s for 30 organic and inorganic amphibian contaminants for which the laboratory tests were conducted at or below concentrations reported from environmental studies.[12] Sparling concluded that many compounds occurred in the environment at higher than toxic concentrations. Chemical formulations also differed in their effects, and the same contaminant could vary in its effect depending on the amphibian species and life stage tested (e.g., larva or adult). Finally, sediment toxicity may be more important than water toxicity, because sediments can have higher chemical concentrations.

We often think of amphibians as superior indicators of environmental quality because various species, and sometimes the same species, live in aquatic and terrestrial habitats and could therefore be exposed to toxins from multiple sources. But Boone and Bridges concluded that, compared to other animals, amphibians can be either more or less sensitive to toxins, depending on the species and study. More research is needed; as of the year 2000, only about 3 percent of vertebrate toxicological studies included amphibians. To gain a sense of the task, Sparling also noted that some 75,000 chemicals are manufactured in the United States each year, not including pesticides or the 3000 high-production, high-volume chemicals (>450 kg produced). Many more data are needed to understand how contaminant exposure affects amphibians. A key step in this process is extending our knowledge of how toxins affect animals in the laboratory to how they affect animals in the field.

A toxicology study is designed to assess the ways in which a chemical might be harmful before release or might cause harm after release. Assessing the latter is especially difficult for chemicals introduced into the environment, because ecosystems have many complex interactions that make it difficult to extrapolate from the laboratory to the field. Ecotoxiocology arose in the 1960s in response to this challenge, and it is the next step for connecting how toxins act in the body of an individual to cause changes in amphibian population sizes.

Ecotoxicology is the branch of toxicology concerned with the effects of pollutants on ecosystem constituents in an integrated context, which means that studies are done in the field or under simulated field conditions.[13] James Burkhart at the National Institute for Environmental Health Sciences in North Carolina and colleagues outlined the steps that are followed.[14] Once an acute effect on an organism is observed in the field, the pathology's expression is described—imagine describing the deformities of the frogs collected by the Minnesota school children. Then the physical and biotic environment is described, including suspected toxins in the site's water, sediment, or surrounding terrestrial habitats. Duplicating the pathology in the laboratory using the suspected chemical is next. A model that integrates the laboratory and field results is then developed. Finally, researchers assess how the number of acutely affected animals (e.g., deformed frogs) influences population size. This last step deserves more discussion here, because the details are important.

Studying how chemicals introduced into the environment affect individuals, populations, and ecosystems is difficult, complex, detailed research.

To conclude that a contaminant is negatively affecting a population, an investigator first must show that ecologically relevant concentrations of the contaminant are a new source of population stress. This means that there is reduced recruitment of new animals or increased deaths beyond those caused by other stressors. Then the investigator must show that these decreases in population recruitment or increases in population losses exceed the usual numbers resulting from predation, competition, parasitism, and other factors.

Here is another way to think of the problem. Each year, a population of frogs or salamanders suffers deaths due to predation, competition, or parasitism. If the number of animals killed by toxins does not exceed the number of these deaths, then we conclude that the toxin is not affecting the amphibians' population size. Under these circumstances, ecologists call the deaths due to toxins "compensatory" deaths, or animals that would have died anyway from natural causes.[15] Ecotoxicologists face the challenge of, first, knowing what the average population size or growth rate is, and then demonstrating that a toxin kills animals above and beyond the deaths from usual causes.

These are tough standards to meet, especially for many amphibians. As we mentioned earlier, populations often fluctuate by orders of magnitude in different years, and the population dynamics of most species are unknown. Indeed, in the same review discussed already, Sparling concluded, "In reality, these criteria are seldom, if ever, met in ecotoxicological studies; thus, only indirect evidence is available to determine the role contaminants may have in amphibian population declines."[16] Indirect evidence includes (1) field data documenting population declines in the presence of environmental contaminants, (2) laboratory studies demonstrating lethal or sublethal responses to contaminants at or below field concentrations, (3) increase in populations after removal or neutralization of contaminants, and (4) laboratory or field studies showing that contaminants interact with other causes of amphibian declines to increase the likelihood of population losses.

Experiments are run regularly with suspected chemicals, introduced alone or as mixtures, to test their effects on one or several species in the laboratory, in field enclosures (cages placed in an aquatic or terrestrial habitat), in mesocosms (which are static, outdoor, above ground stock tanks or children's swimming pools in which the natural habitat is simulated by adding multiple species and then introducing the toxin),[17] or directly in natural or manmade habitats.[18] The latter experiments may be deliberate, as with the introduction of a chemical poison, or inadvertent, as when a pesticide sprayed on crops enters a nearby habitat via wind or runoff. The logical chain connecting toxins and population declines remains incomplete until studies have demonstrated how ecologically relevant concentrations of a contaminant act as a new source of stress to diminish recruitment or to increase losses beyond those typically experienced from natural causes. This chain of causation must be completed to show a connection between inorganic and organic toxins and amphibian population declines.

The two most studied inorganic effects on amphibians are acidification of habitats and nitrogen deposition. Laboratory and mesocosm studies have shown that acidification adversely affects amphibians. However, when Frank Vertucci and Stephen Corn, U.S. Fish and Wildlife Service researchers at Fort Collins, Colorado, reviewed surveys of habitats in which amphibians have declined in the U.S. Rocky Mountains and in the Sierra Nevada in California, there was insufficient evidence to conclude that acidic deposition caused the declines. Therefore, this was a case in which laboratory results did not translate readily to field performance. After reviewing studies conducted across the eastern and western United States, Christopher Rowe and Joseph Freda from the Chesapeake Biological Laboratory concluded that there was no evidence to relate acidification to declines in amphibian populations.[19]

In contrast to acids, elevated levels of nitrogen, often as nitrate fertilizers, negatively affect development and survivorship of amphibians, at least in the United States, Europe, Canada, and Australia. Jeremy Rouse and colleagues with Environment Canada concluded that "nitrate concentrations in some watersheds in North America are high enough to cause death and developmental anomalies in amphibians and impact other animals in aquatic ecosystems."[20] Indeed, for several decades now, nitrogen fertilizers and nitrogen oxides discharged by cars and factories are thought to have had a widespread negative effect on biodiversity and a range of ecosystem processes.[21]

Rick Relyea and Jason Hoverman at the University of Pittsburgh reviewed 319 studies of the effects of pesticides and herbicides alone or together on freshwater ecosystems.[22] Toxins alter the relationships of species in food webs by changing behavior, growth, physiology, life history traits, and even susceptibility to pathogens. A toxin does not have to kill a frog or salamander directly to have a negative effect; it can do so by compromising an animal's ability to escape from a predator, compete, avoid disease, or grow quickly enough to complete development before a pond dries. These important findings add great complexity to ecotoxicology studies. A toxin may also kill the aquatic invertebrates that amphibians eat. Either way, there is a negative effect—directly through the animal's physiology or indirectly through the ecosystem's food web.[23]

The reviewed studies included surveys of natural habitats after contamination with experimental application of toxins, and experiments in which toxins were applied to enclosures in natural habitats and to mesocosms;

20 cases included amphibians. Relyea and Hoverman concluded, "One of the most challenging issues to address is whether pesticide effects on individuals actually affect population dynamics. For example, when a pesticide causes low to moderate mortality in a species or the feminization of males due to endocrine effects (e.g., Hayes et al. 2002), it is an open question whether these effects will affect population growth. . . . We need more work in this area, but the data will not come easily for most species."[24]

We will return later to Hayes' studies on feminization of male frogs, but here is what a hierarchical research strategy tells us so far. Based on laboratory studies with toxin concentrations at or below lethal levels, contaminants can cause amphibian declines. Field experiments with cages in ponds, mesocosms, and experimental ponds bring these studies closer to reality. They demonstrate how a toxin can reduce amphibian abundance and negatively alter traits such as size at metamorphosis, under conditions much closer to those experienced by wild populations. But do toxins applied in the field affect populations? Carlos Davidson at San Francisco State University asked just this question in a series of studies.[25]

Sierra Nevada amphibian populations are downwind of California's Central Valley and its extensive agricultural industry. Davidson tested the hypothesis that historical patterns of pesticide use in the valley might also explain declines of five Sierra Nevada frog species, one *Bufo* and four *Rana*, because the prevailing west-to-east winds carry toxins from the valley and deposit them in montane amphibian habitats. Indeed, as Gillian Daly and Frank Wania from the University of Toronto reported, organic chemical contaminant deposition generally increases with elevation for mountains at middle and high latitudes.[26]

Kimberly Hageman and associates at the U.S. Environmental Protection Agency's Western Ecology Division in Corvallis, Oregon, also studied the atmospheric deposition and fate of semivolatile organic compounds in alpine, sub-Arctic, and Arctic ecosystems in the Western United States.[27] Their correlation analysis revealed that current and historical agricultural practices are largely responsible for the distribution of pesticides in the national parks in the region.

Davidson found a significant association between amphibian declines for the four *Rana* species and total pesticide use in the Central Valley. He also teamed with Roland Knapp to assess the effect of pesticides and fish on the abundance of mountain yellow-legged frogs and three additional ranid species in and around the Sierra Nevada.[28] In chapter 4 we described how Knapp and his colleagues had already documented in detail the negative effect of fish on mountain yellow-legged frogs.

Davidson and Knapp quantified the habitat characteristics, presence or absence of fish and frogs, and pesticide use upwind of 6831 aquatic habitats.

Fish and pesticides together significantly reduced the likelihood of finding frogs, with the landscape-scale effect of pesticides being much stronger than that of fish. So, even in sites without fish, high upwind use of pesticides most likely accounted for the absence of frogs. The fact that the likelihood of finding live frogs increased for sites sheltered from Central Valley winds by topographic features reinforced this conclusion. Their results show, first, that amphibian declines have complex, multiple causes, a conclusion that will be reiterated throughout this chapter. Second, this study demonstrates how investigators are using multiple, complementary approaches to understand the effects of toxins on amphibian populations. Natural systems are complex, and at any time multiple causes may act on a population simultaneously or follow each other historically. Finally, although the data are extensive, the Davidson and Knapp study is still a correlation, and additional studies are needed to confirm that toxins in the habitats of frogs are responsible for the variation in population sizes.

Back to Minnesota

Trematodes and Amphibian Abnormalities

After the children's discovery of frogs with extra limbs, public health officials tested those frog habitats for toxic chemicals but found none. Researchers discovered, however, that the Minnesota incident was not an isolated instance. A key first step in answering a scientific question is placing the observations in context: Has anyone collected abnormal frogs before? And if so, under what circumstances and could the same cause or causes account for Minnesota's abnormal frogs?[29]

Statistics compiled by the North American Center for Amphibian Deformities show that abnormalities such as extra limbs, malformed or missing limbs, and facial malformations have been reported in almost 60 amphibian species from 44 U.S. states.[30] In some populations, 60 percent of the animals observed are abnormal. In a review in 2000, Martin Ouellet at McGill University in Quebec reported that deformities are more common in frogs (67 species) than in salamanders (26 species).[31] Michael Lannoo recognizes 21 kinds of frog abnormalities in 3 classes: craniofacial (e.g., a small or missing eye), limbs (e.g., missing or shortened), and whole body (e.g., swelling).[32]

Stanley Sessions, at Hartwick College in upstate New York, is an expert on amphibian abnormalities. He makes an important distinction between "deformities," which are abnormalities resulting from the response of normal tissue to mechanical forces (trauma), and "malformations," which result from an intrinsic defect in development caused solely by genetic or gene-environment interactions.[33] Use of either term implies knowing the process that is causing the pathology. The cause is typically unknown in such cases, at least initially, which is why the Minnesota case was alarming. Here, we will use the neutral term, "morphological abnormality," or just "abnormality."

As the Minnesota schoolchildren made their discovery, other researchers were also studying abnormal frogs. In the mid-1990s, Martin Ouellet was a graduate student at McGill University. For his doctoral dissertation, he was trying to understand why frogs in Ontario's agricultural areas developed with no hind legs or more than two hind legs (see Figure 5.1). Ouellet found residues of insecticides, herbicides, and fungicides at the affected sites, but he did not find a significant relationship between toxin levels and frequency of abnormalities because of high variation among sites for both variables.[34]

Ouellet's work brought an important perspective to the problem by reminding everyone that abnormalities might be caused by something other than toxins.[35] For example, while trying to catch a frog, a predator might get only part of a limb; the partial limb remaining is a "morphological abnormality." Genetic defects and nutritional deficiencies also cause abnormal development. In fact, as early as 1987, Sessions suggested that parasites could cause frogs to develop abnormally.[36]

Trematodes are small (a bit longer than a pinhead), parasitic "worms" that have a fascinating, complex life cycle. An adult trematode is also called a flatworm or fluke, and the ones of most interest to amphibian researchers live in the digestive tracts of birds. The parasite's eggs are released into aquatic habitats with bird feces and hatch into a stage called a miracidium that infects some, not all, pond snail species. The parasite multiplies again inside the snail and is released into the pond as a free-swimming stage called a cercaria.

In the mid-1980s, Stephen Ruth, at Monterey Peninsula College, contacted Sessions because he had found Pacific treefrogs (*Hyla* [now *Pseudacris*] *regilla*) and long-toed salamanders (*Ambystoma macrodactylum*) in northern California with missing or extra hind limbs. Sessions suspected that a trematode infection might be causing the abnormalities. Specifically, a cercaria infects a tadpole, embeds near the site of the developing pair of hind limbs, and forms a cyst that appears to disrupt normal limb development, leading to a metamorphosed frog with no limbs or more than two hind limbs. Sessions and Ruth tested the hypothesis in the laboratory by implanting inert resin beads, simulating the trematode cysts, into the developing limbs of frogs and salamanders.[37] As predicted, the animals produced more than two hind limbs, but there is no still no agreement on the developmental mechanism that might account for the abnormalities.

In the mid-1990s, a Stanford University undergraduate, Pieter Johnson, and colleagues built on the Sessions and Ruth field work and experiments.[38] They surveyed 14 ponds in northern California and found that up to 45 percent of the frogs in 4 ponds had abnormal limbs. Two kinds of trematodes in the genera *Ribeiroia* and *Alaria* were common. Johnson and colleagues concluded that the trematode studied by Ruth and Sessions was likely *Ribeiroia ondatrae.* If parasites caused the problem, they predicted, then abnormal frogs would have higher parasite levels than frogs without abnormalities. After dissecting frogs from various ponds, they found higher numbers of flatworms in abnormal frogs, and the organisms were mostly clustered near a frog's groin. Then they reasoned that, if parasites were the cause, adding them experimentally to tadpoles would increase abnormalities. Johnson and colleagues collected eggs of the tree frog *Pseudacris regilla* from a California site with no abnormal frogs and divided the hatchling tadpoles into six groups: light, intermediate, heavy, or no exposure to *Ribeiroia,* exposure to *Alaria,* and exposure to both trematode species. The *Ribeiroia* treatment caused missing or multiple legs; the frequency of the abnormality increased with exposure level, and tadpole survivorship decreased with exposure level. Trematodes, it seemed, caused abnormal hind limb development, at least in this species in the laboratory. Abnormal animals are probably easy prey for birds, completing the parasite's cycle from bird to water to snail to water to tadpole/frog to bird. This reasoning represents the parasite hypothesis for amphibian abnormalities.

In a subsequent study, Johnson and others used museum collections to document limb deformities in three frog and three salamander species in seven states from 1946 to 1988.[39] Between 1999 and 2002, they resurveyed those same sites for abnormal animals and found many. *Ribeiroia,* they concluded, has caused abnormal development for decades, which also meant that, although the abnormal Minnesota frogs were startling, they were not new to science: the phenomenon had been seen before. The new research did, however, offer some evidence that parasite-induced malformations had increased, which led the team to conclude that limb deformities were an emerging parasitic disease in amphibians. Pieter Johnson is now at the University of Colorado in Boulder, where he continues to study the causes of abnormal development and its relationship to amphibian population dynamics. He has visited the Minnesota site where the schoolchildren reported the abnormal frogs, and he reported to us that trematode infections most likely caused the abnormalities.[40]

Does all of this mean that toxins do not cause morphological abnormalities? The answer is "No," according to the research of Joseph Kiesecker, then from Pennsylvania State University, Andrew Blaustein at Oregon State University, and Pieter Johnson.[41] Nitrogen fertilizers might be an indirect cause of amphibian abnormalities. They hypothesized that when excess nitrogen fertilizer runs into aquatic ecosystems, the added nutrients increase algal growth and support more snails. More trematodes follow, leading to an increased frequency of frog infections and more abnormalities.

Kiesecker tested the trematode-agricultural runoff hypothesis with laboratory and field experiments using wood frogs, *Rana sylvatica*.[42] Field experiments demonstrated that exposure to trematodes led to abnormal limbs; moreover, the closer the cages were to agricultural areas with runoff containing detectable levels of organochlorine pesticides and organophosphorus compounds (e.g., atrazine, malathion), the greater was the frequency of abnormal frogs. He also exposed frogs in the laboratory to the triazine herbicide atrazine, the organophosphate insecticide malathion, the synthetic pyrethroid insecticide Esfenvalerate, and then the trematodes. For each contaminant, exposure increased the proportion of two trematode species in the genera *Ribeiroia* and *Telorchis* that encysted. Pesticide exposure also correlated with a decrease in circulating leukocytes in the frogs, suggesting a compromised immune system. Laboratory experiments corroborated the association between pesticide exposure and increased infection, with a pesticide-mediated reduction in immune competency as the possible mechanism.

Johnson and colleagues also tested the trematode-agricultural runoff hypothesis with field mesocosm experiments using green frogs (*Rana clamitans*), *Ribeiroia*, and a snail (*Planorbella trivolis*).[43] They added these three species plus algae, phytoplankton, and zooplankton from local aquatic habitats to 1200-L mesocosms at the University of Wisconsin's Trout Lake Station. The design also included ambient and enhanced concentrations of nitrogen and phosphorous to simulate eutrophication of an aquatic system. In eutrophic mesocosms, the added nutrients promoted algal growth, increased snail density, and increased the intensity of infection in the frogs. The increased frog infections resulted from both an increased density of snail hosts and an enhanced production (average number of infectious parasites) by each snail.

Jason Rohr at the University of South Florida and colleagues add to our understanding of this mechanism.[44] In a field study of 18 wetlands north of Le Sueur, Minnesota, they found that among 240 plausible predictors, the herbicide atrazine and phosphate concentrations, both presumably from agricultural runoff, were the best indicators of larval trematode, including *Rebeiroia*, prevalence in northern leopard frogs (*Rana pipiens*). They hypothesized that atrazine increased periphyton levels that then supported larger snail populations; recall that snails are an intermediate host for trematodes. They also demonstrated a suppression of amphibian immunity, which, like Kiesecker, they felt increased susceptibility of larval frogs to infection.

But this is not the story's end. Lannoo's recent literature review makes it clear that abnormalities like missing eyes that are inconsistent with the parasite hypothesis occur frequently in areas without trematodes.[45] Relying on the research of Ouellet and others, he summarized 10 natural causes (e.g., wounding, high tadpole densities) and 7 manmade causes (e.g., radiation, agricultural chemicals) of frog abnormalities. Trematode-induced abnormalities are only one of the natural causes, and they are largely restricted to the upper Midwest and parts of California in the United States.

David Skelly and his students at Yale University are studying Vermont ponds, where they find high frequencies of amphibian abnormalities and no *Ribeiroia*.[46] Nearby agricultural areas are possible risk factors, but urban and suburban areas may be an even greater risk. Excess nitrogen is a possible cause, but other possibilities include insecticides, herbicides, or pharmaceuticals in surface water. Lannoo extended this argument.[47] His review indicated that trematode-associated abnormalities represent a minority of cases. There are many examples of laboratory experiments in which agricultural pesticides induce amphibian abnormalities. Collectively, these analyses led Lannoo to conclude that no single mechanism explains amphibian abnormalities, but he went on to hypothesize that chemical pollutants, perhaps via chromosomal damage, are an underappreciated and probably widespread cause of abnormal development in frogs. As yet, there are no tests of this hypothesis.

Endocrine Disruptors and Amphibian Abnormalities

Modern agriculture is one of the great achievements of the 20th century. Widespread use of fertilizers and pesticides (herbicides, fungicides, and insecticides) has been among the major causes of increased food production, especially after World War II.[48] Atrazine, one of the most widely used herbicides in the world, provides us with a cautionary tale. In a series of studies, Tyrone Hayes, at the University of California, Berkeley, tried to determine whether atrazine might have a role in amphibian abnormalities and declines. Hayes and his students tested atrazine's effects on wildlife, using laboratory studies with African clawed frogs (*Xenopus laevis*) and then laboratory and field studies with American leopard frogs (*Rana pipiens*).[49]

Recalling the hierarchical research strategy, we will first discuss the study with African clawed frogs, which are native to cooler regions of sub-Saharan Africa. Since the 1930s, clawed frogs have been common laboratory animals worldwide, and laboratory-reared *X. laevis* embryos are the model for a 96-hour whole-embryo developmental toxicity screening assay called FETAX—Frog Embryo Teratogenesis Assay–Xenopus.[50] The first important fact about atrazine is that concentrations in rainwater can be as high as 40 parts per billion (ppb), and in agricultural runoff, several parts per million (ppm). In the laboratory, Hayes and his students exposed larval and adult clawed frogs to doses between 0.01 and 200 ppb. These doses ranged from less than to greater than rainwater concentrations, and they were less than agricultural runoff concentrations.

For larvae exposed to atrazine at doses from 0.02 to 200 ppb, there was no effect on survival, time to metamorphosis, or size to metamorphosis. But the larynx of atrazine-treated males at metamorphosis was, on average, smaller than in control males, and reduction in size varied directly with dose. The frogs were abnormal: The higher the dose, the smaller the larynges, which are the organs a male uses to call and attract female frogs. There was no significant effect on the larynges of females, and, importantly, control males not exposed to atrazine had larger larynges than control females. Atrazine-treated males were demasculinized.

The researchers suspected that atrazine was interfering with steroid production, resulting in a differential, and negative, effect in males compared with females. They tested their suspicion by exposing adult frogs to 25 ppb of atrazine for 46 days. As expected, atrazine-treated males had testosterone levels 10 times lower than the levels in control males and not significantly different from those in females. Finally, after metamorphosis, atrazine-treated male larvae had abnormal gonads: There were as many as six testes in one animal, or the animals were hermaphrodites with multiple male and female organs, which was again a result consistent with reduced testosterone levels. Physiologically, a developing animal was responding as both sexes and therefore incapable of properly functioning as either. Controls had no abnormalities. In contrast to the Minnesota frogs, atrazine-treated frogs were internally rather than externally abnormal.

Is there evidence connecting atrazine and deformities in the field? Hayes and his group examined this question by collecting adult, wild leopard frogs at eight sites across the United States where atrazine had been applied in high concentrations. At seven of the eight sites where atrazine concentration in the water exceeded 0.20 ppb, between 10 percent and 90 percent of males had oocytes developing in the testes. At sites with low atrazine levels or no agriculture, males were not hermaphrodites. The similarity of the field and laboratory results led Hayes' group to conclude that atrazine was responsible even though other chemicals might be present.[51] They reasoned that, because atrazine is known to induce the enzyme (aromatase) that converts male hormones to female hormones (androgens to estrogens), it was causing the synthesis of female hormones at the cost of male hormones and thereby affecting sex differentiation.

This outcome is called "endocrine disruption," and it also occurs in fish, reptiles, and mammals. Sheldon Krimsky, in the Department of Urban and Environmental Policy and Planning at Tufts University, described the scientific and social origins of the environmental endocrine hypothesis.[52] The hypothesis is that environmental chemicals can interfere with hormonal messages in living organisms, including humans, affecting reproduction and development and also causing disease. Hayes and colleagues concluded, "Atrazine is a potent endocrine disruptor that both chemically castrates and feminizes male amphibians."[53]

Krista McCoy at the University of Florida also studied how agriculture, endocrine disruption, and amphibian abnormalities might be related.[54] Her research team found that frequency of gonadal abnormalities and frequency of intersex gonads increased with agricultural land use in cane toads, *Bufo marinus,* along a gradient of increasing agriculture in south Florida. Hormone concentrations of males collected in the field were altered by chemicals experienced in the environment in ways that caused them to develop with intersex traits—they were feminized—while females were unaffected. A research group at Environment Canada led by Tana McDaniel also reported potential endocrine disruption in northern leopard frogs (*Rana pipiens*) and green frogs (*Rana clamitans*) from farm ponds and agricultural drains in southwestern Ontario.[55] The proportion of oocytes in males correlated particularly with atrazine and nitrates in the water. Such abnormalities, both groups suggested, could contribute to amphibian declines.

We have now come full circle in discussing contaminants. In *Silent Spring,* Rachel Carson referred to the possibility that chemicals could have hormonal effects, but for amphibians a puzzle piece is missing. For the *Rana pipiens* study, Hayes and collaborators concluded "that the current data strongly suggest a connection between atrazine exposure and intersexuality," and the laboratory and correlative field data supported that conclusion. They went on to say, "Combined with decreases in dissolved oxygen, pH, and available food sources ... caused by atrazine, this common contaminant could be a contributing factor in amphibian declines."[56] Data are still needed to link the effects of contaminants on amphibians in laboratory and field experiments with abnormalities and with population declines and extinctions. The studies of Rohr, McCoy, McDaniel, Boone, Davidson and other researchers move us closer to that goal, but we still need data that quantify adult amphibian population size at one density before intoxication and at a lower density afterward in order to demonstrate a direct connection between toxins and declines or extinction. Alternatively, ecotoxicologists need to employ the analytical method that Raina Plowright at the Consortium for Conservation Medicine and colleagues described as "triangulation:" the process of gathering scientific evidence about a system by combining laboratory/field studies, modeling, and historical investigations.[57] This approach would be an effective way to analyze the problem by employing models to make predictions about population sizes without chemicals that can then be compared to estimates of population sizes when toxins are present.

Although research is needed to complete the connection from laboratory to field, we do have clear evidence that contaminants at environmentally realistic levels kill amphibians in the laboratory and in the field. Intoxication also has important sublethal effects on behavior and development in ways that make individuals vulnerable to predators or even impotent. Toxins kill organisms that larval and adult amphibians eat, asserting an indirect, negative effect through the food web. Collectively, the evidence suggests reason for concern, and reason to suspect that contaminants can reduce numbers of animals to levels that make populations and species vulnerable to extinction.

We need more research to connect laboratory and field experiments and observations to population change. Until this gap is closed, ecotoxicology with amphibian models will be subject to the criticism that what happens in the laboratory or mesocosm does not apply directly to communities outside experimental systems.[58] Still, the ill effects from toxins that amphibians suffer based on laboratory and field experiments and observations has an accumulated dispositive weight. The results implicate toxins as likely playing a role in population declines. It is in the interests of sustaining amphibian biodiversity (as well as ourselves) to understand better how toxins in ecosystems relate to amphibian declines and extinctions. We will return to this thought in our final chapter.

We conclude this section with two studies that signal our next theme. Robert Poulin at the University of Otago in New Zealand compiled a database of studies on the effects of temperature on the emergence of infective stages of trematodes from their snail intermediate hosts. He found that small increases in temperature, such as those forecasted to occur by many recent climate models, increase parasite output from their hosts. Increasing temperatures, he concluded, are likely to influence the geographic distribution of parasites such as the trematodes that infect frogs, as well as the number of infective stages hosts produce.[59] Second, after reviewing the literature on how climate change might be affecting amphibian declines, Stephen Corn concluded in 2005, "Climate change may be a relatively minor cause of amphibian declines, but it may be the biggest future challenge to the persistence of many species."[60] What is the relationship of climate change to amphibian declines and extinctions?

Change in Climate and Related Variables

Earth has been changing in many ways since it formed billions of years ago. Climate in particular has changed relative to the planet's radiation balance.

Orbital changes called Milankovitch cycles happen over some 30,000 years, and with them the amount of solar radiation reaching Earth's latitudes changes. Ice ages with large effects on polar and temperate regions have come and gone over the last 3 million years and are closely tied to these cycles. The Pleistocene Epoch, from about 2 million to 11,000 years ago, was a period of glacial cycles in which regional ice sheets extended from the poles. In the north, some extended south as far as the 40th parallel, covering much of Canada and the northern tier of the United States, Europe, and Asia. Glaciers topped mountain peaks, and remnants remain today at high elevations across the globe. Earth has warmed over the last 10,000 years. During this latest interglacial period, ice sheets have been retreating toward the poles and glaciers have been melting. The big question is, Why has the rate of global warming accelerated recently?

In late January 2007, the Intergovernmental Panel on Climate Change (IPCC), a group established by the United Nations, finally took a big step toward settling the "mystery" surrounding the late 20th century rise in average global temperature. Humans are largely responsible. Increasing greenhouse gases, warmer land and sea temperatures, rising sea levels, and more extreme weather are predicted. The *IPCC Fourth Assessment Report: Climate Change 2007* highlights the important relationship between the physical and biological aspects of climate. Climate changes life on Earth—and life on Earth, in turn, changes climate. The report's central conclusion makes this point clearly: With a 90 percent certainty, humans are the main cause of recent planetary change—not volcanoes or variations in the amount of sunlight reaching Earth, but living organisms. Most of the warming in the last 50 years is attributable to human activities, largely through the increased production of greenhouse gases associated with industrialization.[61]

The IPCC report has a strong message. The interactions at Earth's surface, which include the geosphere, atmosphere, and biosphere, are changing before our eyes. Yellow-bellied marmots in Colorado are emerging from their 8-to-9-month hibernations 38 days earlier than they did 23 years ago. New York and New England rivers have 20 fewer days of ice cover today than in 1936, a change that affects frogs and other animals that spend their winters beneath a frozen blanket. The increased warming will affect climate across the globe and influence the distribution of all organisms.[62] Amphibians are no exception. In the late 20th century, the extinction rate of amphibians increased at least 200 times above the rate of the last 350 million years.[63] Are these amphibian losses related to climate change?

A review of the climate change literature yields two quick lessons. First, Earth's climate varies regionally. Although the planet's atmosphere is on average warming, surface temperatures in some places are warming while other places are cooling; some places are getting drier, and others wetter.[64]

Lesson two is that throughout Earth's history climate change has affected many species both negatively and positively. Current changes will be no exception, and amphibians illustrate both outcomes.

Some Clear Effects of Climate Change

Ecosystems Will Vary in Response to Climate Change

The retreat of the Pleistocene ice sheets uncovered great land areas, especially in the Northern Hemisphere. As ice sheets melted, amphibian populations expanded from their ranges in southern refugia.[65] At present only three species—the wood frog in Canada (*Rana sylvatica*), the Siberian wood frog (*Rana amurensis*), and the Siberian newt (*Salamandrella keyerslingii*)—range above the Arctic Circle (about 66° N); no amphibians live below the Antarctic Circle (about 66° S).

Many shallow ponds across the Arctic are centers of biodiversity and natural amphibian habitats. Species will extend their ranges into more northern latitudes as temperatures warm. The research of David Vieites and colleagues at the University of California in Berkeley illustrates how this process acts over the long term. They discovered that periods of global warming probably promoted diversification of northern hemisphere salamanders along with intercontinental and transcontinental dispersal by the mechanism just discussed: New terrain shortened dispersal routes and afforded new evolutionary opportunities.[66] In particular, the primary groups of plethodontid salamanders in North America today were established during major global warming events during the late Cretaceous (100 to 65 million years ago) and again during the Paleocene-Eocene thermal maximum (55 million years ago). However, a big difference between the time in which those changes occurred and the 20th and early 21st centuries is the rate of global change.

Estimates of the effect of 20th century temperature increases vary from species' range shifts (toward the poles) of some 6 km per decade to 6 km over the last 50 years; the latter estimate includes a corresponding estimated shift of 6 m upward in elevation.[67] Either way, it is clear that species are shifting their ranges. As a result, ecosystems will change quickly and in complex ways. The research of John Smol and Marianne Douglas from Queens University and the University of Alberta gives us some insights into these changing ecosystems. Cape Herschel is on Canada's eastern coast, above the Arctic Circle at 78° N on Ellesmere Island. After monitoring habitats there for 24 years, they report that ponds that had been permanent water bodies for millennia crossed a key "tipping point" in 2006 and are now ephemeral. Smol and Douglas "link the disappearance of the ponds to increased evaporation/precipitation ratios, probably associated with climatic warming."[68]

In Yellowstone National Park species have been protected longer than anyplace else in the world. Yet, even in this western American reserve, amphibian species are declining because of climate change. Sarah McMenamin at Stanford University and her research team demonstrated that over the last 60 years annual precipitation has decreased while temperatures in the warmest months have increased.[69] Compared to 16 years ago the number of permanently dry ponds has increased 4-fold. A survey of 49 ponds revealed that during this time 3 of 4 common amphibian species declined significantly (blotched tiger salamander, *Ambystoma tigrinum melanostictum;* boreal chorus frog, *Pseudacris triseriata maculata;* Columbia spotted frog, *Rana luteiventris*); boreal toads (*Bufo b. boreas*) were present, but less common.

The ponds at Cape Herschel and in Yellowstone hold an important lesson: Climate warming will affect some organisms directly as temperatures exceed species' limits, but in many cases, perhaps most, its effects will be indirect. Predators will move or perish as their prey diminishes. The transition of ponds from permanent to ephemeral or terrestrial habitat illustrates how amphibians, invertebrates, plants, fungi, and microbes that prefer aquatic habitats will be displaced. The losses will occur, not because they cannot tolerate higher temperatures, but because the increasing temperatures will affect other features of their habitat. If the negatively affected organisms cannot shift their ranges, then they will go extinct locally—and, if their range is small, perhaps entirely.[70]

Tracie Seimon and colleagues at Columbia University's Earth Institute have shed additional light on these processes.[71] Between 2000 and 2005, the team studied rapid ecological succession of high Andean plants and animals following ice retreats in the Cordillera Vilcanota, a heavily glaciated range in southern Peru. In recent decades, this region's troposphere has warmed about 0.33° C per decade. An aerial and photographic series since 1931 documents ice recession. In the last 10 years, three frog species—*Pleurodema marmorata, Bufo spinulosus,* and *Telmatobius marmoratus*—have colonized newly formed ponds in recently deglaciated terrain at 5244 m; *Pleurodema* reached 5348 m. All are the highest elevation records in the world for amphibians. A similar vertical expansion of range has happened in *Eleutherodactylus* and *Hyla* species in Ecuador. We will return to the Peruvian case later, because two of the colonizing species also harbor an emerging infectious amphibian disease.

Climate change effects will vary regionally, and in the Andes new amphibian populations will arise as animals colonize habitats exposed by retreating glaciers. But again, things are rarely as simple as they seem at first glance. In late July 2007, the World Bank reported that mountain lakes and wetlands in the Andes were drying, threatening water supplies for major South American cities such as Bogota, Quito, and La Paz. Andean wetland habitat called the paramo, which occurs above the tree line and below the permanent snow line between about 3000 and 5000 m, is especially at risk. Walter Vergara, a World Bank climate change specialist in Latin America, attributed the drying to rising temperatures that are causing Andean clouds to condense at higher altitudes.[72]

Later in this chapter, we will discuss this same elevation phenomenon for Central America. The point is that even as new Andean habitats are created for amphibians by the retreating glaciers and a permanent snow line that is higher in elevation, old habitats at lower elevations are lost to drying wetlands. Amphibian populations at the upper and lower elevations will be in flux. Most climate change research focuses on the atmosphere. We need research that links a changing atmosphere with the biosphere and geosphere before we can reliably forecast the consequences of climate change, the effects of which will vary regionally and by species. In most instances, the positive or negative effects on individual organisms and species will not occur because of increased temperatures, per se, but through the influence of temperature on abiotic habitat variables such as water availability or on other organisms in the ecosystem. One thing for certain is that species will vary in their responses.

Species Will Vary in Response to Climate Change

Most species have a seasonal cycle, and at a particular time each year they mate, give rise to young animals or seeds, and move through other stages of their life cycle. The study of the timing of such recurrent natural phenomena is known as phenology. Records of the time each year when various species flower is an example of the kinds of data researchers collect. Wilfried Thuiller, at the Alpine Ecology Laboratory of the Université Joseph Fourier in Grenoble, reported that in the last 50 years seasonal phenomena advanced, on average, 2.3 to 5.1 days per decade.[73] Amphibians are changing both more slowly and faster than these limits.

James Gibbs and Alvin Breisch, at the State University of New York's College of Environmental Science and Forestry, analyzed a 100-year record (1900–1912 to 1990–1999) of the earliest dates each year when males of six frog species near Ithaca, New York, started calling to attract mates.[74] By the end of the 20th century, four frog species called 10 to 13 days earlier than they did at the century's start, and the other two species were unchanged. A temperature increase of 1.0° to 2.3° C during the calling season accompanied this shift. Trevor Beebee, at the University of Sussex, found that two frog species and three salamander species in the

United Kingdom were breeding earlier by some 1 to 3 weeks per decade over 16 years from 1978 to 1994; one frog species had not changed.[75] Andrew Blaustein and students summarized additional evidence for four frog species from the northwestern and midwestern United States showing no changes.[76] Taken together, these studies reveal a correlation suggesting some species are responding to climate change by shifting their breeding season, whereas others are not. The same species may also shift in one part of its range and not another. Climate change varies regionally, and species vary in physiology and life history, so it is not surprising that species respond differently. In contrast, it is precisely the fact that two major groups, amphibians and reptiles, showed similar population changes in Costa Rica over the last 25 years that makes a recent study so interesting.

The La Selva field station of the Organization for Tropical Studies is in an evergreen wet forest in the Caribbean lowlands of Costa Rica. For decades, graduate students contributed to a field project in which a standard method was used to estimate population densities of amphibian and reptile species living in the leaf litter of undisturbed forest and nearby abandoned cacao plantations. Steven Whitfield of Florida International University asked the question, Have population sizes of leaf litter amphibian and reptile species at La Selva changed relative to climate? He took advantage of the graduate students' long-term data set to find an answer.[77]

Whitfield's team found that daily minimum temperature increased by about 1.0° C between 1980 and 2005; rainfall did not increase from 1970 to 2005, but the proportion of days with no rain decreased. La Selva got warmer and wetter; that is, dry season length decreased. Since 1970, all species of terrestrial amphibians declined by 75 percent, and all common reptile species declined as well. In primary forest, the population densities of 12 frog species and 4 lizard species decreased significantly, at annual rates of about 4 percent each; 4 frog species and 1 lizard species did not change significantly. In contrast, in nearby abandoned cacao plantations, amphibian and reptile densities increased by annual averages of 4 percent and about 3 percent, respectively—six frog species declined significantly and one increased, whereas one lizard species declined significantly and three increased. Notably, even species that increased in density in plantations declined in primary forest.

The researchers argued that climate change, specifically changes in temperature and rainfall, caused the decline in primary forest densities and the differences between habitats, each a subset of the total community of reptiles and amphibians. As La Selva got warmer and wetter, they postulated, leaf litter depth in primary forest decreased due to increased decomposition rates, and this negatively affected habitat availability and resources for frogs and lizards. In contrast, litter accumulation was greater in the cacao plantations, where trees produce new leaves and drop old ones several times a year. Unfortunately, the annual student exercise did not include quantification of leaf litter depth, so that mechanism is untested. Likewise, no data were reported for insect densities. Insects are a key prey item for both groups, and their availability is expected to vary with both rainfall and litter depth.[78] Still, the population declines are clear, and the fact that they occurred in two very different groups of animals is a compelling argument for a cause, such as rainfall variation, that could affect two major groups independently. Climate change was a readily suspected cause because of this potential for influencing a broad spectrum of species and the fact that its effects would be felt over wide areas, both directly and indirectly.

Regions Will Vary in Response to Climate Change

When researchers first realized that amphibians were declining, especially in protected areas, climate change was among the first hypotheses tested. Cynthia Carey at the University of Colorado, Ronald Heyer from the Smithsonian Institution's National Museum of Natural History, and others worked with the U.S. National Aeronautics and Space Administration (NASA) to organize a workshop to explore the issue. In 2001, the journal *Conservation Biology* published the group's results (Table 5.1); these will be discussed further in the next chapter.

In a summary paper, Carey and colleagues compiled data from 120 temperate and tropical localities with and without declines.[79] They gathered data from satellites, airplanes, and direct ground samples and used them to evaluate temporal variation in temperature, precipitation, wind direction, ultraviolet-B (UV-B) radiation, and concentrations of contaminants. They sought unusual changes in environmental variables in advance of or concurrent with amphibian declines. Their conclusion: "The variation in certain environmental variables documented by others [see Table 5.1] appears unlikely to have directly caused amphibian deaths."[80] The group left open the possibility that interactions among environmental variables and indirect causes might be at work. They reached this overall conclusion even though ultraviolet radiation, specifically UV-B, had increased significantly from 1979 to 1998 at Central and South American sites with declines. Differences regarding the UV-B dose that actually reached organisms on the ground in their native habitats were the basis for the discrepancy. And differences over the UV-B dosage reaching amphibians on the ground or in aquatic habitats is one of the factors contributing to the controversy surrounding the relationship of climate to amphibian declines.

TABLE 5.1 Remote Sensing and Environmental Correlates of Declines

Study Areas	Variables Measured	Methods	Conclusion	Source
Puerto Rico	Forest cover; annual mean, minimum, and maximum daily temperature; annual rainfall; rain and stream chemistry; atmospheric-dust transport	Satellite imagery and air-chemistry samples from a single NASA aircraft flight	"...unusual climate, as measured by regional estimates of temperature and precipitation, is unlikely to be the direct cause of amphibian declines, but it may have indirectly contributed to them."	Stallard (2001)
Central and South America	Trends in solar UV-B radiation exposure (280–320 nm)	Total Ozone Mapping Spectrometer satellite data	"UV-B exposure ... increased in both regions. ... results should justify further research"	Middleton et al. (2001)
Colorado (USA), Puerto Rico, Costa Rica–Panama, Queensland (Australia)	Temperature and precipitation in the Americas; wind patterns in Australia	Observations from airplanes, land stations, satellites, ships, and weather balloons with output from a weather-forecast model; temperature and precipitation measured with thermometers and rain gauges at fixed locations	"...unusual climate, as measured by regional estimates of temperature and precipitation, is unlikely to be the direct cause of amphibian declines, but it may have indirectly contributed to them."	Alexander and Eischeid (2001)

NASA, National Aeronautics and Space Administration; UV-B, ultraviolet B light.

Cases in Which the Effects of Climate Change and Related Variables Are Hypothesized, Less Clear, or Debated

Utraviolet Radiation and Amphibians

Scientists divide the continuously varying spectrum of UV radiation into three discrete classes—UV-A, UV-B, and UV-C, in order of decreasing wavelengths. Most UV does not reach Earth's surface because of atmospheric ozone. But some UV-A and UV-B reaches the surface, and in humans it can cause sunburn and even increased risk of skin cancer. As noted in the previous paragraph, there is evidence that UV-B levels at Earth's surface increased during the last decades of the 20th century.

Andrew Blaustein suspected that UV-B changes might be connected to amphibian declines.[81] Joining with students at Oregon State University and colleagues, he conducted field and laboratory experiments to test how UV-B levels might affect amphibians. No effect was demonstrated in some species, but in others there was a range of lethal and sublethal effects that included embryo mortality, morphological abnormalities, reduced larval growth, and altered behavior. Other researchers have reached similar conclusions for amphibians as well as other species. Blaustein and his students also argued that climate-induced reductions in water depth at oviposition sites in Oregon caused mortality of embryos by increasing their exposure to UV-B radiation and, consequently, their vulnerability to infection by a pathogenic oomycete or water mold called *Saprolegnia ferax*.

But making the transition from evidence that UV-B negatively affects amphibians under experimental conditions to demonstrating that UV-B causes declines in wild populations is just as difficult as supporting the same logical sequence for contaminants. At present, there is no evidence relating negative experimental effects to long-term population declines or extinction. An experiment by Sara Broomhall and colleagues at the University of Sydney offers some support.[82] In the Snowy Mountains of southeastern Australia, they compared survivorship of embryos and tadpoles of *Litoria verreauxii alpina* (Alpine tree frog), a declining species, with that of *Crinia signifera* (common froglet), a nondeclining species. Artificial pools at three elevations (1365, 1600, and 1930 m) were the experimental units, and at each elevation they used three UV-B treatments: ambient, unshielded, and shielded (UV-B excluding). Both species always survived better when UV-B was excluded. At the two highest elevations, *L. v. alpina* had a significantly higher probability of dying for a given UV-B treatment than did *C. signifera*. The team concluded that UV-B is most likely a contributing factor to the disappearance of *L. v. alpina* at higher elevations.

In a review on AmphibiaWeb of how UV-B affects amphibians, Blaustein concluded: "We do know that UV-B radiation is harmful to many species of amphibians. However, we do not know how it, or for that matter, any other agent affects amphibians long-term at the population level. Long-term studies of the effects of UV-B radiation are warranted for a fuller understanding of how UV-B affects amphibian populations."[83]

There are two challenges to understanding UV-B's effects. The first is assessing how much radiation actually reaches an egg, larva, or adult. Many amphibians are active in shade, at night, or they live underground, which precludes a major or any UV-B effect. For this reason, Carey and colleagues concluded that, even though UV-B had increased near Central and South American sites of population losses, there was insufficient evidence to implicate UV-B as a cause. Many temperate amphibians breed in aquatic habitats, and their young develop in water. Stephen Corn concluded that boreal toads (*Bufo boreas*) in the Rocky Mountains near Fort Collins, Colorado, used breeding ponds at a time in the season that minimized UV-B exposure.[84] In a subsequent study in Glacier National Park, Montana, Blake Hossack from the Aldo Leopold Wilderness Research Institute and others wrote, "We found no support for the hypothesis that UV-B limits the distribution of populations (of boreal toads) in the park. ... Instead, toads were more likely to breed in water bodies with higher estimated UV-B doses."[85]

Corn's conclusion was reinforced by researchers from the U.S. Geological Survey, the U.S. Environmental Protection Agency, and the University of Washington, who sampled 136 potential amphibian breeding sites in the Olympic Mountains of Washington and the Cascade Mountains of Oregon and measured UV-B penetration of the water.[86] Organic matter dissolved in the ponds absorbed UV-B and protected 85 percent of the sites sampled. Levels of dissolved organic matter were high enough to protect most populations from UV-B levels known to harm amphibians. However, in 2004, in a series of papers in the journal *Ecology*, all of these researchers differed on how to interpret each other's findings.[87] Although this may seem chaotic or confusing, it is just the way science is supposed to work. A hypothesis is proposed and tested; others try to confirm it. If results are inconsistent, revised hypotheses and more tests follow until differences are resolved.

The second challenge is separating the effects of UV-B from those of other causes. Carlos Davidson teamed with Bradley Shaffer from the University of California at Davis and Mark Jennings from the California Academy of Sciences to study how pesticides, UV-B, and climate change affected eight frog species and one salamander species in California.[88] All were declining based on state or federal listing or other research. Land use change that destroyed or modified aquatic habitats was the main factor causing declines in one salamander (*Ambystoma californiense*) and two frogs (*Scaphiopus hammondi, Bufo californicus*). Windborne agrochemicals most affected five frog species (*Bufo canorus, Rana aurora draytonii* [now *R. draytonii*], *Rana boylii, Rana cascadae,* and *Rana muscosa*). Patterns of decline were not consistent with UV-B as a cause, nor were they consistent with climate warming. In a related study using 237 historic locations for *R. draytonii,* they reached similar conclusions, except that there was a greater effect of habitat loss through urbanization and partial support for an effect of UV-B.[89]

Across a range of studies, no evidence indicates that UV-B is a cause of extinctions, and there is conflicting evidence regarding its role in declines. Overall, without major new evidence regarding UV-B, the results to date suggest that this mechanism can be removed from among the hypothesized causes of global amphibian extinction.

Modeling Climate Change Effects on Amphibians

The effects of climate warming vary from none to positive or negative, depending on the species. We will have more to say about climate change, infectious disease, and declines in the chapters that follow. For the moment, however, the most recent IPCC report makes it clear that climate warming is expected far into the 21st century. How will projected changes affect amphibians? At the Museo Nacional de Ciencias Naturales in Madrid, Miguel Araújo and colleagues study climate warming and range distributions. For European amphibians and reptiles they recently asked, What proportion of species will lose and gain suitable climate space in the future? Do projections vary with taxonomic, spatial or environmental properties? And what climate factors drive projections?[90]

Their models indicated that most species will expand their distributions if dispersal is not limited—for example, by barriers such as large urban areas; most ranges diminish if expansion is limited. Dry conditions will increase in southwestern Europe, which means a decrease in suitable habitat space as regions such as the Iberian Peninsula become more similar in climate to current conditions in North Africa. Warming in currently cooler, northern Europe will allow some species to expand ranges, echoing our earlier discussion of range expansion in Canada or after glacial retreat. For amphibians and reptiles, climate cooling would actually be more deleterious than warming—ignoring possible effects of toxins, habitat loss, disease, and invasive species.

The forecasts of Araújo and colleagues will need revision, especially for amphibians, if projected decreases in water materialize (recall our earlier discussion of Canadian and Yellowstone ponds). As Earth warms, freshwater availability is likely to be the real limiting variable for amphibians and many other species. The authors also recognized that, especially for Europe, "no dispersal," or at least limited dispersal, is a more realistic assumption. Human developments of various sorts already block movement of all kinds of organisms, not just amphibians. These developments will only increase for the foreseeable future. On balance, they concluded, "most European species of amphibians and reptiles modeled in this study would be projected to lose suitable habitat by 2050."[91] Overall, future climate change is expected to cause amphibian declines in Europe, with some indication that salamanders might be at more risk than frogs. A recent U.S. study reinforces this result.

Joseph Bernardo, from the Southern Appalachian Biodiversity Institute, teamed with James Spotila at Drexel University for a more fine-grained look at the effects of climate change.[92] They used two terrestrial salamander species, *Desmognathus carolinensis* and *Desmognathus ocoee,* as models for montane amphibians and tested the capacity of species to respond to warming temperatures. Both species are distributed over broad elevations (900 to >2000 m) in North Carolina. In laboratory experiments, individuals of the two species collected across the elevation range showed significant depression of metabolism at temperatures within the natural thermal range experienced by populations at low elevation. Populations at the low elevation were also near their limits of physiological tolerance. Thus, even if dispersal is possible, warming temperatures will place many montane, cool-adapted species beyond the limits of physiological tolerance. The conclusion assumes limited evolutionary response, and the authors argued that this is reasonable, because climate change will create warm, dry regions that fragment populations, reduce effective population sizes (with a correlated loss of genetic variability), and reduce dispersal ability and the potential for infusion of new genetic diversity.

Christopher Raxworthy at the American Museum of Natural History led a team that studied how climate change was affecting amphibians and reptiles in tropical montane habitats in Madagascar.[93] The island nation is warming: from 1993 to 2003 they found an average shift of 19 to 51 m upslope in the ranges of 30 species of amphibians and reptiles. The group is cautious in its conclusions, but by one projection, "a temperature increase of 1.7° C would be sufficient to lead to complete habitat loss for three species" (two frogs and one lizard).[94] They go on to note that these extinction displacement threats are not unique to Madagascar.

Finally, living in a thermally stressful environment may also compromise the ability of amphibians to resist infectious disease. The role of pathogens as a cause of declines and extinctions is the next topic we will examine.

Amphibian Diseases

The IPCC report concluded that climate change will affect some species negatively and others positively. Amphibian populations are already exhibiting both outcomes. But in cases in which amphibians do migrate successfully to occupy new ranges, they will also bring along their pathogens. As the three species of Peruvian frogs studied by Seimon and colleagues colonized new habitats after glacial retreat, two of the species harbored an emerging pathogenic chytrid fungus, *Batrachochytrium dendrobatidis,* that causes an infectious amphibian disease called chytridiomycosis.[95] Increasing evidence suggests that emerging infectious disease is a threat that might explain a number of enigmatic amphibian losses.

Epidemiologists distinguish endemic infectious diseases that are already established in an area from emerging infectious diseases. An emerging infectious disease may be one that is newly recognized by biologists; it may be a known disease that recently appeared in a population; or it may be an established disease that is rapidly increasing in incidence, virulence, or geographic range. Some infectious diseases show emergent and endemic characteristics, depending on the pathogen, the population structure, and how the disease is transmitted.[96]

Acquired immunodeficiency syndrome (AIDS), severe acute respiratory syndrome (SARS), and hepatitis C are all diseases that have "emerged" in human populations in the past 50 years or so. West Nile virus emerged in New York City birds in 1999; as it spread across the United States, it caused population declines in seven bird species. American crow populations declined by some 45 percent.[97] Amphibians do not suffer from these diseases, but they are affected by others. The two major emerging pathogens of amphibians are a fungus, *Batrachochytridium dendrobatidis,* and a class of viruses known as iridoviruses. The tale of these diseases and amphibian declines begins in the 1990s.

Macroparasitic Amphibian Diseases

All species harbor parasites. Small parasites such as viruses are microparasites. Fittingly, large parasites such as the trematodes that cause abnormal development in frogs are macroparasites. Recall that Pieter Johnson and colleagues concluded that the trematode *Ribeiroia* is a longtime cause of malformations in amphibians and suggested that parasite-induced malformations have increased. Limb abnormalities, they concluded, were an emerging disease in amphibians. Still, they offered no evidence that amphibian population size decreased with frequency of abnormal individuals, and so far no one has provided evidence that macroparasites cause amphibian extinctions. Microparasites are a different story.

Microparasitic Amphibian Diseases

Protozoa, bacteria, viruses, and fungi are microparasites that cause infectious diseases in amphibians and many other organisms. All may kill frogs or salamanders under ideal conditions, but only protozoa, viruses, and fungi are implicated as causes of declines and extinctions. We have not mentioned caecilians, because we know very little about infectious disease in that amphibian order.

Protozoa

As we already discussed, Blaustein and his students Joseph Kiesecker and Lisa Belden at Oregon State University tested *Saprolegnia ferax,* a water mold, to determine whether it was a pathogen causing population declines in frogs—*Bufo boreas, Rana cascadae,* and *Pseudacris regilla*—in Oregon's Cascade Mountains. Water molds look like fungi and were once classified as such, but they are now classified as Protozoa closely related to brown algae and diatoms. In one experiment, the Blaustein team used plastic swimming pools in lakes as experimental units.[98] Control treatments had only *Saprolegnia;* in the experimental treatments, they exposed developing frog eggs to UV-B alone or UV-B plus the pathogen. Embryo mortality rates with *Saprolegnia* plus UV-B exceeded that with either treatment alone. In a related study with only *B. boreas,* field experiments assessed *Saprolegnia,* UV-B, and water depth interactions.[99] Embryo mortality was greatest when the pathogen and UV-B were present and water depth was low. The team concluded that recent high levels of embryonic mortality in local populations correlated with climate-induced reductions in water levels of Oregon ponds caused by changes in the El Niño/Southern Oscillation Cycles. But to date no data directly link *Saprolegnia* infections to declines or extinctions of local free-living amphibian populations.

Viruses

Eight viral groups infect amphibians—adenoviruses, caliciviruses, flaviviruses, parvoviruses, retroviruses, togaviruses, herpesviruses, and iridoviruses—but only the last two are known to be pathogenic.[100] Herpesviruses cause renal tumors in frogs, but they are rare and not implicated in population declines or extinctions; iridoviruses are common and involved in amphibian epidemics.

Iridoviridae is a large family distributed worldwide. First described in insects, the name derives from the fact that an array of virus particles, or virions, in an infected insect can cause patches of iridescence. The family has four genera: *Iridovirus* infects invertebrates, mainly insects; *Chloriridovirus* infects mosquitoes; *Lymphocystivirus* infects fish; *Ranavirus* infects salamanders, frogs, fish, and reptiles.[101]

At least eight *Ranavirus* strains are reported from amphibians and may infect multiple species or only one. Bohle iridovirus (BIV), for example, infects amphibians, fish, and reptiles.[102] Frog virus 3 is reported from almost a dozen frog species and one salamander species. So far Ambystoma tigrinum virus (ATV) is reported only from amphibians[103], and it is the strain we know the most about.[104] ATV-infected amphibians slough a lot of skin while also developing external polyps and small lesions on skin and viscera. Experiments by Jesse Brunner, Amy Greer, and Danna Schock while graduate students at Arizona State University under one of us (JPC) showed that infected salamanders clear ATV, remain chronically infected, or die within weeks.[105] The likelihood of an infected individual's dying depends on the viral strain and dose, host vigor, life history stage, and species.

Ranaviruses cause epidemics in native frogs in Europe, South America, and Australia; in frogs and salamanders in North America; and in aquaculture colonies of frogs in Asia.[106] Alex Hyatt, a world authority on iridoviruses, manages and directs the diagnostic electron microscopy laboratory at the Australian Animal Health Laboratory in Geelong. He and his colleagues are examining whether iridoviruses, or their genes, can be used in a biocontrol program to stop the spread of exotic cane toads in Australia.[107]

Iridoviruses are large (150–200 nanometers in diameter) compared to most viruses. The genome is a single molecule of linear, double-stranded DNA that can encode about 100 polypeptides/proteins. Iridovirus viral particles have a protective protein coat called an icosahedral capsid—think of a geodesic dome, or more precisely a geodesic sphere—that encloses the genome. A soccer ball is also an icosahedron. Hand, foot and mouth disease; poliomyelitis; rhinovirus (which causes the common cold); and hepatitis are icosahedral viruses that infect humans.

In the late 1980s, David Pfennig at the University of North Carolina was a postdoctoral associate in JPC's laboratory. We tested whether infectious disease might explain the variation in frequency of cannibalistic tiger salamanders in Arizona.[108] JPC had observed that where epidemics were common, cannibalism was rare; conversely, if epidemics were rare, cannibals were common. We confirmed the hypothesis and attributed the epidemics to a bacterium, *Aeromonas hydrophila*. Jamie Sedon, an undergraduate in the laboratory, followed up this research and showed that *Aeromonas* infected and killed salamanders at very high doses and only if a larva were injured. But epidemics were sufficiently common and involved high enough numbers of animals to suggest that something other than *Aeromonas*, which required relatively specialized conditions for infection, might be involved.

James Jancovich, a Masters student in JPC's laboratory and Elizabeth Davidson's laboratory, investigated and showed that indeed a new *Ranavirus* strain, which he named Ambystoma tigrinum virus, was the primary pathogen, and *Aeromonas* was most likely a secondary infection.[109] Jancovich's research included completing Koch's postulates: he isolated the virus from sick animals, grew it in pure culture in vitro using fish cells, injected the virus into healthy salamanders with controls, and reisolated the virus from animals that became sick with symptoms identical to those of the original group of ill animals. Control salamanders had no disease.

There is no evidence that ranaviruses are causing species extinctions. Research by Greer and Brunner for their dissertations indicated that ranaviruses in salamanders most closely fit a density-dependent model of host-pathogen dynamics, and under these conditions, as we will learn in chapter 7, extinction is not expected.[110] Indeed, tiger salamander populations in Arizona experience epidemics every few years that reduce population sizes and may even cause local extinctions, but there is no evidence that ranaviruses cause species extinctions. So far, this fact distinguishes viral and protozoan infections in an important way from chytrid fungi.

Fungi

Mushrooms, rusts, smuts, puffballs, truffles, morels, molds, and yeasts are common fungi.[111] Many microscopic fungi are less well known. For example, a yeast-like fungal cell, *Pneumocystis jirovecii,* causes pneumocystis pneumonia in humans with suppressed immune systems. The fungus *Coccidioides immitis* grows in soils in areas of low rainfall, high summer temperatures, and moderate winter temperatures, such as in the southwestern United States. If humans inhale airborne *Coccidioides* spores, valley fever, or coccidioidomycosis, may result. The decline and extinction of frogs in Australia, Central America (salamanders also), Europe, Africa, the Caribbean, and North America are associated with a chytrid fungus called *Batrachochytrium dendrobatidis* (Bd).

For decades, herpetologists have followed in the footsteps of the late Margaret Stewart of the State University of New York at Albany (now the University at Albany) and studied Puerto Rico's coqui frogs in the genus *Eleutherodactylus.* Stewart's most recent followers are Patricia Burrowes and her husband Rafael Joglar from the University of Puerto Rico. Burrowes and Joglar joined with David Green from the U.S. Geological Survey's National Wildlife Health Center to study the decline and disappearance of eight *Eleutherodactylus* species in Puerto Rico. As mentioned in chapter 1, *Eleutherodactylus karlschmidti, Eleutherodactylus jasperi,* and *Eleutherodactylus eneidae* are now most likely extinct; eight populations of six other species have declined at elevations above 400 m in Puerto Rico's mountains.[112]

Just as at La Selva,[113] the daily minimum temperature in the Puerto Rican study sites increased by about 1.0° C between 1970 and 2000. In contrast to La Selva, mean monthly precipitation was not significantly different, but the number and length of dry periods each year increased over this same period. These researchers also reported for the first time in the Caribbean a pathogenic chytrid fungus known to kill amphibians. A study of museum specimens revealed chytrids on frogs since at least 1976. Burrowes and her team hypothesized that drought-stressed frogs infected by chytrid are more susceptible because the pathogen compromises their ability to absorb water.[114]

If disease played a role in the extinction of Puerto Rico's amphibians, this would be an important discovery, because pathogens are not usually a suspected cause of extinction. An exception to this generalization is the Polynesian snail, *Partula turgida,* extinct because of a pathogen infection in the only remaining captive colony.[115] In three other cases a pathogen is the suspected cause of extinction: 1) the sharp-snouted day frog in Australia along with many other amphibians; 2) some 13 Hawaiian land bird species; and 3) the possibility that a "hyperdisease" caused the extinction of large Pleistocene mammals—a point we will discuss in the final chapter.[116] In chapter 7, we examine in detail why pathogens are not generally thought of as causing extinction and why the chytrid fungus is such a striking exception.

Chytrids or chytrid fungi are in the most basal phylum of Fungi, called Chytridiomycota, which means that they show the traits of ancestral fungi; members of this phylum typically degrade organic matter.[117] The group has motile gametes (zoospores) with a posteriorly directed flagellum that makes them move and look like human sperm; only fungi in this phylum have zoospores. Sexual reproduction and a life stage resistant to stressful environments, typically called a spore stage, are not known for Bd, but are suspected by several investigators. Other Chytridiomycota species have sexual reproduction, resulting in a resistant life stage. Resting stages are important mechanisms for survival during periods when a host is absent, and they often play a role in long distance dispersal.

Many of the approximately 1000 described chytrid species parasitize algae, other fungi, vascular plants, protozoa, and invertebrates, but Bd is the first chytrid known to parasitize a vertebrate host. Chytrids occur in most aquatic and terrestrial habitat types—and even in the rumen and hindgut of large herbivores—from the tropics to the Arctic. The only chytrid parasitizing a vertebrate group is Bd, and the only vertebrate group parasitized is amphibians. All evidence to date indicates that Bd lives only on amphibian skin. Bd transmits easily between conspecific and heterospecific individuals in laboratory environments, and researchers suspect the same is true in the field. Bd is reported from all continents except Antarctica, which has no amphibians.[118]

The amphibian chytrid has a simple vegetative tissue called a thallus, with a reproductive body (zoosporangium) and branched, thread-like rhizoids, which look like tiny roots but have no nuclei. Zoospores are the only known reproductive stage, and they need water to disperse. In amphibian hosts, thalli (7–15 μm) can occur in large numbers within the epidermis, resulting in a thickening of the skin that may be the proximate cause of death. The thick skin may compromise an animal's ability to maintain a proper physiological balance of salts.[119]

Louise Rollins-Smith and her students at Vanderbilt University have evidence that proteins secreted from frog skin can kill Bd.[120] Reid Harris at James Madison University has evidence that the community of microbes on the skin of the four-toed salamander, *Hemidactylium scutatum,* and the red-backed salamander, *Plethodon cinereus,* both of which have wide distributions in the eastern United States, have antifungal properties and may protect against Bd.[121] If microbial communities differ among species or populations, these results could explain why some species are susceptible to Bd and others are not; they also offer hope for protecting frogs and salamanders in zoos and conservation facilities from infectious disease.

Megan Johnson and Richard Speare, of The School of Public Health and Tropical Medicine at James Cook University, showed that zoospores can survive apart from an amphibian host in tap water and in deionized water for 3 and 4 weeks, respectively.[122] In lake water, Bd zoospores survived and were active 7 weeks after introduction. In all three cases, they used sterilized water. They reasoned that sterile water afforded the best opportunity for survival and recovery of Bd because microbial competitors were absent, and this proved true. Their lake water included dead, sterilized microscopic algae, arthropods, protists, and plant debris. Chytrid zoospores introduced to the experimental flasks in the laboratory attached to this organic material, developed, and released new zoospores only in the lake water, not in the tap or deionized water.

Karen Lips and colleagues from Southern Illinois University, working in Panama, have added to this picture.[123] They used a molecular method that detects Bd DNA and recovered Bd evidence from rocks previously occupied by frogs as well as rocks downstream from these spots, but they could not say for sure that the specimens collected were alive. Recovering only DNA is not the same as recovering viable cells, just as nonviable DNA evidence in human cases may be recovered years after a crime.

We need a consistently reliable method to recover Bd in natural environments to help us understand how it survives apart from amphibian hosts and how it moves among habitats. In the first moves in this direction Julie Kirshtein at the U.S. Geological Survey and Susan Walker at Imperial College have partnered with other researchers to develop a molecular method for detecting Bd in water and sediment samples from the field.[124] Does Bd move only by amphibian-to-amphibian contact, or by amphibians contacting birds or flying insects that can then disperse over several kilometers? Or perhaps Bd can be picked up by the wind and dispersed as naked zoospores in droplets of water in mist, fog, or clouds. Perhaps Bd disperses from aquatic habitats attached to leaves, twigs, or debris. If there is a resting stage or spore that aids survival in stressful environments, this is a more likely dispersal stage than unprotected zoospores.

Two major questions still remain unanswered: How can Bd drive species extinct? And where did Bd originate?

Several research teams are working to answer these questions, and we will meet them in chapter 7. For the moment, however, we will backtrack and meet some of the scientists and examine in detail the steps that led to the conclusion that Bd was indeed a major cause of enigmatic amphibian declines and likely extinctions.

6

Unraveling the Mystery

We ended the last chapter with a brief description of an emerging disease caused by the chytrid fungus *Batrachochytrium dendrobatidis* (Bd), a pathogen associated with many amphibian population declines around the world. Whether Bd is the direct and only cause of certain declines, whether it acts only when catalyzed by or associated with some other environmental factor, or whether it comes into play only when amphibians are stressed will be discussed in chapter 7. The answer differs among species and geographical locations.

Our current understanding of the chytrid scenario results from more than a decade's worth of "detective work" by scientists worldwide. These scientists have diverse expertise, from ecology to pathology. They have used various methodologies, including empirical work and field observations of amphibians, laboratory and field experiments, rigorous hypothesis testing, and searching for correlations between factors. Each scientist built on previous work by others, either expanding a line of investigation or proposing alternative ideas.

In this chapter, we present three of these "scientific detective stories." Each focuses on an amphibian assemblage or species that occupies a limited geographical area. The studies involve stream-dwelling frogs from northeastern Queensland, Australia; the boreal toad (*Bufo boreas*) from the Colorado Rocky Mountains; and montane frogs from Costa Rica and Panama. In each case, the declines initially were enigmatic, in part because habitat destruction or modification was not an obvious cause. Each story illustrates how international scientific collaboration helped unravel the mystery.

These "detective stories" are fascinating in themselves, but they also illustrate how science works when the causes of complex phenomena cannot be observed easily. They also reveal that we have not reached the final page of the detective story. New discoveries, or recognition of flaws in our current thought, surely will continue to enlighten us.

Stream-dwelling Frogs from Northeastern Queensland

The tropical rain forest of northeastern Queensland, Australia (Figure 6.1), used to be a special place to see a diverse array of endemic stream-dwelling frogs. It still is a special place, but you will see fewer frogs there now than you would have 2 decades ago. Here is the story of the frogs' disappearances and how scientists identified the cause. See Figure 6.2 for photographs of

Figure 6.1 Australian wet tropics, Birthday Creek, northeastern Queensland. Photograph by Ross Alford.

Figure 6.2 Six species of frogs whose populations have declined or disappeared from northeastern Queensland. *Top left*: green-eyed treefrog, *Litoria genimaculata*. *Top right*: waterfall frog, *Litoria nannotis*. *Middle left*: common mist frog, *Litoria rheocola*. *Middle right*: mountain mist frog, *Litoria nyakalensis*. *Bottom left*: sharp-snouted dayfrog, *Taudactylus acutirostris*. *Bottom right*: Australian lace-lid, *Nyctimystes dayi*. Photographs of green-eyed treefrog, waterfall frog, and common mist frog by Ross Alford; photographs of mountain mist frog, sharp-snouted dayfrog, and Australian lace-lid by Stephen Richards.

six of these species and Table 6.1 to match the common and scientific names and ecological information.

Early Observations

Stephen Richards at James Cook University in Townsville, Australia, and two colleagues first reported, in 1993, major declines in six species of endemic stream-dwelling frogs.[1] Their data came from monitoring studies of 4 sites between 1989 and 1992 and from quantitative surveys of 47 sites at four streams during 1992. Most declines occurred at high elevations. The field crews failed to find two species, the mountain mist frog (*Litoria nyakalensis*) and the northern tinker frog (*Taudactylus rheophilus*), at any site during their 1991–1992 survey. These two species are restricted to upland rain forest. Four species were absent from most high-elevation sites south of the Daintree River: waterfall frog (*Litoria nannotis*), common mist frog (*Litoria rheocola*), Australian lace-lid (*Nyctimystes dayi*), and sharp-snouted dayfrog (*Taudactylus acutirostris*). Curiously, three of the four (all except the sharp-snouted dayfrog) persisted at most lowland sites surveyed.

Richards and his colleagues analyzed the streams' water chemistry but found no evidence for water pollution or elevated acidity levels. They ruled out habitat destruction. Most of the area is protected under World Heritage legislation, and delayed effects of earlier forestry and mining activities seemed unlikely. They eliminated lack of rainfall during the wet season, overcollecting, and road kills. Presence of feral pigs seemed the only possible contributing factor. The investigators found only one dead frog during their surveys, but because mortality in nature is often hard to observe, they did not rule out disease as a potential cause for the declines and disappearances.

Subsequent studies documented further declines from the same general area of the wet tropics of northeastern Queensland. Between July and November 1993, Michael Trenerry, then a student at James Cook University, and two colleagues censused tadpoles at 62 upland sites along six stream systems in northeastern Queensland rainforest.[2] They focused on tadpoles, because that stage occurs year-round and therefore is more readily and reliably detected than adult frogs. Their results supported and extended the 1993 report. In 1996, Jean-Marc Hero at Griffith University Gold Coast in Queensland, reported that a field crew had searched with mixed success for previously missing and declining frogs in rain forest areas, including northeastern Queensland.[3] The data reinforced earlier surveys but also revealed that the sharp-snouted dayfrog had disappeared from the rain forest of northeastern Queensland. Clearly, something was wrong. What was causing the declines?

A Hypothesis

William Laurance, then at the CSIRO Tropical Forest Research Centre in Atherton, Queensland, and two colleagues published a paper in 1996 suggesting that a rapidly spreading epidemic disease may have caused the catastrophic decline of at least 14 species of endemic rain forest frogs in eastern Australia.[4] They speculated that the pathogen was a water-borne virus that affected stream-dwelling frogs. Potential vectors for the disease might be migrating fish, exotic aquarium fish released into the wild, the introduced cane toad (*Bufo marinus*), aquatic birds, aquatic insects, or hikers' shoes. Laurance and his colleagues offered eight lines of evidence that they suggested collectively supported their epidemic disease hypothesis. Briefly, these lines of evidence and selected salient points are as follows:

- *Ecological similarities of vulnerable species.* The affected species all breed in and live near streams or adjoining seepages.
- *"Extinction wave" pattern.* There is an apparent south-to-north pattern in the declines of frog populations in Queensland. Over a period of 15 years, the frog declines spread northward across 1500 km of Queensland at a mean rate of 100 km/year. Such a wave-like pattern of decline, they argued, is typical of an epidemic involving a highly virulent infectious agent spreading through populations lacking immunity.
- *Rapid declines.* In all known cases, populations declined rapidly, supporting the hypothesis of infection by a highly virulent pathogen.
- *Pathological evidence.* Pathologists examined 24 moribund frogs collected in 1993 from declining populations on Big Tableland, near Cooktown in northeastern Queensland. Histology of infected tissues suggested a viral infection. The frogs' kidneys, livers, skin, and other organs had widespread necroses. The frogs were lethargic and exhibited motor dysfunction.
- *Experimental evidence.* The authors referred to an unpublished 1994 report of a translocation experiment in which larval and adult sharp-snouted dayfrogs and common mist frogs were held in field aquaria fed by stream water at three sites in northern Queensland. As a control, sharp-snouted dayfrogs were held in an aquarium fed with rainwater. The control population persisted for over 10 months. In contrast, each of the 13 stream-fed aquaria populations died in less than 3 months. The sick and dying frogs exhibited lethargy, motor dysfunction, and skin necroses—external appearances indistinguishable from those of the moribund frogs collected from Big Tableland in 1993.

- *Persistence of lowland populations.* Lowland populations of several affected species persisted, whereas upland populations disappeared. Animal viruses often have narrow optimal thermal ranges. This particular virus may be more virulent at colder temperatures. Furthermore, low temperatures might limit the upland frogs' antibody response, reducing their ability to combat disease.
- *Selective declines.* The fact that unaffected species of frogs occurred in the same areas as declining or disappearing species was consistent with a disease hypothesis, because many viruses exhibit host specificity. The unaffected species might even serve as reservoirs for the virus and transmit it to vulnerable species.
- *Absence of plausible alternatives.* Laurance and his colleagues ruled out natural population fluctuations, large-scale habitat modification, water pollution, increased exposure to ultraviolet B (UV-B) radiation, unusual weather, predation, and overcollecting as possible causes.

This disease hypothesis generated considerable discussion among herpetologists. Had the cause of Australian, or perhaps even worldwide, declines finally been identified? Many biologists argued it was too soon to know. The hypothesis required testing, and the pathogen—if indeed there were one—needed to be isolated and identified.[5]

A Breakthrough

Meanwhile, other scientists focused on the cause of mass mortality of frogs observed in 1993 from Big Tableland, Australia. Andrew Cunningham at the Institute of Zoology in London, Chair of the Disease and Pathology IUCN Working Group, recounted in the December 1998 issue of *Froglog* the series of events that led to isolation of the disease-causing pathogen.[6] The team of Keith McDonald (herpetologist, Conservation Strategy Branch, Queensland Department of Environment), Alex Hyatt (virologist, Australian Animal Health Laboratory), Richard Speare (veterinarian, James Cook University), and Lee Berger (veterinary pathologist, James Cook University) detected the same parasite in the skin of almost every carcass examined from Big Tableland. Microscopically, the parasite resembled either a fungus or a protozoan. Needing further expertise, they sent samples to a British protozoan taxonomist, Peter Daszak, then at Kingston University, who later identified the pathogen as a chytrid fungus—not a protozoan.

In what Cunningham referred to as "one of those fortuitous accidents of history," Karen Lips, then a graduate student at the University of Miami, discovered dead and dying frogs in Panama during 1992–1997 (discussed later in this chapter). In early 1997, Lips contacted David Green, a veterinary pathologist then at the Maryland Animal Health Laboratory. Green reported histological findings in the skin of the dead Panamanian frogs similar to those described from the Australian frogs. The *New York Times* published a photograph of the chytrid and an article describing the amphibian die-off in Panama. When the Australian biologists read the article, they strongly suspected their fungal pathogen was the same.

During this same time period, Donald Nichols, a veterinary pathologist at the National Zoological Park in Washington, D. C., was studying an outbreak of a skin disease from the zoo's captive frogs.[7] Nichols and his colleagues, Allan Pessier (National Zoological Park) and Joyce Longcore (University of Maine), eventually identified the culprit as a chytrid fungus. Nichols heard about the disease-associated declines in Australia and Central America and made contact with the scientists involved in those regions; they all suspected that they were dealing with the same pathogen.

In short order, scientists from three continents—Australia, Europe, and North America—gathered for a week-long meeting in Champaign, Illinois, funded by George Rabb, then Director of the Brookfield Zoo. Participants included Daszak, Green, Berger, Cunningham, Lips, Longcore, Nichols, Pessier, and others. Comparative study suggested that the pathogen found on the diverse frogs was indeed the same.

In 1998, Lee Berger and 13 collaborators published their findings in a landmark paper entitled "Chytridiomycosis causes amphibian mortality associated with population declines in the rain forests of Australia and Central America."[8] Instead of a virus as first hypothesized, a fungus was the culprit, a member of the phylum Chytridiomycota. How did chytridiomycosis emerge as a disease of wild frogs on the two continents? The authors suggested alternate possibilities. First, the fungus might be an introduced pathogen spreading through populations previously unexposed to the chytrid. Second, the chytrid might be a widespread organism that only recently became pathogenic to frogs, either because it increased in virulence or because frogs as a group are now more susceptible to it. Environmental changes or other factors may have increased susceptibility. In 1999, Longcore and her collaborators named the chytrid *Batrachochytrium dendrobatidis* (Bd), a new genus and species.[9]

As news of the amphibian chytrid broke, journalists enthusiastically announced the discovery to the public. Ian Anderson's headline in *New Scientist* read, "A Great Leap Forward." The first sentence stated, "At long last, zoologists may know what is killing the world's amphibians."[10] That was indeed a great leap—from a possible cause of specific population declines to an explanation for worldwide declines that were not due to obvious habitat destruction or degradation.

Tim Halliday, at the time International Director of the Declining Amphibian Populations Task Force and Professor at The Open University in England, responded to Anderson's article in the journal *Nature*.[11] He referred to isolation of Bd as "a very exciting discovery," but warned that "claims in the popular media that the answer to the declining amphibian conundrum has been found are premature. It [the discovery] does not solve the puzzle, but it does raise further questions." Halliday wisely warned that, until we know more about the pathogen's distribution and ecology, we do not know how significant the fungus is as a cause of declines. There may be more to the story than simply a pathogen infecting otherwise healthy frogs. We might discover that frogs' immune systems are being compromised by environmental factors such as climate change or chemical contamination. Halliday warned that it would be a mistake for scientists to stop investigating other potential causes of declines in the belief that Bd is the only answer.

Halliday was not the only biologist skeptical of assuming that Bd explained worldwide enigmatic amphibian declines. Many biologists argued that if we focused entirely on Bd, we would almost certainly paint a biased picture of the causes for worldwide declines, and we might overlook interactions among factors.

There was good reason for the skepticism. In 1998, scientists had no prior model of an infectious disease causing such havoc. No other global pathogen had ever affected so many species. In fact, at the time, there had never been a report of extinction of a free-ranging animal species caused by infectious disease. The ensuing debate was healthy. After all, science is a process of testing alternative hypotheses.

Research Continues

Meanwhile, Australian frog populations kept declining, and local biologists continued to survey and document those declines. In 1999, Keith McDonald and Ross Alford (James Cook University) reported that populations of eight species of stream-breeding frogs from northeastern Queensland had declined or disappeared—the same six reported by Richards and colleagues in 1993, plus two additional species, the armoured mist frog (*Litoria lorica*) and the green-eyed tree frog (*Litoria genimaculata*). In contrast, apparently none of the terrestrial-breeding, direct-developing microhylids that occur in these same areas had declined.[12] This made sense, because Bd was thought to be a water-borne pathogen.

Because investigators often found dead and dying frogs infected with Bd during cooler seasons and at high elevations, biologists speculated that

- 96 percent of the trade was commercial, mainly for pets and food
- Few commonly traded species were monitored or regulated at the species level
- Most of the trade involved individuals from nine frog families and two salamander families[39]
- The United States represented about 15 percent of the world market made up of all other CITES party nations, but there was no global database or monitoring program for trade in non-CITES species between countries outside the United States
- The analysis did not include animals collected in the United States for the domestic or black markets.

Schlaepfer and colleagues concluded that the data "do not provide conclusive evidence of widespread, unsustainable collections. They do, however, reveal that the volume of animals taken from the wild (for the U. S. market alone, let alone globally) is large enough to potentially extirpate populations or species."[40] Clearly, more research is needed to understand when and at what harvest levels the taking of wild-caught animals is sustainable.

We know of no case in amphibians where a joint analysis of the population biology of the species, estimated sustainable yields, and expected economic costs and profits preceded establishment of a market. Exploitation of amphibian populations is usually an empirical experiment; that is, animals are harvested and sold, and collecting stops when the population size becomes so low that harvesting is no longer rewarding economically.[41] Schlaepfer's team made two recommendations to address this deficiency: (1) improve estimates of how many animals can in theory be harvested from wild populations, and (2) quantify with greater precision how many animals are removed. We agree. Such data are required to develop the mathematical models needed to predict when populations can support sustained collecting.

The ability to assess the effects of collecting will increase in importance as new and easier trade routes unite previously unconnected markets, linking species to ecosystems far beyond where they evolved. A recent report from Japan highlights this reality. In 2002, Bryan Stuart, a researcher at the Field Museum of Natural History in Chicago, described the Laos warty newt (*Paramesotriton laoensis*), a strikingly colored salamander species from Southeast Asia.[42] Following standard scientific methods, he described the new species and where he collected it. In March, 2006, Laos warty newts appeared for sale by Japanese pet traders for US $170 each. Apparently, a dealer noted the description of the new species in the technical scientific literature and used the report to locate animals for collection and sale. The tactic alarmed scientists because of how the research literature was used for commerce and because the species' range is unknown, so it is unclear whether it needs protection. The dealer justified his actions by observing that there were "lots of them [salamanders]" and, in fact, "local people eat them." As the world behaves more and more like one market, research is needed to assess the effects of globalization on biodiversity in terms of increased trade and the increased rate at which new species can be brought to commercial use.

It is likely that indigenous amphibians of the right size for human consumption have been hunted for food for centuries. Under the right conditions, harvesting could be sustainable, but the red-legged frog in California was hunted to unsustainable levels by the late 1800s. Land use changes and introduced bullfrogs combined with unsustainable harvesting to diminish populations even more.

We suspect that two broad classes of amphibian populations are being harvested. In the first class, animals are harvested from largely intact, native communities. We predict that in these cases commercially successful yields will rarely be sustainable, and native populations will decline to levels that threaten survival as small populations become vulnerable to other forces, including land use change, introduced species, diseases, and contaminants. The red-legged frog fits this scenario.

In the second group, animals are harvested from greatly modified native communities, such as tropical marshlands converted to rice paddies. Under these conditions, amphibian biodiversity is reduced because species are lost as complex tropical and subtropical habitats are converted to simpler agricultural communities, but, for the amphibian species that can coexist with agriculture, population numbers will stay high enough to support sustainable yields. American bullfrogs harvested from rice fields in Cuba seem to fit this model.

Collecting Amphibians for Bait

Amphibians are a minor part of the bait trade in comparison to artificial lures or other live bait such as fish or invertebrates. Still, amphibians are used as bait throughout the United States, where the commercial trade activities can be divided into midwestern/north-central, southeast, and western regions.[43]

Frogs are the preferred bait in the midwestern and north-central region, where there are harvest seasons, size restrictions, and collecting limits. There is a limited conservation concern for these species, but, as we will learn, collecting and moving these animals is likely also moving their pathogens. There is a smaller bait industry in the southeastern part of the country, where salamanders (genus *Desmognathus*) are sold as "spring lizards." Most states lack regulations for salamanders.

In the western United States, millions of tiger salamanders (*Ambystoma tigrinum*) of several subspecies are moved annually from the central plains to the intermountain west for sale as bait. As in the southeast, the western U.S. bait industry developed in response to fishing, mostly for sunfish, bass, and catfish indigenous to the eastern United States. The fish are introduced into ponds, lakes, rivers, streams, and large reservoirs. A 1970 study concluded that almost 2.5 million salamanders, valued at approximately US $500,000 (equivalent to US $2,766,489 in 2005 dollars[44]) were sold annually along the Colorado River bordering Arizona and California. Arizona tiger salamanders (*Ambystoma tigrinum nebulosum*) are collected in Arizona and sold within the state. Barred tiger salamanders (*Ambystoma tigrinum mavortium*) have been imported as bait into Arizona, Colorado, and California and deliberately or inadvertently released into native habitats and livestock ponds. Importing of salamanders into California and southeastern Arizona is presently prohibited.

The barred tiger salamander is now an established, nonindigenous subspecies that breeds with native races, altering the gene pools of California (*Ambystoma californiense*) and Sonora (*Ambystoma tigrinum stebbinsi*) tiger salamanders.[45] Both taxa are on the U.S. Endangered Species list. Fifty years ago, Bernard Martof, a University of Georgia herpetologist, warned of just this kind of alteration of gene pools. The introduced subspecies is causing the decline of native races, but not by killing animals or displacing them from habitats; rather, the genetic integrity of the native populations is altered by a mechanism that is subtle and not obvious to casual investigation, and which can be discovered only by sophisticated genetic analyses.[46]

Pathogens also move with bait animals, raising the possibility that native populations will be affected. Angela Picco, a doctoral student at Arizona State University now with the U.S. Fish and Wildlife Service, used amphibians as a model system to study how pathogens are moved in commerce.[47] Andrew Cunningham, Head of Wildlife Epidemiology at the Institute of Zoology in London, and his colleagues coined the term "pathogen pollution" to describe this anthropogenic movement of pathogens, often associated with commerce.[48] Picco studied Ambystoma tigrinum virus (ATV), an ideal candidate for pathogen pollution. These viruses cause die-offs of amphibian populations, and the potential for moving them outside their native range is high due to the trade in amphibians as fishing bait. Her results indicate that 85 percent of Arizona bait stores sold ranavirus-infected tiger salamander larvae during the 2005 fishing season.

James Jancovich from Arizona State University and Andrew Storfer from Washington State University led two teams conducting molecular analyses of ATV pathogens and tiger salamander populations throughout the western United States.[49] Except for a few populations, the salamanders and the virus showed clear patterns of co-evolution. The exceptional populations appear to be cases in which salamanders had been moved from their original population, probably as bait, and introduced into a new population. There is evidence, therefore, that ATV is moved via the tiger salamander bait trade, but the impact of this pathogen movement on possibly naïve hosts is unknown.

As yet, we have no example in which commerce alone has decreased amphibian population sizes to extinction—but the potential is there. Matthew Fisher and Trent Garner at Imperial College and the Institute of Zoology in London reported that 28 species of introduced frogs and salamanders carried the amphibian chytrid, which is a lethal pathogen that we will discuss in the next chapter.[50] They concluded that there is evidence that trade in amphibians is contributing to the spread of the pathogen. In general, concerns are raised by the large number of amphibians collected for food, the pathogens that move with exotic species, and the collecting for the pet trade of species distributed over small areas.

Commercial use of amphibians often affects small areas of the landscape. Humans, of course, also manipulate larger areas of the landscape in what is the greatest cause of amphibian declines—habitat alteration.

Land Use and Land Cover Changes

When Mark Twain published his short story, "The Celebrated Jumping Frog of Calaveras County" in the November, 18, 1865, edition of *The New York Saturday Press,* that California county was known for its gold mining camps in the foothills of the Sierra Nevada. Twain set the story in Angels Camp, and the California red-legged frog was at the story's center—"the celebrated frog." Fast-forward almost 150 years to a medium Twain never imagined, and on May, 1, 2006, the red-legged frog is a headline on the MSNBC News Web site: *"Habitat or homes? Frog jumps into debate. Species made famous by Mark Twain is being squeezed out."*[51] The point of the story is clear from its title and lead sentence: "The national debate over protecting fragile species comes to life here [Livermore, California], where upscale housing developments push ever deeper into the rumpled blanket of grassy hills at the eastern edge of the San Francisco Bay area." In 1996, the red-legged frog was listed as threatened under the U.S. Endangered Species Act, and about 1.6 million hectares were designated as critical habitat to ensure the species' survival by limiting development. Subsequent litigation reduced the area set aside as critical habitat to 182,000 hectares, allowing private landowners to develop areas that otherwise could serve as frog habitat. Red-legged frogs now occupy about 30 percent of their former range, and the California debate is one example of how land use and land cover changes affect amphibians and many other species.

Three points are useful to make before we discuss how land change affects amphibians. First, the natural world changes constantly and land can change as a result of marginal, indirect, or no human influence. The mechanisms can be rapid and violent, as in the eruption of the volcanic island of Krakatoa, or gradual, as when grassland slowly changes to forest after invasion by shrubs and then fast-growing trees, leading finally to the establishment of large, mature tree species typical of an old growth forest. Animal, fungal, and microbial species change with the plants. Ecologists call this gradual community change "succession," and it provides an important perspective, standard, or baseline for calibrating the rapid land changes that humans can initiate.

Second, when humans alter the way in which land is used, it is called simply land use change. Land cover change is alteration by humans of the physical or biotic nature of a site (e.g., conversion of a forest to grassland).[52] These processes can occur together or separately, and we will follow Peter Vitousek at Stanford University and label both "land use change."[53] Land use change can facilitate local and, eventually, regional extinction of populations and species by killing organisms, removing or transforming habitat, or preventing access of animals to breeding sites.

Third, except for marine environments and the Antarctic, most of Earth's freshwater and terrestrial biomes have amphibian species. Amphibians not only inhabit wetlands, but they also occupy deserts and high mountains. Some species live their entire lives in water, whereas others never venture near ponds, lakes, streams, or marshes. Instead, they have a life cycle in which eggs are laid and the embryos develop directly into miniature adults that then grow to maturity. Even in species that require water for development of larvae, the juvenile and adult metamorphosed stages often live away from water in terrestrial habitats. The point is that loss of aquatic habitats often negatively affects amphibian species, but most species also require associated terrestrial habitats. Land use change, therefore, includes aquatic and terrestrial habitats.

We will set the stage for understanding the relationship between land use change and population or species losses by first summarizing what would be expected if humans were not affecting things. Would changes still be expected? Indeed they would be, and it is important to recognize that amphibian populations can decline, even to local extinction, as a result of natural changes around them. Then we will review what we know about the transformation of landscapes by human actions and how that affects amphibian populations.

Monitoring Changes in Amphibian Population Sizes

Drawing firm conclusions about long-term population trends is not easy. Estimating a population's size is technically difficult, laborious, and complicated by the fact that many adult amphibian populations have "boom and bust" population cycles in which numbers may vary some 20 times over only a year or two.

As we discussed in chapter 1, a general problem in studying amphibians is separating the "noise" caused by annual variation in population size from the "signal" that is the long-term trend in population size (increasing, decreasing, or no change).[54] Declines in some years are expected even without negative pressures such as introduction of an exotic species, clear cutting of a forest, or spread of a toxin. After analyzing the data for hundreds of frog populations, Ross Alford and Stephen Richards at James Cook University reached the important conclusion that, on average, amphibian populations decrease in more years than they increase.[55] We might think, then, that "boom" years are key points for population persistence. David Green at McGill University also analyzed data for many populations.[56] He pointed out that average amphibian population size is not a function of relatively rare years of high recruitment offset by intervening years of gradual decline, so that the years with declines would outnumber years with increases without negative effect. It is more complicated than that, and here is why.

For many species of amphibians, breeding and recruitment are strongly influenced by water availability. In any year, the "decline" of a particular population may be nothing more than a year of reduced recruitment that is part of the expected ups and downs typical of many amphibian species in response to water availability. But the occasional "boom" year alone is not sufficient to offset years of reduced recruitment. Something important, but not well understood, happens in the mix of years of decreasing and increasing population size that is necessary to ensure persistence in the long term. Part of the explanation probably rests with the fact that we have a poor understanding, for most species, of how variation at one location relates to changes in the larger metapopulations in which many amphibians live. Specifically, population biologists speak of populations that are "sources" or "sinks." Some habitats, such as a pond or lake, are sources of individuals that colonize other ponds or lakes; the recipient populations are "sinks" in that fewer animals leave the habitat than enter it. In general, monitoring of the dynamics of local populations absent from the web of interconnecting habitats that constitute the metapopulation is not the best level of analysis for understanding processes at the regional or larger scales that control biodiversity.

These observations lead to a clear question: What is the best way to study populations that vary so significantly over years? Deciding whether a habitat is occupied by a species can be a challenging question, because not all individuals or even all species may be active when the location is sampled. The likelihood of detecting a species varies with the techniques used, the species, and the environmental variables that affect the species' activity. Long-term datasets increase our confidence in concluding that a species does or does not occupy a habitat, because observations over time increase the likelihood of sampling under just the right conditions needed for a species to be active and recorded. Long-term datasets that include a large number of habitats are the best way to detect meaningful trends for populations that vary greatly in number of individuals between years. Repeated sampling of populations, perhaps over decades depending on the amount of annual variation, may be needed to reach statistically meaningful conclusions.

A 40-year study of change in 14 species of frogs and salamanders at the E. S. George Reserve in Michigan exemplifies the long-term record needed for understanding variation in amphibian population sizes. Most importantly, the study documents the increases and decreases expected in systems not affected by humans through the introduction of exotic species, commercial use, or major land use changes.

The E. S. George Reserve in southeastern Michigan is a protected game reserve that has been maintained since 1930 by The University of Michigan. It comprises about 600 hectares of moraine and basin topography. Lowlands are marshes and bogs filled with a rich diversity of aquatic vegetation, including sedges, cattails, red maple, birch, and tamarack trees. Uplands are old fields and oak-hickory woodlands. There are some 50 permanent, semipermanent, and temporary aquatic habitats: marshes, swamps, bogs, and ponds. Between 1967 and 1974, one of us (JPC) and Henry Wilbur, who is now at the University of Virginia, surveyed annually all of the amphibians living in each habitat on the Reserve.[57] From 1988 to 1992, David Skelly, who is now at Yale University, and colleagues resurveyed all of the habitats and determined the net change in the number of breeding populations of 14 frog and salamander species in 37 ponds between the two surveys.[58]

Forty population colonizations and 34 population extinctions occurred between the two surveys. The number of amphibian populations stayed about the same, but the mix of species in each habitat changed. Why? Over time, especially small, shallow ponds and marshes filled with decaying vegetation and silt, making them shallower and less likely to hold water for long periods; put technically, the hydroperiod shortened. Likewise, open, sunny woodlands became shady over time as trees grew and filled the light gaps in the canopy. Ecological succession occurred as young, second-growth forests developed into older, mature forests. Skelly's group concluded that, as a result of succession leading to altered pond hydroperiod and canopy cover, the average amphibian species experienced about five colonizations and five extinctions between the two surveys. Said differently, over these 25 years in southeastern Michigan, a species of frog or salamander in an area protected from all but natural processes would appear as a "new" species in about five habitats and disappear from about the same number.

A long-term study like this one teaches us that amphibian populations decline even to local extinction as a result of natural habitat changes, often over relatively long time periods, such as decades. This is an important perspective to have as we review conditions in which humans play a major role in species declines through land use changes (clearing of forests, draining of wetlands) that can occur much faster—over days, weeks, or a year. We will start at the local and regional levels to get a sense of how land use change affects individuals and populations. Then we will assess how large forces extending over a greater scale affect populations and species.

Human Transformation of Landscapes and Its Effect on Amphibian Populations

In this section, we present examples of how land use change in various countries and regions has affected amphibians.

Starting in 1992, Stephen Hecnar and colleagues at Lakehead University began studying the variables controlling amphibian communities across 42,962 km^2 in southwestern Ontario, Canada.[59] Wetlands covered 69 percent of the southern and 23 percent of the northern part of the study area in presettlement times (the 1700s). Drainage of wetlands for agriculture began in the 1800s, and wetlands (often artificial agricultural ponds) came to cover 3 percent of the southern and 10 percent of the northern portion. Now, moving from south to north, there is increased forest cover, decreased human population density, and increased richness in amphibian species. Hecnar concluded: "The most important factor that has affected amphibian diversity in southwestern Ontario has occurred historically. The massive deforestation and wetland drainage of the 1800s destroyed most amphibian habitat. While the building of artificial ponds may have been beneficial (those not stocked with fish), the magnitude of habitat loss and change resulted in less diverse communities."[60]

The amount of regional woodlands, predation by fish, habitat isolation, and perhaps agricultural chemical use interact at local and regional scales to control amphibian diversity. Overall, however, "The primary process affecting local amphibian richness in southwestern Ontario appears to be historic habitat loss.... Continued deforestation in the more northern regions of southwestern Ontario will likely result in less diverse amphibian communities consisting mainly of green frogs [*Rana clamitans*], leopard frogs [*Rana pipiens*], and American toads [*Bufo americanus*]."[61] Historically, this region supported 13 species of frogs and salamanders. This fact leads to the more general concern that land use change that eliminates native woodlands and wetlands selects against endemic amphibian taxa and favors species that can adapt to human-dominated landscapes.

The negative effects of land use change on amphibian populations were at work in Ontario before the 1980s. Habitat conversion over more than a century had diminished amphibian biodiversity in southwestern Ontario across some 43,000 km^2. Habitat conversion is an issue in tropical areas as well: Between 1990 and 1997, an average of about 60,000 km^2 of humid tropical forest were converted to agricultural landscapes each year, and an average of about 23,000 km^2 were degraded.[62]

Gentile Ficetola and Fiorenza Bernardi at the University of Milano studied 84 wetlands in the Lombardy region surrounding Milan in northern Italy and showed how human-driven land use change negatively affected amphibians.[63] In the late 19th and early 20th centuries, there were nine amphibian species in the region, and all but one were abundant. At present, for the most common species, the edible frog (*Rana escuelenta*) and the Italian tree frog (*Hyla intermedia*), metapopulations remain intact because individuals move across the landscape using canals and hedgerows. Newts and toads are rare because they are sensitive to habitat alteration, which is common as a result of agricultural and industrial development. Because they do not move readily across the landscape, subpopulations of newts and toads are isolated. The authors concluded that "degradation of wetland ecological features (like pollution and fish release) and landscape alteration (like increased wetland isolation) cause the extinction of sensitive species from many patches."[64]

Tim Halliday at the Open University worked with one of us (JPC) and summarized data for the United Kingdom. Over the last 50 years, 20 percent of 350,000 actual or potential amphibian habitats were lost, at a rate of 1 and 3 percent annually (Table 4.4).[65]

Under the editorship of Michael Lannoo at the Indiana University School of Medicine, 215 scientists contributed to a 2005 summary of the conservation status of the 289 U.S. amphibian species, comprising 103 frog species and 186 salamander species, each with a species account that also described potential threats.[66] David Bradford analyzed most of these accounts and concluded that land use change—habitat loss, alteration, or fragmentation—was the most frequently cited cause of population declines and losses (about 77 percent of 91 frog species and about 91 percent of 176 salamander species analyzed).[67] The most-cited land changes

TABLE 4.4 Loss of Ponds (Actual or Potential Amphibian Habitats) at Selected Localities in the United Kingdom over the Last 50 Years

Location	% Lost	Annual Rate	Source
All UK	20% of ~350,000 (1958–1988)	0.7%	Swan & Oldham 1993
Sussex downland (rural)	18% of 33 (1977–1996)	0.9%	Beebee 1997
Huntingdonshire (rural)	99% (over 40 years)	2.5%	Nature Conservancy Council 1982
Milton Keynes (urban)	33% of 126 (1984–1994)	3.3%	Barnes & Halliday 1997
Geneva Basin *Triturus cristatus* populations (rural/urban area)	68% of 22 (1975–1997)	3.1%	Arntzen & Thorpe 1999

Source: Collins and Halliday, "Forecasting changes in amphibian biodiversity," 312.

were conversion of native landscapes to agriculture, urban development, and timber/silviculture. Exotic species were the second most frequently implicated factor for frog declines and the third most common for salamanders. Chemical contamination was the second most-cited factor for salamanders and the third for frogs. As we already discussed, the authors of the 2004 Global Amphibian Assessment reached a similar conclusion.[68] By far the greatest threat to all amphibians worldwide is habitat loss. Land use change is a major cause of reduced biodiversity globally, and amphibians are no exception.

Loss of native habitats, especially wetlands, has typified the 19th and 20th centuries, and it has every prospect of continuing. Amphibian populations have been declining for centuries as a result of land use changes, and the losses have gone largely undocumented.[69] It should be noted that land use change, especially for agriculture, exposes frogs and salamanders to toxins, a point we will discuss in the next chapter. Kenneth Dodd, who at the time was with the U.S. Geological Survey, and Lora Smith, also with the U. S. Geological Survey, made the important point that amphibian habitats must be thought of as all of the elements that affect their life history—both water and terrestrial sites; as they wrote, "The biotic integrity of amphibian habitats needs to be ensured."[70] Land use change is but one of the many ways in which humans change the Earth.

We are in the midst of significant global change, much of which is initiated and driven by humans. The patterns we reviewed are being seen worldwide, and things are likely to get worse before they get better. Michael Lannoo teamed with colleagues at the U.S. Geological Survey and the

University of Wisconsin-Milwaukee to analyze global land use change and amphibian distributions.[71] They showed that amphibian species diversity is highest in terrestrial and freshwater habitats in warm, moist regions such as Central America, northern and central South America, central Africa, and Southeast Asia. Before the 20th century, human population growth and spread was slower and occurred mainly in temperate zones, away from regions of high amphibian species richness. Throughout the 20th century, however, human population growth has accelerated together with its accompanying land use changes, mainly for agriculture, and especially in regions favored by amphibians. Overall, regions with the highest species richness of amphibians are undergoing the highest rates of landscape modification and will continue to do so for the immediate future. A clear prediction is that more amphibian declines are expected in the immediate future, as humans continue transforming Earth's ecosystems.

Habitat change, amphibian extinctions, and our incomplete knowledge of amphibian biodiversity were brought home by a 2002 study in Sri Lanka. Madhava Meegaskumbura at Boston University and Sri Lanka's Wildlife Heritage Trust led a research team in a project that we first discussed in chapter 1.[72] Surveys performed primarily in Sri Lanka's rain forest revealed some 140 frog species on the island—100 of which were previously unknown. In addition, a check of 19th century museum collections uncovered approximately 100 additional new species that were not recorded in the recent survey, presumably because they had become extinct. The authors concluded that this is not surprising, because the island has lost more than 95 percent of its rain forest habitat. Seventeen of Sri Lanka's frog species disappeared within the past decade—representing 50 percent of the 34 confirmed amphibian extinctions occurring worldwide in the past 5 centuries. Land use change is a major cause of extinctions in amphibians.

If exotic species, commercial use, and land use change were the only causes of amphibian declines, the issue would not have drawn the attention of scientists and the public as it did in the late 1980s and 1990s, because these factors are threatening global biodiversity of many taxa other than just frogs and salamanders. Something else was driving the interest in amphibian declines: the fact that species had disappeared in the midst of protected areas. By 2004, these losses came to be called the "enigmatic declines," and their unknown causes were the real driver of the acute interest in amphibian declines.[73] Untangling this mystery is the next step.

5

Contaminants, Global Change, and Emerging Infectious Diseases

In the late 1980s and 1990s, we were hearing reports of some amphibian species going extinct in only a few years; the search for the answer to the question, Why are they gone? was becoming paramount. But as we have seen already, the reality in science, as in any good detective novel, is complicated.

In the last chapter, we learned that historical explanations point to such causes as competition with exotic, introduced species or predation by the same; harvesting of wild animals for food or pets; and changes in land use. These mechanisms accounted for most of the damage to amphibian populations for much of the 20th century, and even today. These historical causes also typically leave clues—some as obvious as the mark of a chainsaw—from which an investigation can begin.

However, at least some of the amphibian declines and extinctions in the late 20th century left few, if any, clues. Animals went missing in national parks and nature reserves, where obvious causes such as habitat destruction did not apply; somehow, habitat protection, historically the best way to ensure a species' survival, was failing to safeguard some amphibians.

The realization that the standard historical explanations could not solve the mystery led researchers to consider explanations based on recent changes. Three suspects emerged: toxic chemicals in the environment; global change, particularly global warming or increased ultraviolet radiation; and emerging—in some sense, new—infectious disease. Each hypothesis had its champions, and given the complexity of ecological interactions, it was likely none would act alone. In addition, these suspected causes could have still undiscovered accomplices.[1]

Contaminants

As mentioned briefly in chapter 2, Minnesota schoolchildren from Le Sueur took a field trip to a farm pond in the summer of 1995. While collecting frogs, they found many that lacked legs or even had extra legs (Figure 5.1).[2] The deformed animals were surprising and unsettling. Were these animals signaling something about the water in which they lived and the environment surrounding the pond? And did these animals relate to global amphibian declines? The children's discovery evoked memories of Rachel Carson's *Silent Spring:* Were harmful chemicals being introduced into the environment?[3]

Environmental contamination is the introduction of manmade organic and inorganic chemicals into the environment, a field large enough to warrant many books. We will focus only on contaminants related to amphibians.

Why think about contaminants? Minnesota's frogs attracted a lot of media attention and a book by the journalist William Souder—*A Plague of*

Figure 5.1 Deformed green frogs from Quebec. (After Souder, *A Plague of Frogs: The Horrifying True Story*; photograph by Martin Ouellet.)

Frogs. The Horrifying True Story.[4] The deformed frogs were a mystery because everything else about the pond seemed okay: There were no obvious disturbances in the area, no incriminating pipe discharging an unknown fluid, and no hazardous waste dump in sight. All of this suggested that something might be entering the area unnoticed, and one possibility was a contaminant—a toxin draining into the pond undetected or wafting in on air currents—that was interfering with normal frog development. But why should a contaminant come to mind?

The publication of *Silent Spring* in 1962 sensitized researchers and the general public to the influence of chemicals on wildlife. A classic example was the effect of the organic insecticide DDT (dichlorodiphenyltrichloroethane) on developing bird embryos, especially among predators high on the food chain. Top predators, such as bald eagles, accumulated DDT in quantities that led to thin eggshells and dead chicks. Extensive research connected DDT to declines in populations of eagles and other U.S. birds. In the closest thing to a large field experiment, U.S. raptor populations rebounded after the federal government restricted DDT usage.

Environmental toxins also affect humans. Japanese Minamata disease was recognized in the 1950s as a neurological syndrome caused by mercury poisoning. Deformities can result, especially in a congenital form of the disease that affects fetuses. The release of methyl mercury, an inorganic toxin, in the industrial wastewater of a Japanese chemical factory caused the disease. The mercury became concentrated in marine organisms in Minamata Bay, Japan, poisoning people who ate the animals.

These are two well-known cases involving an organic and inorganic toxin. It is easy to see how such examples might lead to a suspicion that toxins caused the Minnesota frog deformities. And knowing how DDT caused raptor populations to decline raised the possibility that deformed frogs, toxins, and amphibian declines might be connected. But there is a big difference between suspecting that a toxin causes deformities or population changes and demonstrating either result.

Studying the effects of chemicals in the environment and in organisms is difficult. When a chemical such as DDT enters a terrestrial or aquatic ecosystem, it retains its original chemical structure or separates into parts. The original form and its parts usually affect organisms differently, and separate analyses are required to isolate them. In some cases, the components are more powerful than the parental compounds.[5] Also, a chemical such as DDT is usually not delivered in its pure form, but instead another chemical with its own properties is used as a solvent or transporter. Research shows that the transporting chemical—in some cases, a surfactant that allows easier spreading—can be more hazardous than the pure chemical.[6]

When more than one chemical in combination enters an ecosystem, it is called a "chemical cocktail." Identifying the harmful chemical or chemicals within a cocktail is difficult, and in some cases there is a synergistic effect in which it is the mixture itself that is harmful and not one single chemical.[7] Chemicals originate from point sources, for example the chemical factory in Japan, or from nonpoint sources, of which the Central Valley of California is an example.

Thousands of metric tons of insecticides and herbicides, some 25 percent of the amount used in the United States, are applied annually in the Central Valley. These chemicals enter the atmosphere directly, as planes apply the chemicals to fields or when they adhere to soil particles and disperse as dust. Either way, the chemicals are transported on prevailing winds. At one scale, the valley is a discrete geological feature or a "point source"—imagine the valley viewed from outer space. But for the purposes of describing the transport of chemicals in Earth's atmosphere, scientists and policy makers treat such cases as nonpoint sources, meaning that the chemical is deposited kilometers downwind, and the "origin" is within a large volume of prevailing wind currents that include chemicals from the Central Valley and perhaps other sources. Distinguishing point sources and nonpoint sources is important, because the former are easier to manage than the latter. We will return to this example later.

Finally, once a chemical enters an organism's body, its effects vary. The same chemical acts differently depending on the species, the organism's age or developmental stage, environmental factors such as temperature, and the amount ingested. The latter factor may seem obvious: The more chemical ingested, the more lethal the toxin in a direct linear effect. In fact, for chemicals such as the widely used herbicide dioxin, low doses in some species produce beneficial physiological effects, whereas high doses are lethal in a classic nonlinear response that toxicologists call hormesis.[8]

Isolating a cause amid such complexity is difficult. Add the fact that each year many new chemicals are developed and introduced into the environment, and we have some idea of how hard it is to identify a chemical cause of wildlife deaths or deformities from among many candidates. Detection requires specialized analyses, and there is no single test to apply to a sample that will lead to the conclusion that a "toxin" is present. An investigator must often know or at least have a strong suspicion about what she or he is seeking. A test may require a particular tissue type (liver, brain, blood, muscle) that may be in short supply if only a few, small specimens are available. In addition, the amount of toxin applied varies, duration varies (e.g., a pulse or prolonged application), and persistence varies from minutes to days to years. It takes significant planning, skill, and often site-specific plus toxin-specific knowledge to identify contaminants in the field. The next section describes how an investigation often proceeds.

A Hierarchical Research Strategy

A general strategy used to study how chemicals affect amphibians, and other organisms in general, is what Michelle Boone and Christine Bridges at the University of Missouri called a hierarchical approach.[9] Researchers expose individual organisms to chemicals to understand basic physiological responses and then integrate these findings with studies of individuals exposed to chemicals under diverse environmental conditions. Ideally, a conclusion regarding a chemical's effect integrates all outcomes as a test for consistency of response. A researcher then wants to connect an understanding of how toxins act in an individual's body with changes in population sizes of amphibians in the field.

Step 1 is often assessing a chemical's effect on performance or development of organisms in the laboratory. Toxicology is the study of how chemical substances and their doses adversely affect traits such as survival, development, reproduction, likelihood of genetic mutations, or behavior. Researchers typically study one species under controlled laboratory conditions. Ideally, variation in the toxin's effects is assessed among individuals in a population and in multiple populations, among genetically different families within one population, and among species.

After exposing a sample of organisms to a test chemical, a typical measure of outcome is the LD50, or time until 50 percent of the test population dies. Thousands of such studies have been reported for vertebrates and invertebrates. The U.S. Environmental Protection Agency maintains a database of ecological toxicity records dating from 1926 that summarizes the effects on organisms of organic and inorganic toxic substances.[10] For amphibians, there are tests assessing chemical effects relative to temperature, pH, and ultraviolet (UV) light levels. For example, in nine species of frogs in the genus *Rana,* time to death after exposure to a lethal dose of the pesticide carbaryl varied from 5 hours in wood frogs to 35 hours in northern red-legged frogs. Pesticide tolerance varies a good deal among frog species, within species, and within populations of a single species.[11]

In a 2003 review, Donald Sparling, who was with the U.S. Geological Survey, summarized LD50s for 30 organic and inorganic amphibian contaminants for which the laboratory tests were conducted at or below concentrations reported from environmental studies.[12] Sparling concluded that many compounds occurred in the environment at higher than toxic concentrations. Chemical formulations also differed in their effects, and the same contaminant could vary in its effect depending on the amphibian species and life stage tested (e.g., larva or adult). Finally, sediment toxicity may be more important than water toxicity, because sediments can have higher chemical concentrations.

We often think of amphibians as superior indicators of environmental quality because various species, and sometimes the same species, live in aquatic and terrestrial habitats and could therefore be exposed to toxins from multiple sources. But Boone and Bridges concluded that, compared to other animals, amphibians can be either more or less sensitive to toxins, depending on the species and study. More research is needed; as of the year 2000, only about 3 percent of vertebrate toxicological studies included amphibians. To gain a sense of the task, Sparling also noted that some 75,000 chemicals are manufactured in the United States each year, not including pesticides or the 3000 high-production, high-volume chemicals (>450 kg produced). Many more data are needed to understand how contaminant exposure affects amphibians. A key step in this process is extending our knowledge of how toxins affect animals in the laboratory to how they affect animals in the field.

A toxicology study is designed to assess the ways in which a chemical might be harmful before release or might cause harm after release. Assessing the latter is especially difficult for chemicals introduced into the environment, because ecosystems have many complex interactions that make it difficult to extrapolate from the laboratory to the field. Ecotoxiocology arose in the 1960s in response to this challenge, and it is the next step for connecting how toxins act in the body of an individual to cause changes in amphibian population sizes.

Ecotoxicology is the branch of toxicology concerned with the effects of pollutants on ecosystem constituents in an integrated context, which means that studies are done in the field or under simulated field conditions.[13] James Burkhart at the National Institute for Environmental Health Sciences in North Carolina and colleagues outlined the steps that are followed.[14] Once an acute effect on an organism is observed in the field, the pathology's expression is described—imagine describing the deformities of the frogs collected by the Minnesota school children. Then the physical and biotic environment is described, including suspected toxins in the site's water, sediment, or surrounding terrestrial habitats. Duplicating the pathology in the laboratory using the suspected chemical is next. A model that integrates the laboratory and field results is then developed. Finally, researchers assess how the number of acutely affected animals (e.g., deformed frogs) influences population size. This last step deserves more discussion here, because the details are important.

Studying how chemicals introduced into the environment affect individuals, populations, and ecosystems is difficult, complex, detailed research.

To conclude that a contaminant is negatively affecting a population, an investigator first must show that ecologically relevant concentrations of the contaminant are a new source of population stress. This means that there is reduced recruitment of new animals or increased deaths beyond those caused by other stressors. Then the investigator must show that these decreases in population recruitment or increases in population losses exceed the usual numbers resulting from predation, competition, parasitism, and other factors.

Here is another way to think of the problem. Each year, a population of frogs or salamanders suffers deaths due to predation, competition, or parasitism. If the number of animals killed by toxins does not exceed the number of these deaths, then we conclude that the toxin is not affecting the amphibians' population size. Under these circumstances, ecologists call the deaths due to toxins "compensatory" deaths, or animals that would have died anyway from natural causes.[15] Ecotoxicologists face the challenge of, first, knowing what the average population size or growth rate is, and then demonstrating that a toxin kills animals above and beyond the deaths from usual causes.

These are tough standards to meet, especially for many amphibians. As we mentioned earlier, populations often fluctuate by orders of magnitude in different years, and the population dynamics of most species are unknown. Indeed, in the same review discussed already, Sparling concluded, "In reality, these criteria are seldom, if ever, met in ecotoxicological studies; thus, only indirect evidence is available to determine the role contaminants may have in amphibian population declines."[16] Indirect evidence includes (1) field data documenting population declines in the presence of environmental contaminants, (2) laboratory studies demonstrating lethal or sublethal responses to contaminants at or below field concentrations, (3) increase in populations after removal or neutralization of contaminants, and (4) laboratory or field studies showing that contaminants interact with other causes of amphibian declines to increase the likelihood of population losses.

Experiments are run regularly with suspected chemicals, introduced alone or as mixtures, to test their effects on one or several species in the laboratory, in field enclosures (cages placed in an aquatic or terrestrial habitat), in mesocosms (which are static, outdoor, above ground stock tanks or children's swimming pools in which the natural habitat is simulated by adding multiple species and then introducing the toxin),[17] or directly in natural or manmade habitats.[18] The latter experiments may be deliberate, as with the introduction of a chemical poison, or inadvertent, as when a pesticide sprayed on crops enters a nearby habitat via wind or runoff. The logical chain connecting toxins and population declines remains incomplete until studies have demonstrated how ecologically relevant concentrations of a contaminant act as a new source of stress to diminish recruitment or to increase losses beyond those typically experienced from natural causes. This chain of causation must be completed to show a connection between inorganic and organic toxins and amphibian population declines.

The two most studied inorganic effects on amphibians are acidification of habitats and nitrogen deposition. Laboratory and mesocosm studies have shown that acidification adversely affects amphibians. However, when Frank Vertucci and Stephen Corn, U.S. Fish and Wildlife Service researchers at Fort Collins, Colorado, reviewed surveys of habitats in which amphibians have declined in the U.S. Rocky Mountains and in the Sierra Nevada in California, there was insufficient evidence to conclude that acidic deposition caused the declines. Therefore, this was a case in which laboratory results did not translate readily to field performance. After reviewing studies conducted across the eastern and western United States, Christopher Rowe and Joseph Freda from the Chesapeake Biological Laboratory concluded that there was no evidence to relate acidification to declines in amphibian populations.[19]

In contrast to acids, elevated levels of nitrogen, often as nitrate fertilizers, negatively affect development and survivorship of amphibians, at least in the United States, Europe, Canada, and Australia. Jeremy Rouse and colleagues with Environment Canada concluded that "nitrate concentrations in some watersheds in North America are high enough to cause death and developmental anomalies in amphibians and impact other animals in aquatic ecosystems."[20] Indeed, for several decades now, nitrogen fertilizers and nitrogen oxides discharged by cars and factories are thought to have had a widespread negative effect on biodiversity and a range of ecosystem processes.[21]

Rick Relyea and Jason Hoverman at the University of Pittsburgh reviewed 319 studies of the effects of pesticides and herbicides alone or together on freshwater ecosystems.[22] Toxins alter the relationships of species in food webs by changing behavior, growth, physiology, life history traits, and even susceptibility to pathogens. A toxin does not have to kill a frog or salamander directly to have a negative effect; it can do so by compromising an animal's ability to escape from a predator, compete, avoid disease, or grow quickly enough to complete development before a pond dries. These important findings add great complexity to ecotoxicology studies. A toxin may also kill the aquatic invertebrates that amphibians eat. Either way, there is a negative effect—directly through the animal's physiology or indirectly through the ecosystem's food web.[23]

The reviewed studies included surveys of natural habitats after contamination with experimental application of toxins, and experiments in which toxins were applied to enclosures in natural habitats and to mesocosms;

20 cases included amphibians. Relyea and Hoverman concluded, "One of the most challenging issues to address is whether pesticide effects on individuals actually affect population dynamics. For example, when a pesticide causes low to moderate mortality in a species or the feminization of males due to endocrine effects (e.g., Hayes et al. 2002), it is an open question whether these effects will affect population growth. . . . We need more work in this area, but the data will not come easily for most species."[24]

We will return later to Hayes' studies on feminization of male frogs, but here is what a hierarchical research strategy tells us so far. Based on laboratory studies with toxin concentrations at or below lethal levels, contaminants can cause amphibian declines. Field experiments with cages in ponds, mesocosms, and experimental ponds bring these studies closer to reality. They demonstrate how a toxin can reduce amphibian abundance and negatively alter traits such as size at metamorphosis, under conditions much closer to those experienced by wild populations. But do toxins applied in the field affect populations? Carlos Davidson at San Francisco State University asked just this question in a series of studies.[25]

Sierra Nevada amphibian populations are downwind of California's Central Valley and its extensive agricultural industry. Davidson tested the hypothesis that historical patterns of pesticide use in the valley might also explain declines of five Sierra Nevada frog species, one *Bufo* and four *Rana*, because the prevailing west-to-east winds carry toxins from the valley and deposit them in montane amphibian habitats. Indeed, as Gillian Daly and Frank Wania from the University of Toronto reported, organic chemical contaminant deposition generally increases with elevation for mountains at middle and high latitudes.[26]

Kimberly Hageman and associates at the U.S. Environmental Protection Agency's Western Ecology Division in Corvallis, Oregon, also studied the atmospheric deposition and fate of semivolatile organic compounds in alpine, sub-Arctic, and Arctic ecosystems in the Western United States.[27] Their correlation analysis revealed that current and historical agricultural practices are largely responsible for the distribution of pesticides in the national parks in the region.

Davidson found a significant association between amphibian declines for the four *Rana* species and total pesticide use in the Central Valley. He also teamed with Roland Knapp to assess the effect of pesticides and fish on the abundance of mountain yellow-legged frogs and three additional ranid species in and around the Sierra Nevada.[28] In chapter 4 we described how Knapp and his colleagues had already documented in detail the negative effect of fish on mountain yellow-legged frogs.

Davidson and Knapp quantified the habitat characteristics, presence or absence of fish and frogs, and pesticide use upwind of 6831 aquatic habitats.

Fish and pesticides together significantly reduced the likelihood of finding frogs, with the landscape-scale effect of pesticides being much stronger than that of fish. So, even in sites without fish, high upwind use of pesticides most likely accounted for the absence of frogs. The fact that the likelihood of finding live frogs increased for sites sheltered from Central Valley winds by topographic features reinforced this conclusion. Their results show, first, that amphibian declines have complex, multiple causes, a conclusion that will be reiterated throughout this chapter. Second, this study demonstrates how investigators are using multiple, complementary approaches to understand the effects of toxins on amphibian populations. Natural systems are complex, and at any time multiple causes may act on a population simultaneously or follow each other historically. Finally, although the data are extensive, the Davidson and Knapp study is still a correlation, and additional studies are needed to confirm that toxins in the habitats of frogs are responsible for the variation in population sizes.

Back to Minnesota

Trematodes and Amphibian Abnormalities

After the children's discovery of frogs with extra limbs, public health officials tested those frog habitats for toxic chemicals but found none. Researchers discovered, however, that the Minnesota incident was not an isolated instance. A key first step in answering a scientific question is placing the observations in context: Has anyone collected abnormal frogs before? And if so, under what circumstances and could the same cause or causes account for Minnesota's abnormal frogs?[29]

Statistics compiled by the North American Center for Amphibian Deformities show that abnormalities such as extra limbs, malformed or missing limbs, and facial malformations have been reported in almost 60 amphibian species from 44 U.S. states.[30] In some populations, 60 percent of the animals observed are abnormal. In a review in 2000, Martin Ouellet at McGill University in Quebec reported that deformities are more common in frogs (67 species) than in salamanders (26 species).[31] Michael Lannoo recognizes 21 kinds of frog abnormalities in 3 classes: craniofacial (e.g., a small or missing eye), limbs (e.g., missing or shortened), and whole body (e.g., swelling).[32]

Stanley Sessions, at Hartwick College in upstate New York, is an expert on amphibian abnormalities. He makes an important distinction between "deformities," which are abnormalities resulting from the response of normal tissue to mechanical forces (trauma), and "malformations," which result from an intrinsic defect in development caused solely by genetic or gene-environment interactions.[33] Use of either term implies knowing the process that is causing the pathology. The cause is typically unknown in such cases, at least initially, which is why the Minnesota case was alarming. Here, we will use the neutral term, "morphological abnormality," or just "abnormality."

As the Minnesota schoolchildren made their discovery, other researchers were also studying abnormal frogs. In the mid-1990s, Martin Ouellet was a graduate student at McGill University. For his doctoral dissertation, he was trying to understand why frogs in Ontario's agricultural areas developed with no hind legs or more than two hind legs (see Figure 5.1). Ouellet found residues of insecticides, herbicides, and fungicides at the affected sites, but he did not find a significant relationship between toxin levels and frequency of abnormalities because of high variation among sites for both variables.[34]

Ouellet's work brought an important perspective to the problem by reminding everyone that abnormalities might be caused by something other than toxins.[35] For example, while trying to catch a frog, a predator might get only part of a limb; the partial limb remaining is a "morphological abnormality." Genetic defects and nutritional deficiencies also cause abnormal development. In fact, as early as 1987, Sessions suggested that parasites could cause frogs to develop abnormally.[36]

Trematodes are small (a bit longer than a pinhead), parasitic "worms" that have a fascinating, complex life cycle. An adult trematode is also called a flatworm or fluke, and the ones of most interest to amphibian researchers live in the digestive tracts of birds. The parasite's eggs are released into aquatic habitats with bird feces and hatch into a stage called a miracidium that infects some, not all, pond snail species. The parasite multiplies again inside the snail and is released into the pond as a free-swimming stage called a cercaria.

In the mid-1980s, Stephen Ruth, at Monterey Peninsula College, contacted Sessions because he had found Pacific treefrogs (*Hyla* [now *Pseudacris*] *regilla*) and long-toed salamanders (*Ambystoma macrodactylum*) in northern California with missing or extra hind limbs. Sessions suspected that a trematode infection might be causing the abnormalities. Specifically, a cercaria infects a tadpole, embeds near the site of the developing pair of hind limbs, and forms a cyst that appears to disrupt normal limb development, leading to a metamorphosed frog with no limbs or more than two hind limbs. Sessions and Ruth tested the hypothesis in the laboratory by implanting inert resin beads, simulating the trematode cysts, into the developing limbs of frogs and salamanders.[37] As predicted, the animals produced more than two hind limbs, but there is no still no agreement on the developmental mechanism that might account for the abnormalities.

emergence of chytridiomycosis might be associated with cool temperatures. An experiment was in order. Douglas Woodhams, at the time a doctoral student at James Cook University, and two colleagues raised a set of infected and a set of uninfected juvenile Australian red-eyed tree frogs (*Litoria chloris*) in each of four different thermal environments in the laboratory:[13]

1. Naturally fluctuating temperatures ranging from 13.5° to 23.2° C
2. Constant 20° C
3. Naturally fluctuating temperatures with two 8-hour periods at 8° C
4. Naturally fluctuating temperatures with two 8-hour periods at 37° C.

Only 1 of the 40 uninfected frogs died, but many infected frogs died. Never being exposed to warm temperatures (environment 2) proved deadly for the infected frogs: All of them died within 55 days. In contrast, infected frogs thrived in the warmest temperatures: All 10 infected frogs exposed to environment 4 survived at least another 5 months after the experiment's end, and they had no trace of fungus. Exposure to high temperatures actually cured the frogs.

In 2005, Woodhams and Alford updated the status of the 10 species of stream-dwelling frogs from the wet tropics of northeastern Queensland (Table 6.1).[14] The only two species that were still common at high elevations and at that time had not at least temporarily declined were the stoney-creek frog (*Litoria lesueuri*) and the great barred frog (*Mixophyes shevilli*). The green-eyed tree frog declined at high elevation sites during 1990–1994 but later recovered. The Australian lace-lid and the waterfall frog disappeared above 400 m, but low-elevation populations were stable. The common mist frog disappeared above 400 m but was "possibly recovering." The armoured mist frog, mountain mist frog, sharp-snouted dayfrog, and northern tinker frog all disappeared throughout their ranges.

Some frog species appear more susceptible than others to chytridiomycosis, and some sites seem to be harder hit. Within a population, only some individuals succumb. So what makes a given frog vulnerable to Bd?

Woodhams and Alford tackled this question by examining the effect of selected variables on the probability of Bd infection on northeastern Queensland frogs.[15] The biologists surveyed for infection at three upland and three lowland rainforest sites between August 2000 and March 2003. Of the frogs surveyed, they found at least some individuals of all six species infected with Bd, and they found infected frogs at each site. Infection prevalence was higher during the May-to-September cool, dry winter season than during the warm, wet season, and prevalence was higher at high elevations (750–800 m) than at lowland sites (40–160 m). The other variables

TABLE 6.1 Status of 10 Species of Stream-Dwelling Frogs from Northeastern Queensland, Australia

Common Name (Scientific Name)	Elevational Distribution[a]	Decline[b]	Current Status[c]	Decline Period[d]
Green-eyed tree frog (*Litoria genimaculata*)	Upland and lowland	Present at most upland and lowland sites surveyed	Recovered from declines, currently common	1990–1994
Stoney-creek frog (*Litoria lesueuri*)	Upland and lowland	Not discussed	Common	No declines
Armored mist frog (*Litoria lorica*)	Upland	Not discussed	Disappeared from throughout range	1991
Waterfall frog (*Litoria nannotis*)	Upland and lowland	Absent from most upland sites south of Daintree River, present at most lowland sites surveyed	Disappeared above 400 m throughout range, stable at lower elevations, some recovery or recolonization	1989–1993
Mountain mist frog (*Litoria nyakalensis*)	Upland	Not found at any site surveyed 1991–1992	Disappeared from throughout range	1990
Common mist frog (*Litoria rheocola*)	Upland and lowland	Absent from most upland sites south of Daintree River; present at most lowland sites surveyed	Disappeared throughout range above 400 m, stable at lower elevations, some recovery or recolonization	1989–1994
Great barred frog (*Mixophyes shevilli*)	Upland and lowland	Present at most upland and lowland sites surveyed	Common	No declines
Australian lace-lid (*Nyctimystes dayi*)	Upland and lowland	Absent from most upland sites south of Daintree River, present at most lowland sites surveyed	Disappeared throughout range above 400 m, stable at lower elevations	1989–1994
Sharp-snouted dayfrog (*Taudactylus acutirostris*)	Upland	Absent from most upland sites south of Daintree River	Disappeared from throughout range	1989–1994
Northern tinker frog (*Taudactylus rheophilus*)	Upland	Not found at any site surveyed in 1991–1992	Disappeared from throughout range	1989–1991

Source: Woodhams and Alford (2005).

[a]Upland is defined as 300 m and higher.
[b]Information from Richards et al. (1993).
[c]McDonald and Alford (1999); Woodhams and Alford (2005); R. A. Alford, pers. comm. (November 2006).
[d]From Woodhams and Alford (2005).

they examined were not associated with infection prevalence. These variables included:

- Population density
- Presence of metamorphosing tadpoles and juveniles (tadpoles may serve as reservoirs for Bd's infectious zoospores)
- Mean body condition (proportion of weight to body length)
- Level of fluctuating asymmetry of hind limb length (an estimate of environmental and genetic stress)

Frogs at high elevations seemed to be more susceptible to Bd than their lower elevation counterparts, and these high-elevation frogs were hit hardest during the winter.

There were other questions on many biologists' minds: Are frogs doomed if they come into contact with Bd? Do frogs have any natural defenses? In 2006, Woodhams and collaborators published data suggesting that the natural antimicrobial peptides produced by granular skin glands might protect some species from contracting the disease.[16] The investigators focused on the difference in potency between antimicrobial peptides in species that had experienced declines versus those in nondeclining species. They examined the same five stream-dwelling species from northeastern Queensland that were the focus of previous studies—the Australian lace-lid (declining), the common mist frog (declining), the waterfall frog (declining), the green-eyed tree frog (declined, but since recovered), and the stoney-creek frog (populations stable). Woodhams and colleagues found the least potent peptides from the Australian lace-lid and the common mist frog—both declining species. The authors emphasized, however, that peptide defenses are only one factor that may influence susceptibility to Bd infection. As shown in the Woodhams and Alford 2005 study, environmental conditions (presumably temperature) may affect susceptibility to Bd. Other factors might include age of the host and behavior.

A study carried out in east-central Queensland suggested that Bd does not affect all species within a community equally. During 1985–1986, one of the gastric-brooding frogs, *Rheobatrachus vitellinus,* and the Eungella torrent frog, *Taudactylus eungellensis,* disappeared suddenly from rainforest streams in Eungella National Park. Biologists still have not found any gastric-brooding frogs, but they subsequently discovered a few remnant populations of the torrent frog. Had Bd been the cause of the declines? We do not know. No specimens are available from that time period for testing. The first record of Bd in the national park is from a moribund frog collected in 1995. Yet Eungella torrent frogs persist in the habitat. Has Bd disappeared since then? No. In 2004 Richard Retallick at James Cook University and two colleagues published a paper demonstrating that the remaining populations of Eungella torrent frogs now persist with stable infections of Bd. The investigators analyzed toe clips from frogs collected in the park between 1994–1998. Eungella torrent frogs had an overall infection rate of 18 percent. One sympatric species that appeared not to have declined had an infection rate of 28 percent. None of the other four species showed any sign of infection.[17]

Where Are We Now?

In 2006, Lisa Schloegel at the Consortium for Conservation Medicine in New York City and five co-authors published a review of the evidence that Bd caused the extinction of sharp-snouted day frogs (*Taudactylus acutirostris*).[18] They proposed that this anuran is the first animal known to have gone extinct in the wild due to infection—that is, where disease (in this case Bd) had acted as both the proximate and the ultimate cause of extinction. Three co-authors on this paper were also part of the Australian group (referred to in the Preface) who characterized the impact of infectious disease on frogs as "the most spectacular loss of vertebrate biodiversity due to disease in recorded history."[19] Their statement was warranted. It was not hyperbole.

In November 2006, while visiting one of us (MLC), Ross Alford reported that the waterfall frog and several other species of frogs had reappeared at many sites in northeastern Queensland where they were previously missing. Yet Bd is still present in these areas.

If Bd is a native pathogen, perhaps something about the environment in these northeastern Queensland sites has changed. Perhaps the exact combination of environmental factors that caused the fungus to go into outbreak phase and resulted in die-offs does not exist currently, making the pathogen less virulent. We do not know.

Based on observations of outbreaks in Australia and elsewhere, however, it seems likely that Bd is an invader. An exotic pathogen might better explain the massive die-offs of numerous species in a short period of time, if it were more virulent than a native pathogen with which hosts have co-evolved for a long time. If this were the case, once frog populations experience Bd-associated declines, how quickly could they co-evolve with the new pathogen? Perhaps there has been some change in either the frogs or the fungus in northeastern Queensland that would explain their reappearance despite the continued presence of Bd. The frogs' antimicrobial peptide secretions may have changed. Frogs exhibit considerable variability in skin secretions within species, so perhaps there has been rapid natural selection along these lines. Individuals with the most effective secretions might live to reproduce and pass on this trait to their offspring. Or, perhaps, the naturally occurring bacteria or fungi on frogs' skin have evolved to outcompete Bd. For now we can only speculate.

In either case, whether the pathogen is native or exotic, after Bd reduces frog populations to low levels, the host-pathogen dynamics may change. Since the initial infections 2 decades ago, the frogs may have changed their behavior in some way that reduces their susceptibility to Bd. In addition, as a result of natural selection, some frogs may now be immune to Bd or at least able to tolerate its presence.

Are frogs seeking out warmer microhabitats in their natural environments? Results from the laboratory experiment performed by Woodhams and his colleagues suggest that such behavior by wild frogs would be prudent! One of Alford's graduate students at James Cook University, Jodi Rowley, recorded how much time rain forest frogs of northeastern Australia spent in sites where they could elevate their body temperatures, such as basking in sunny spots. She found a correlation between behavior and susceptibility to Bd.[20] Individuals of species less vulnerable to the chytrid tended to spend time in warmer, drier sites, and they did not frequently contact others of their same species. In contrast, individuals of the species most vulnerable to Bd tended to spend time in wet, cold sites. These frogs frequently contacted other individuals because they aggregated in rock crevices near streams. Contact presumably increased fungal transmission. Rowley's data show that observations of frogs in their natural habitat are just as valuable as laboratory experiments. We will come back to Rowley's work in chapter 7.

Clearly, we have learned much about frog declines in northeastern Queensland since the early 1990s. There is no doubt that chytridiomycosis is associated with certain declines and disappearances. We still have much to learn, however, including understanding interactions between the pathogen and environmental factors and the ecology of Bd.

Boreal Toads from the Colorado Rocky Mountains

The second ongoing "detective story" involves not a complex assemblage of frogs, each with its unique traits, distribution, and history, but rather a single species that once was familiar to naturalists, ranchers, and hikers across much of the mountainous west of the United States: the boreal toad (*Bufo boreas*), named for the ancient Greek god of the North Wind, Boreas.

Sick and Dying Toads

Boreal toads used to thrive in the cool summers and survive the harsh winters of the high mountains over much of the western United States

Figure 6.3 Boreal toad, *Bufo boreas*, from Colorado. Photograph by Cynthia Carey.

Figure 6.4 Typical habitat of *Bufo boreas* in Colorado. Photograph by Cynthia Carey.

(Figure 6.3). In the early 1970s, boreal toads were abundant in the West Elk Mountains and other Colorado ranges (Figure 6.4). Because snow covers the ground 6 to 8 months of the year in this area, the toads spent at least half their lives hibernating. Once the snow melted, they emerged and congregated at lakes, beaver ponds, glacial kettle ponds, and small temporary pools formed by snowmelt, where they mated and laid eggs.

Cynthia Carey, a Ph.D. graduate student in physiological ecology at the University of Michigan in the early 1970s, yearned to spend her summers in the Colorado mountains. She had previously worked with birds but decided that the ideal focus for her dissertation would be the abundant, easy to capture, and easy to transport boreal toads. She could spend her summers in Colorado doing fieldwork, then take toads back to Michigan and conduct physiological experiments.

Carey began fieldwork in the summer of 1971, working out of the Rocky Mountain Biological Laboratory in the ghost town of Gothic, near Crested Butte. All went well until the summer of 1973, when she began finding toad carcasses. Many dead toads had their back legs stretched out, as though convulsed by muscle spasms just before death. At the end of that field season, as in previous years, Carey brought apparently healthy toads back to Michigan. But that year, the toads sickened and died one by one. She took some ailing toads to a veterinarian, who diagnosed the problem as a bacterial disease known as "red-leg." The veterinarian suggested putting pennies in the water with the toads, because of the antibacterial qualities of copper. The toads still sickened and died. Another veterinarian suggested tetracycline. Carey tried the antibiotic, but her toads continued to die. She raced against time to gather enough data to complete her studies before all her animals died.

She made it and completed her doctoral dissertation on the physiological ecology of boreal toads. Colorado still beckoned, and in 1976 she joined the faculty at the University of Colorado in Boulder. When she returned to her former field sites in the West Elk Mountains, however, she found few toads. What to do? As a new faculty member, she had to "publish or perish." She switched her research to the physiological ecology of bird eggs.

Declines of Boreal Toads Continue

While Carey focused on bird eggs from the late 1970s to the early 1990s, populations of boreal toads continued to decline and disappear. Surveys on the western slope of Colorado completed in the late 1980s to early 1990s revealed that the toads persisted at only 1 of 377 historical sites.[21] Stephen Corn, then with the U.S. Fish and Wildlife Service in Fort Collins, and his colleagues found no boreal toads at 49 of 59 known localities on the eastern slope, in the Front and Park Ranges of Colorado, and in the Medicine Bow Mountains in Wyoming.[22]

Corn and Frank Vertucci at Colorado State University wondered if the cause of these declines and disappearances could be acid snow that accumulates over the winter and then melts and forms ponds the toads use for breeding. They studied the effects of acidic precipitation on six species of amphibians, including the boreal toad, from the Rocky Mountains of Wyoming and Colorado.[23] The habitats where amphibians bred in these areas—vernal pools, beaver ponds, semidrainage lakes, and drainage lakes—were not sufficiently acidic to kill developing embryos. Furthermore, boreal toads do not breed until May to mid-July, much later than the acid flush that would occur during early spring with melting snow. Acidic deposition appeared not to be the cause of declining populations of boreal toads. Some other factor was responsible.

Carey was working with bird eggs when, in 1989, she attended the First World Congress of Herpetology and heard talk of declining and disappearing amphibians. She had assumed that the boreal toad decline in Colorado was just a unique, local phenomenon—her bad luck. After the meeting, she read up on "red-leg" disease. Since the 1970s, when she had talked with the veterinarians in Michigan, much had been learned about the disease. One of the pathogens that causes "red leg" is *Aeromonas hydrophila,* a bacterium that is ubiquitous in fresh water and omnipresent in the digestive tract and on the skin of healthy amphibians. It infects the animals when they become immunosuppressed or when their immune systems lose the ability to prevent infection. They stop eating, septicemia (blood poisoning) develops, and eventually the skin hemorrhages, which yields the characteristic reddish color of infected animals.

Carey pondered what had stressed the immune systems of her toads and why their natural defenses had failed. In 1993, she published a working hypothesis for the disappearance of 11 populations of boreal toads from the West Elk Mountains between 1974 and 1982.[24] She suggested that some environmental factor, or a combination of factors, had stressed the toads. This stress in some way suppressed the toads' immune system, and they succumbed to infection by normally congenial *Aeromonas hydrophila* bacteria. She offered her hypothesis "as a means of stimulating research into the causes of amphibian declines." And indeed it did.

In 1996, Carey received a grant from the National Science Foundation to study possible environmental factors that might stress toads and affect their immune systems. She exposed toads to various stresses in the laboratory—pesticides, acidified water, and rapid temperature changes. Nothing emerged as a culprit. Imagine her surprise when she learned about the chytrid, *Batrachochytrium dendrobatidis.* She now suspected that the same fungus associated with declines in Australia and Central America might have infected her boreal toad populations.

We will return to Carey's research in a bit, but first we will follow a different line of investigation from the 1990s—another attempt to explain boreal toad declines.

Ultraviolet-B Radiation?

Andrew Blaustein and associates at Oregon State University published an intriguing paper in 1994 reporting on the effect of UV-B radiation on three species of amphibian embryos from Oregon.[25] One of these was the boreal toad. Embryos of both boreal toads and Cascade frogs (*Rana cascadae*), species whose populations have declined, had lower hatching success in field enclosures exposed to full sunlight (ambient levels of UV-B radiation) than in enclosures fitted with UV-B filters. Embryos of Pacific tree frogs (*Hyla regilla*, now *Pseudacris regilla*), whose populations are not known to have declined, showed no difference in hatching success when shielded from UV-B radiation, compared with exposure to natural UV-B levels. The investigators found differences among the three species in the activity of photolyase, an enzyme involved in repair of photo-damaged DNA. Boreal toad and Cascade frog embryos had low photolyase activity. Embryos of Pacific tree frogs had the highest levels of photolyase activity. The investigators suggested that differential sensitivity to UV-B radiation among species of amphibians might contribute to population declines.

Could exposure to high levels of UV-B radiation have caused population declines of boreal toads in the Colorado Rocky Mountains? In contrast to Blaustein and colleagues' research, other published studies had not demonstrated a convincing link between increased levels of UV-B radiation and mortality in amphibian embryos. Stephen Corn repeated the Oregon experiments with boreal toad embryos from two breeding sites in Rocky Mountain National Park.[26] In contrast to the Oregon study, however, there was no difference in hatching success related to UV-B exposure. Why the opposite results? Corn suggested several possibilities, including different environmental conditions at the sites and different experimental enclosures. Whatever the reason, the results of Corn's study did not support the hypothesis that increased levels of UV-B radiation caused declines of populations of boreal toads in the southern Rocky Mountains.

Back to Bd—And Has Climate Change Played a Role?

The year after Corn published his research, the October 2, 1999, issue of *Science News* reported dead boreal toads collected in May of that year from Clear Creek County, west of Denver. David Green, the veterinary pathologist who had examined the diseased frogs from Panama, identified Bd on the boreal toads.

That same year, Carey and two collaborators published a paper addressing an immunological perspective on amphibian declines.[27] They speculated that fungal infection may have been the primary cause of death of boreal toads and other amphibian species that experienced mass die-offs in the 1970s. A preserved specimen of *Rana pipiens* (leopard frog) collected from the Colorado Rockies carried Bd on its skin, revealing that the pathogen was present in Colorado as early as 1974. It is likely that we will never know for sure what caused those early boreal toad population die-offs, but circumstantial evidence now points to Bd.

Other scientists also had directed their research toward boreal toads. Erin Muths at the U.S. Geological Survey Biological Resources Division in Fort Collins and her colleagues began monitoring populations of boreal toads at four sites in Rocky Mountain National Park in 1991.[28] These populations were relics that had somehow escaped the declines of the 1970s and 1980s. In 1996, the researchers began finding fewer animals. In 1998, they found dead toads and lethargic individuals with sloughing skin. Subsequent histology revealed Bd.

But what about climate change? Could that have affected boreal toads in the southern Rocky Mountains? As we discussed in the last chapter, two papers published in a 2001 issue of *Conservation Biology* suggest that the answer is "No." Michael Alexander and Jon Eischeid, both at the Climate Diagnostics Center in Boulder, analyzed variability in temperature and moisture data from four areas experiencing frog die-offs: northeastern Queensland, Puerto Rico, Colorado Rocky Mountains, and montane regions of Costa Rica and Panama.[29] They concluded that unusual climatic factors did not cause any of these declines directly but suggested that climatic variables might have contributed indirectly. Likewise, Carey and 10 colleagues examined environmental correlates coincident with mass mortalities at these same four sites and concluded that unusual climate just before or during mass mortality was unlikely to have caused deaths directly.[30] They emphasized, however, the need to investigate possible synergistic interactions among environmental variables and factors such as food supply, movement of pathogens, and susceptibility to infectious disease. The authors recommended that research focus on predicting where future mass die-offs might occur, with the goal of preventing them.

The time seemed ripe for experiments. Carey and eight collaborators experimentally exposed boreal toads to Bd to document pathogen/host interactions.[31] Because the toads are listed as endangered or threatened throughout their southeastern range, the investigators used toadlets raised at the Native Aquatic Species Restoration Facility (Colorado Division of Wildlife) from eggs or tadpoles collected from five localities in the state. They found that the infection must reach a certain threshold of zoosporangia before the victims die. (Recall from the previous chapter that zoosporangia are tiny structures housing the fungal asexual spores called zoospores.) The investigators housed toadlets at either 12° C or 23° C and found that temperature did not affect survival time. Smaller toads, however, survived fewer days. Toadlets exposed to the pathogen behaved differently from control animals not exposed to the fungus. Whereas control toadlets sat in the water and rarely elevated their undersides, Bd-exposed individuals held their bodies out of water as much as possible and even climbed onto the container walls.

Carey and her colleagues found that toadlets did not need to contact each other directly to transmit or contract the pathogen. Individuals sitting in water in which infected toadlets had been housed contracted chytridiomycosis and died. Previous work by other investigators had shown that the pathogen can remain viable in sterilized water for up to 7 weeks.[32] For boreal toads, this could mean that, if an infected individual shed zoospores in a breeding pond in June, individuals in that pond could be exposed throughout much of the summer, even if there were no infected toads at the pond during that time. Whether a toad exposed in the wild develops chytridiomycosis depends in part on the quantity of zoospores to which it is exposed. Males may be more vulnerable than females because they spend more time in the water at the breeding sites. Furthermore, males presumably attempt to breed every year, whereas females generally do not breed every year.[33]

When one of us (MLC) talked with Carey in mid-December 2006, she believed that Bd alone was responsible for declines of boreal toads in the southern Rocky Mountains, pointing out that "in the laboratory the fungus kills the toads without any other factors. You don't have to evoke global climate change. In the field, once the chytrid infects a formerly chytrid-free population of toads, the toads die whether it's a warm year, cold year, dry year, or wet year."

The Future

What is the prognosis for boreal toads? Carey and five colleagues addressed this question for southern Rocky Mountain populations.[34] Certain life history characteristics, such as clutch size, size at reproductive maturity, and wet egg mass, are similar between boreal toads and other species of lower-elevation *Bufo*. Boreal toads, however, must deal with a short growing season and cold summer temperatures. This combination greatly influences certain other life history characteristics, restricting their ability to recover from major population declines. For example, boreal toads have a much shorter active season in which to accumulate energy for breeding, growth, and maintenance than toads at lower elevations. Boreal toads begin breeding 1 to 3 months later in the year, and they spend 1 to 3 months longer in hibernation than their lower-elevation counterparts do. Cold temperatures prolong larval development. In years when freezing temperatures arrive early in the fall, larvae may not metamorphose in time. Individuals that do metamorphose grow slowly before reaching reproductive maturity. Boreal toads may well have an uphill battle.

In 1995, the Colorado Division of Wildlife formed a Boreal Toad Recovery Team to protect the few remaining boreal toad populations in the southern Rocky Mountains.[35] Since then, personnel have spent considerable time and money to restore population sizes and expand the toads' geographic distribution. Unfortunately, attempts to repatriate eggs, tadpoles, and metamorphosed young into historical sites have been unsuccessful. The Native Aquatic Species Restoration Facility still raises toads for research purposes and to maintain genetic variability, but because so many repatriated toads died, scientists have given up trying to reintroduce boreal toads to the wild.

As of January 2008, there were about 33 known populations of boreal toads remaining in Colorado.[36] Most of the northern populations are infected with Bd, whereas most of the southern populations are still free of the pathogen. Will the fungus spread south?

Carey and other scientists are pondering some profound questions. Are toads that survive an outbreak of chytridiomycosis resistant to the fungus? Will natural selection alter gene frequencies and produce populations of resistant toads? If so, how long will it take? Who will win the evolutionary arms race: boreal toads or *Batrachochytrium dendrobatidis?*

The ancient Greeks believed that Boreas, god of the North Wind and harbinger of winter, was strong and determined. And he had a temper—he got his way. The ancients believed Boreas lived in a place where inhabitants had extraordinarily long lifespans. Perhaps not coincidentally, boreal toads live longer than do toads at lower elevations. We can only hope that boreal toads will fight off the amphibian chytrid fungus and regain their position as salient members of their cold, windy montane environments.

Montane Frogs From Costa Rica and Panama

Our third "detective story" takes us to Central America, where amphibians have declined within diverse assemblages at high-elevation sites. As with the preceding two stories, research on the causes of declines in Central America has been at the forefront of solving the mystery of worldwide declines.

The First Recorded Declines

"Marty, the golden toads are out! You've got to come see them," urged Wolf Guindon, field coordinator for the Monteverde Cloud Forest Reserve in northern Costa Rica. Guindon stood at Marty Crump's door in a driving rain and shouted over the howling wind that it was a sight not to be missed. Golden toads (*Bufo periglenes*) lived underground all year long, except for a few weeks in March through June when they emerged to breed in temporary pools. The next morning, Marty and Guindon hiked uphill through a constant drizzle to the elfin forest near the continental divide in the Cordillera de Tilarán.[37] Rounding a bend on the path, they saw more than 100 bright orange toads gathered around small pools at the bases of stunted trees (Figure 6.5). The toads sparkled like jewels in the otherwise dark forest on that early April day of 1987. Over the next 10 days, the toads laid eggs, but the cool misty weather ended and the breeding pools dried. Most embryos died before hatching. Another bout of rain in May brought the toads to the surface once more, but the refilled pools again dried before many eggs hatched.

The following year, Marty found no toads until a few days after a hard rain on May 21. One male golden toad sat next to a pool. Marty monitored the breeding sites every day through early June but found no additional toads. She assumed weather conditions had not been right. Surely the toads would appear next spring. She worried, however, because harlequin frogs (*Atelopus varius*) also had declined precipitously in the Monteverde cloud forest as well as from her study site 4 km west at the Río Lagarto (Figure 6.6).

Figure 6.5 Golden toads, *Bufo periglenes*, at breeding pool, Monteverde, Costa Rica. Photograph by Martha Crump.

Figure 6.6 Harlequin frog, *Atelopus varius*, on boulder at Río Largarto, Costa Rica. Photograph by Martha Crump.

By June, she could find no harlequin frogs at either site. She facetiously wrote in her field journal that perhaps she had inadvertently transmitted a virus, fungus, or bacterium lethal to frogs at both study sites.[38] She recalled that in 1986 Guindon had asked for her advice on what to feed the captive golden toads kept at the reserve station for ecotourists to see. He said that for years they had kept toads without a problem, but that recently the toads were dying. Now she wondered whether "her" study toads had been sick also.

Guindon monitored the breeding sites throughout March 1989 but found no toads. Marty hiked up to the elfin forest every day in April. Although the depressions filled with water, she found no toads. The leaf litter *Eleutherodactylus* that had been so common in the reserve in past years were missing, as were glass frogs (centrolenids), formerly common along the Río Guacimal. Something was definitely wrong. That summer, Frank Hensley, one of Marty's graduate students at the University of Florida, found the last known golden toad—sitting on the ground within 3 m of the one Marty had seen the previous year.

Three months later, Marty attended the First World Congress of Herpetology in Canterbury, England. She shared her story of golden toads and harlequin frogs with colleagues and also heard their woes of disappearing amphibians.

Marty, Hensley, and Kenneth Clark (then a graduate student at the University of Florida) documented the decline of golden toads in 1992.[39] They speculated that the toads might still be underground, where they normally lived because of unfavorable conditions. They noted, however, that in the late 1980s many other species of amphibians from the area were either absent or infrequently encountered. That observation led them to suggest an alternative hypothesis: Golden toads and other anurans in the region might have experienced mass mortality, perhaps from a pathogen outbreak.

Meanwhile Alan Pounds, Resident Scientist at the Tropical Science Center's Monteverde Cloud Forest Reserve, also was trying to figure out why golden toads had disappeared. He reanalyzed rainfall data, focusing on the 1986–1987 El Niño event. Pounds divided the year, not by months or traditional seasons, but by four components of the yearly cycle based on the amount of rain and cloud mist: late wet season (July–October), transition into dry season (November–December), dry season (January–April), and post-dry or early wet season (May–June). Then he examined 20 years of rainfall data for Monteverde from the perspective of these four annual components. An aberration appeared. The 1986–1987 weather cycle, the one that included the last breeding bouts of the golden toads, had had abnormally low rainfall during each of the four components. The total precipitation for July–October 1986, the time when wet-season rains usually replenish groundwater, was the lowest recorded—almost 40 percent below the 20-year average. The following May, total precipitation was 64 percent below the 20-year average. The major impact of reduced precipitation may have been on the water table. Because golden toads live underground, they may have experienced unusually dry conditions during 1986–1987.

Pounds and Marty collaborated on a 1994 publication addressing the disappearance of golden toads from the Monteverde Cloud Forest Reserve and harlequin frogs from the Río Lagarto in light of Pounds' reanalysis of the weather data.[40] In it, they suggested that the 1986–1987 weather played a significant role in the species' declines, but that the unusual moisture-temperature conditions may have interacted synergistically with some unknown factor. They offered alternative hypotheses. Their "climate-linked epidemic hypothesis" suggested that, as animals exposed to drier conditions and warmer temperatures aggregated in suitable microhabitats, they might have been more vulnerable to microparasites such as bacteria or protozoans. Their "climate-linked contaminant pulse hypothesis" suggested a transient increase in the concentration of atmospheric toxic chemicals. During a normal dry season, these contaminants might not be lethal to amphibians, but an unusually severe dry period might prompt a deadly pulse of toxins. During prolonged dry periods, such as 1986–1987, less water would dilute pesticides or other contaminants that blow into the Monteverde area. Amphibians might absorb these chemicals concentrated in the soil.

Subsequent Declines Farther South and East

In July 1991, Karen Lips began her fieldwork on the population biology of the tree frog *Hyla calypsa* at Las Tablas, a remote cloud forest site in southern Costa Rica. She surveyed other amphibians at the site from 1991 to 1996. Over the years, Lips observed population declines for many species, including *Atelopus chiriquiensis* and *Hyla calypsa*, both stream-breeders.[41] She found dead and dying individuals of six species of frogs and salamanders in September 1992 and between June and August 1993. One of Lips' important contributions was catching these die-offs in the field, as they were happening. By observing simultaneous multispecies population crashes associated with mass mortality within short time periods, she sounded an alarm based on empirical data. Lips sent some of the preserved dead animals to the University of Florida's School of Veterinary Medicine for analysis. The results proved negative for bacteria and virus. In 1994, Lips found *Hyla calypsa* tadpoles that lacked the normal keratinized beaks surrounding the mouths. Because bacteria and fungi are known to degrade keratin, she suggested that the absence of keratinized mouthparts signaled infection by a pathogen.[42]

Lips ruled out habitat destruction, introduced fishes, researcher disturbance, UV-B radiation, habitat acidification, and alteration of rainfall patterns as possible direct causes for the Las Tablas declines. She concluded that "environmental contamination (biotic pathogens or chemicals) or a combination of factors (environmental contamination plus climate change) may be responsible for declines in the amphibian populations at this protected site."[43] Lips noted that the spread of declines from Monteverde to Las Tablas corresponded to a movement of about 250 km in 6 years, or about 42 km/year. She suggested that the marked declines from the two areas were consistent with a pathogen outbreak. The fact that the species most affected were those with aquatic eggs and tadpoles suggested a waterborne agent. Her paper was published in 1998, two years after the paper by Laurance and colleagues hypothesizing a wave of epidemic disease in Australia.[44]

Lips had also been working in Panama, because she suspected that a lethal pathogen was moving from one mountain range to another, in a north-to-southeast direction in Central America. Beginning in 1993, she monitored streamside anurans at the Reserva Forestal Fortuna, in western Panama. Frogs were abundant from 1993 to 1995. Lips returned to Fortuna in December 1996 because she had predicted, based on the epidemic disease wave hypothesis, that population crashes would occur at that site. Her hunch was right. She found frogs at only a few streams—and 54 dead or dying individuals (15 percent of total captures) of 10 species at four streams.[45] She froze or preserved the bodies of sick or dead frogs. Subsequent examination revealed that 18 of 18 animals were infected with the amphibian chytrid fungus, *Batrachochytrium dendrobatidis.* These were the animals included in the 1998 paper by Berger and colleagues revealing that the frogs from Australia and western Panama had been infected with what appeared to be the same pathogen.[46] Of the 106 tadpoles Lips collected from five streams, 12 percent—13 individuals of four species in all five streams—had either partial or complete loss of keratinized mouthparts.

Lips returned to Fortuna during July–August 1997. She sampled nine streams 37 times and found only six frogs. Her observations at Fortuna provided the third report of mass amphibian declines at a remote, protected high-elevation reserve in Central America. Now on the faculty at Southern Illinois University in Carbondale, she published her results in 1999.[47] Lips hypothesized that the cause of the sudden and sequential declines at all three sites—Monteverde, Las Tablas, and Fortuna—was Bd. Las Tablas is about 250 km southeast of Monteverde, and Fortuna is about 75 km east-southeast of Las Tablas (Figure 6.7). The observations suggested directional movement of the pathogen, similar to that hypothesized by Laurance and colleagues in Queensland, Australia.

At the end of her 1999 paper, Lips referred to unpublished data from Erik Lindquist, then a graduate student at Ohio State University. Lindquist's field observations from Panama supported her hypothesis. Between February and June 1994, he had found a dead *Atelopus chiriquiensis* and a dead *Rana warszewitschii* in streams on the northern and eastern slopes of

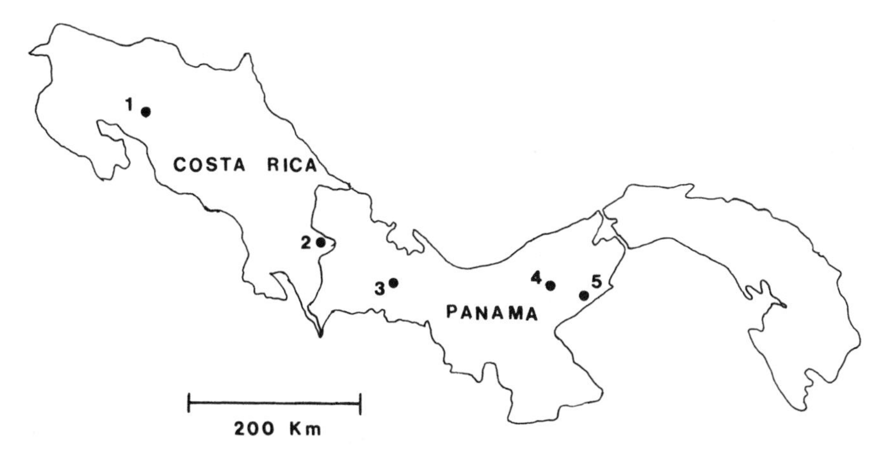

Figure 6.7 Map of Costa Rica and Panama showing sites with major amphibian declines: (1) Monteverde, Costa Rica; (2) Las Tablas, Costa Rica; (3) Fortuna, Panama; (4) El Copé, Panama; (5) El Valle de Antón, Panama. Redrawn from Lips et al., 2003b.

Cerro Pando, a site geographically intermediate between Las Tablas and Fortuna. The timing of the discovery was also intermediate between the declines Lips observed at her study sites. Subsequent examination of five *Atelopus* from the Cerro Pando area revealed Bd in four individuals. By 1997, Lindquist could find no amphibians along upland streams at Cerro Pando.

After the discovery of Bd on the Panamanian frogs, Lips and two colleagues reexamined four specimens of dead or dying amphibians (three frogs, one salamander) and seven seemingly healthy frogs Lips had collected from Las Tablas, Costa Rica, in 1993. All three of the sick or dead frogs and one of the seemingly healthy frogs had epidermal Bd infections.[48] She now had stronger evidence to implicate the fungus in the previously unexplained die-offs and population declines at Las Tablas.

Could Bd have wiped out the harlequin frogs and golden toads from Monteverde in 1987? Robert Puschendorf at the University of Costa Rica examined Costa Rican specimens of *Atelopus varius* from the Museum of Zoology at the University of Costa Rica and reported that an individual collected in 1986 in San Ramón, Sarapiquí, was infected.[49] The San Ramón reserve is a few kilometers south of Monteverde. Unfortunately, no one had preserved specimens from Monteverde during 1986–1987. Although a direct test for Bd was impossible, it certainly seemed likely that Bd was the culprit.

Lips and two colleagues analyzed ecological traits of declining species from four sites in Central America and found several similarities: aquatic habitats, limited elevational ranges, and large body sizes.[50] Their "aquatic index," which estimated the degree of dependence on stream habitats over a species' lifespan, was the most significant factor associated with declines at all sites. The results of their analysis provided a model that allowed them to forecast the likelihood of decline for species at the then unaffected site of El Copé in central Panama: They predicted that 33 species were likely to decline, 9 species were somewhat likely to decline, and 10 species were unlikely to decline. El Copé is geographically and ecologically similar to other sites in Costa Rica and Panama where declines had occurred. The wait began to see whether El Copé would experience amphibian declines.

Effect of Climate?

Golden toads and harlequin frogs were not the only anurans that declined or disappeared from the Monteverde cloud forest. As mentioned in chapter 1, Alan Pounds and a field crew surveyed the forest for frogs from 1990 to 1994 and reported that 20 species disappeared during that period—40 percent of the known anuran fauna in the area.[51] Populations of some other species from the region declined but did not disappear. Pounds continued to study the possible association between climate and amphibian declines in the Monteverde cloud forest.

Pounds and two colleagues analyzed patterns of precipitation, stream flow, air temperatures, and sea surface temperatures to investigate whether changes in abundance and distribution patterns of birds, reptiles, and amphibians were associated with climate change.[52] They noted that there is increasing evidence that the strong warming of tropical oceans since the mid-1970s has altered the climates of tropical mountains. For example, atmospheric temperatures have warmed. As moisture-laden trade winds meet the Caribbean slope of the Cordillera de Tilarán and flow upward, they cool and produce a large cloud bank. The authors hypothesized that cloud formation heights have shifted upward, reducing inputs of tradewind-conveyed precipitation, and that reduced moisture might adversely affect organisms because of less exposure to critical mist and cloud water during the dry season.

Pounds' group reported a dramatic decline in mist frequency since the mid-1970s and increased variability of daily rainfall.[53] From 1973 to 1998, the number of dry days during the dry season increased; 1987 was one of four years with the greatest number of dry days. The annual minimum stream flow decreased from 1965 to 1995. The daily minimum temperature increased relative to the daily maximum, yielding a smaller diurnal temperature range.

The authors suggested that certain biological patterns followed their model's predictions. Many species of premontane breeding birds had invaded lower-montane habitats. Some lower-montane species likewise had moved higher in elevation. Furthermore, wet-season abundance of anoline lizard populations that had declined or disappeared was strongly negatively correlated with the number of dry days in the preceding dry season—the more dry days, the fewer the anoles. The investigators also found that anuran declines were associated with mist-frequency patterns. Amphibian populations collapsed immediately after the driest precipitation extremes recorded for the Monteverde area.

Pounds and colleagues concluded that these changes in abundance and distribution of various birds, reptiles, and amphibians were all associated with the same climatic patterns and occurred at the same time. They calculated that the probability that all three events—birds moving higher in elevation, lizard declines, and collapse of frog populations—would, by chance alone, correspond to these climatic extremes was equal to or less than 0.008. In other words, at 8 chances in 1000, this was a highly statistically significant result. They suggested that the patterns imply "a broad response to regional climate change which crossed an important threshold in the late

1980s."[54] Furthermore, the authors suggested that, because climatic factors affect host-parasite relationships, climatic changes may have "set the stage" for outbreaks of Bd in high-elevation forests.

Recall that Lee Berger and colleagues suggested, in their 1998 paper, alternative hypotheses regarding the sudden virulence of Bd.[55] One hypothesis was that the fungus might be an introduced pathogen spreading through populations previously unexposed to it. The other was that Bd might be a widespread native organism that had only recently become pathogenic to amphibians, either because Bd itself had increased in virulence or because environmental changes or other factors had increased amphibians' susceptibility to the fungus. The idea that climate change could "set the stage" for chytrid outbreaks might be a stronger possibility if Bd were a native pathogen rather than an exotic invader, although in the latter case, climate change could influence both the spread of Bd and its virulence. We will learn in the next chapter how climate change can also facilitate the spread of frogs infected with Bd into communities without the pathogen.

In an essay entitled "Clouded Futures," Pounds and Puschendorf offered some ideas about how climate change might hamper amphibian defenses against Bd.[56] As Pounds had reported earlier, widespread increases in cloud cover are evident on mountain slopes because of increased water vapor associated with increasing temperatures. And we know that Bd survives in the laboratory at temperatures between 6° to 28° C but dies at higher temperatures. Experiments carried out on Australian red-eyed tree frogs revealed that frogs can rid themselves of the fungus if they can elevate their body temperatures.[57] Pounds and Puschendorf wrote that, taken together, these bits of information suggest that increased cloud cover might reduce the ability of amphibians to bask or seek out warm microenvironments, resulting in their becoming more vulnerable to Bd. Dry conditions could compound the problem by constraining the frogs' movements and behavior. These ideas suggested how Bd outbreaks might be linked to climate change.

To test whether such a link existed, Pounds and 13 colleagues analyzed patterns in the timing of disappearances of harlequin frogs (*Atelopus*) in relation to large-scale climate changes in sea surface and air temperatures. The climate-linked epidemic hypothesis predicted that amphibian declines should occur in unusually warm years, because temperature shifts or related variables can influence disease dynamics. For their analysis, they examined *Atelopus* extinctions from Costa Rica to Peru, including 100 species for which data were available indicating the last year the species was observed in the wild.[58] They estimated that 67 percent of the total recognized species of *Atelopus* had disappeared. About 80 percent of the species that disappeared were last seen soon after a relatively warm year.

The authors concluded that large-scale warming was a key factor in the disappearances.

And the mechanism? Pounds and his coauthors suggested that temperatures at many high-elevation sites are shifting toward the optimum for Bd's growth and reproduction. Although the global warming trend is caused by greenhouse gases, changes in cloud cover might also influence changing temperatures. Cloud cover at night can reduce outgoing long-wave radiation, resulting in warmer nights. Daytime cloud cover can diminish warming and block radiant heating, resulting in cooling of microhabitats. The authors suggested that both nighttime warming and daytime microhabitat cooling might improve conditions for Bd. They referred to their hypothesis as the chytrid-thermal-optimum hypothesis, which suggests that "daytime cooling (local or microscale) and night time warming accelerate disease development." They further stated that "The impacts at night may explain the association with warm years and thereby resolve the climate-chytrid paradox."[59] This hypothesis currently is being tested, as we will see in the next chapter.

The paper by Pounds and associates did not "resolve the climate-chytrid paradox." In fact, it added fuel to an already hot controversy. In March 2008, Karen Lips and three collaborators reviewed the evidence for the role of climate change in triggering outbreaks of Bd.[60] They concluded, "Available data support the hypothesis of multiple introductions of this invasive pathogen into South America and subsequent spread along the primary Andean cordilleras." They found no evidence to support the hypothesis that climate change has been driving outbreaks of Bd. The two groups of investigators—those led by Pounds and by Lips—question each other's data, statistics, and data interpretation.

Bd in Costa Rican Specimens

Robert Puschendorf and his colleagues continued to search for Bd in Costa Rican frogs preserved before the amphibian declines at Monteverde.[61] They examined 202 frogs of 30 species from a 1986 collection of preserved specimens from Braulio Carillo National Park. Frogs had been collected along an altitudinal transect running from 100 to 2600 m elevation. The investigators found the fungus in nine species of frogs from the sample collected 1 year before the first documented declines at Monteverde. The frogs had lived about 75 km east of Monteverde. Almost 6 percent of the specimens examined had Bd in their skin. The highest frequency of infected individuals occurred at elevations between 100 and 399 m, but infections were also high at higher elevations. Puschendorf and his collaborators found Bd in the skin of species that later declined as well as those that did not. In addition, they examined 18 frogs collected in 2002 from two lowland Caribbean sites. Most of these frogs also had the fungus in their skin, suggesting that Bd might be widespread in Costa Rica.

To date, mass mortality of Bd-infected frogs has not been observed at any low-elevation site in Costa Rica. But the fact that Bd is found in lowland sites and the infected species there are not declining—or are experiencing declines that are less rapid, less severe, or less obvious—suggests that such amphibians might serve as reservoirs for the disease. The pathogen is not restricted to aquatic habitats but also infects terrestrial species such as *Craugastor fitzingeri,* which is a direct-developer, and *Dendrobates pumilio,* which has terrestrial eggs, with tadpoles developing in water-filled bromeliads.

There is still no direct evidence that an outbreak of Bd caused the local extinction and population declines of many amphibians from Monteverde. Most recently, Lips and three colleagues histologically examined 64 preserved anurans collected from the vicinity of Monteverde between 1979 and 1984—before the documented declines there in 1987.[62] The sample represented 18 species; of these, 11 have been shown to be infected with Bd elsewhere in their range. None of the 64 specimens, including the 1 *Atelopus varius* examined, was infected with Bd. If Bd was the cause of amphibian declines in the area, it may not have been in the environment as early as 1984.

Predictions

As mentioned earlier, Karen Lips and her colleagues predicted, on the basis of ecological traits, which species of frogs would be most likely to decline from El Copé in central Panama once Bd arrived. Before their work in El Copé, we knew that chytridiomycosis was associated with declines in at least 43 species of amphibians in Latin America and 93 species worldwide. But we did not know the status of Bd before any die-offs or declines had been observed at affected sites. El Copé was ahead of the postulated epidemic wave. By sampling the site before Bd arrived, the investigators hoped to describe the complete sequence of events, from arrival of the pathogen to its effect on the amphibian assemblage.

For 6 years (1998–2004), Lips and her coworkers looked for but did not find Bd at the high-elevation site of El Copé, based on examination of 1566 individuals of 59 species.[63] Then, on September 23, 2004, they found the first infected frog. On October 4, they found the first dead frog. Sampling during October revealed that more than 10 percent of the frogs were already infected. Between October 4 and February 15, they found 346 dead frogs of 38 species and 5 dead salamanders. All but 3 of 318 amphibians examined had moderate to heavy Bd infections; the other 3 were too badly decomposed to diagnose infection. Another nine species of live frogs tested positive for Bd. Affected frogs represented 70 percent of all frog species and represented all seven families known from El Copé. All of these infections happened in less than 5 months during the time the extensive fieldwork was carried out, although infections and mortality continued afterward as well (Figures 6.8, 6.9, and 6.10).

The prevalence of the pathogen had increased rapidly from zero to high, suggesting that Bd had invaded the region. The arrival of the pathogen supported the hypothesis that Bd was moving southeastward. Lips and her colleagues had predicted it would hit El Copé, and it did. Populations of all the species they predicted to decline did decline, some more rapidly than others. Even completely terrestrial species contracted chytridiomycosis, and their populations declined, though more slowly than those of more aquatic species. Furthermore, frogs from all habitats sampled—terrestrial, canopy, and aquatic—were affected, both in areas along streams and in the forest. Importantly, the study documented that most species within this amphibian assemblage contracted chytridiomycosis. Most populations of infected species declined, but not all went extinct. And all of this happened at the same time!

Lips and her colleagues predicted that many more amphibian populations will decline and disappear from montane areas directly east of El Copé. They suggested that, given the extent of this disease-driven extinction, we should refer to "global amphibian extinctions" rather than "global amphibian declines." As we will see later, they were right again.

What to Do?

If Bd is an introduced pathogen that is spreading in a "wave-like" pattern, and if scientists are correct in their forecast of the direction of this virulent pathogen's movement in Central America, is there something we can do if we cannot keep the pathogen from moving? Yes, and it has already started: Move amphibians out. Lips and her colleagues predicted that the pathogen would reach El Valle de Antón in central Panama in 2006. El Valle, a town and national park in the bowl of the inactive volcano, houses a rich amphibian fauna including the endangered golden frog (*Atelopus zeteki*) (Figures 6.11 and 6.12). Horrified by the decline at El Copé, Joseph Mendelson (at Zoo Atlanta) and Ronald Gagliardo (at Atlanta Botanical Garden) and a team of Zoo Atlanta volunteers traveled to El Valle before Bd arrived at the site. They collected 600 frogs of 35 species and airlifted them back to Atlanta. As quoted in *The New York Times*, June 6, 2006, Mendelson said: "When you can make predictions with respect to catastrophic population declines and

Figure 6.8 Dead or dying frogs from El Copé, Panama. *Top*: *Colostethus* sp. *Middle*: *Eleutherodactylus* new species. Photographs by Forrest Brem. *Bottom*: *Phyllomedusa lemur*. Photograph by Roberto Figa Brenes.

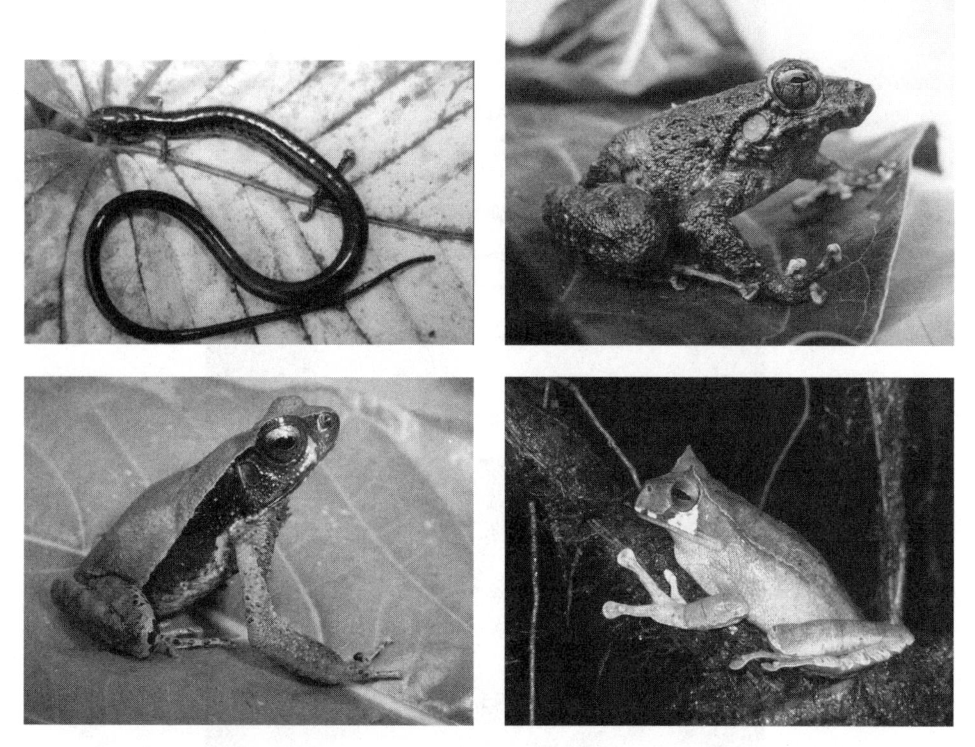

Figure 6.9 Four amphibian species no longer found at El Copé, Panama. *Top left*: *Oedipina collaris*. *Top right*: *Eleutherodactylus punctariolus*. Photographs by Karen Lips. *Bottom left*: *Bufo haematiticus*. Photograph by Roberto Figa Brenes. *Bottom right*: *Gastrotheca cornuta*. Photograph by Forrest Brem.

extinctions, we all agreed you have a moral and ethical responsibility to do something about it."[64] We will return to this heroic airlift in chapter 9.

Meanwhile, Bd arrived in El Valle in the spring of 2006. Frogs were now dying from chytridiomycosis. Had Mendelson, Gagliardo, and their team of volunteers not taken the 600 frogs out of El Valle, the fungus most likely would have killed them. The emergency airlift was worthwhile. But how long can we sustain the effort? And how do we choose which species to rescue, which areas to save? Unfortunately, die-offs of amphibians are likely to continue throughout the region for the foreseeable future.

Although moving animals out of their natural environment and maintaining and breeding them in captivity is a temporary stopgap against extinction, it is a limited option. The devastating and widespread effects of chytridiomycosis make it impossible to save all affected species. Furthermore, the goal is not to ensure that species survive in perpetuity in zoos. To be a conservation strategy, the goal of removing amphibians from

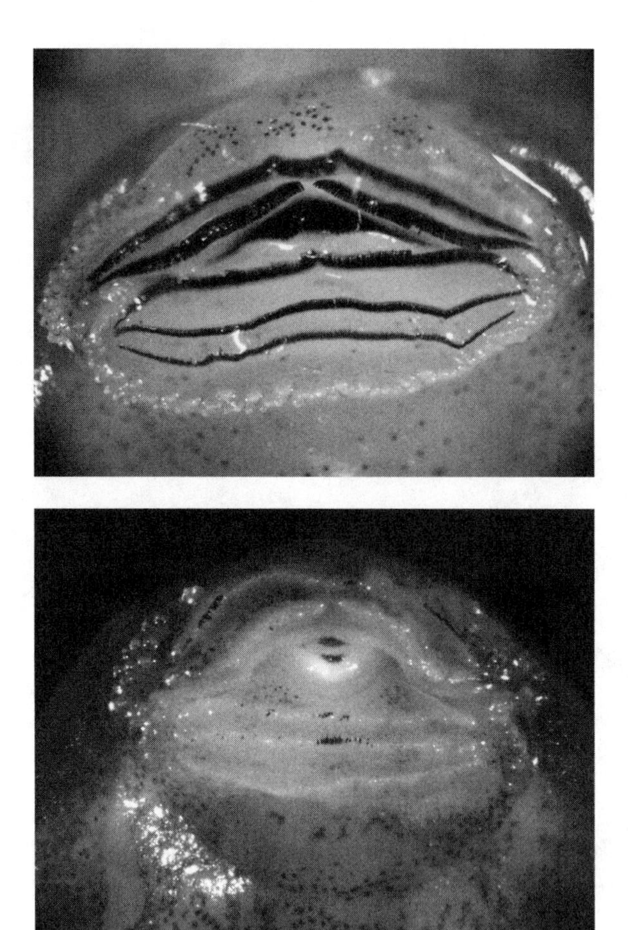

Figure 6.10 Mouths of *Colostethus panamensis* tadpoles from El Copé, Panama. *Top*: normal mouth. *Bottom*: mouth missing about 90 percent of its teeth and jaw sheet. Photographs by Roberto Figa Brenes.

the wild and housing them in captivity should be their eventual return to the wild. But how long will it be before the world is "Bd safe?" By itself, such removal simply leaves zoos with more animals to care for and the natural world just as depauperate. We will return to ethical, policy, scientific, and conservation issues regarding removal of amphibians from the wild in chapter 9.

The culprit in all three scientific detective stories described in this chapter is Bd, the amphibian chytrid fungus. Opinions differ regarding some aspects of the stories, however. In part, the differences revolve around whether the basic premise is that Bd is a widespread, native pathogen or an exotic invader. Furthermore, our understanding of the ecology of

Figure 6.11 Golden frog, *Atelopus zeteki*, from Panama. Photograph by Dante Fenolio.

Figure 6.12 Cloud forest in El Valle, Panama. Photograph by Dante Fenolio.

amphibians and Bd may be clouded because some studies have been carried out in the laboratory and others in the field. A consistent observation in the Bd puzzle is that the fungus tends to be more virulent and to cause mortality more predictably in the laboratory than in the field. We maintain amphibians in the laboratory under "ideal frog conditions"—high humidity and temperatures between 15° and 25° C. These conditions also happen to be ideal for growth and development of Bd. Laboratory studies might give us a biased picture. In the field, many amphibians spend time basking under conditions that are not ideal for Bd, which might reduce susceptibility to infection.

Another controversy is the role of climate change in Bd-caused declines. Evidence suggests that microenvironmental variation in temperature and relative humidity influence infection rate, prevalence, mortality rate, and transmission factors. But beyond this, some scientists argue for a strong link between climate change and outbreaks of chytridiomycosis. Warming temperatures might increase the stress level of amphibians and compromise their immune systems, increasing susceptibility to the ever-present pathogen. Increased cloud cover might reduce amphibians' ability to bask or seek out warmer microhabitats. Nighttime warming and daytime cooling might be improving conditions for growth and reproduction of Bd. Even if Bd is an exotic pathogen, climate change could affect both rate of spread and virulence. And, as we learned from the case of infected frogs colonizing new habitats after the melting of high-elevation glaciers, climate change is affecting the spread of amphibian hosts with Bd (chapter 5). At the other extreme, some scientists argue that Bd itself plays the major role in population declines caused by the pathogen, with climate change playing only a minor role in the outcome of individual populations.

Scientists also disagree on whether stress facilitates outbreaks of chytridiomycosis. If Bd is ever-present, the fungus may be the proximate cause of population declines when individuals have been stressed by elevated levels of UV-B radiation, chemical pollution, warmer temperatures, or some other environmental variable. Other scientists argue that the most parsimonious explanation for simultaneous mass mortality of many species living in one area is an exotic pathogen entering a formerly "clean" area. The victims do not need to have been stressed by some other factor; the pathogen is sufficient.

The debate concerning these differing interpretations is healthy and is generating considerable research. That is what science is all about.

In the following chapter, we will look in more detail at Bd, now recognized as a major culprit in certain amphibian declines and extinctions worldwide.

7

Amphibian Chytrid Fungus as a Cause of Declines and Extinctions

In ecology or conservation biology textbooks, infectious disease is usually discussed as a stressor that can cause a population to decrease but eventually recover, or perhaps stabilize at a smaller size after an epidemic. Rarely is infectious disease discussed as a primary cause of extinction. A well-known exception is chestnut blight in the northeastern United States: An introduced fungus eliminated American chestnuts as major forest trees in urban and native landscapes, although even in this case the species did not become extinct. American chestnuts can still be found, but the species is no longer a dominant canopy tree.

The chytrid fungus, *Batrachochytrium dendrobatidis* (Bd), is regularly associated with amphibian declines and extinctions. If the fungus does have a role in causing declines as well as extinctions, several questions are raised:

- How can this fungus drive species extinct? In particular, are there exceptional features of its biology that make it a likely agent of extinction? Does the biology of the amphibian-chytrid association have features of a relationship consistent with what we know in theory about the traits of a host-pathogen system that might under some conditions cause host extinction?
- What is known about the history of Bd as a pathogen? In particular, where did it originate, and how might it spread across widely separated geographical regions?

Batrachochytrium's Role in Extinction

Infectious diseases are not seen as important agents of extinction for several reasons. For one thing, it appears that species have rarely gone extinct due to infectious disease. Katherine Smith and colleagues at the University of Georgia and U.S. Geological Survey used data from the 2004 IUCN Red List to assess infectious disease as a cause of extinction.[1] Infectious disease explained fewer than 4 percent of species extinctions since 1500 (833 plants and animals), and it contributed to a species' status as endangered in fewer than 8 percent of cases (2852 critically endangered plants and animals). Amphibians, however, were an exception: 30 percent of critically endangered species are amphibians, and more than 75 percent of critically endangered species threatened by infectious disease are amphibians. Said differently, a significant fraction of endangered species are frogs and salamanders, and most species that are endangered as a result of infectious disease are amphibians. Smith and colleagues found that Bd was a risk factor for 97 percent of the critically endangered amphibians threatened by infectious disease. There is a caveat to keep in mind in interpreting these results. As Smith et al. noted, diseases were not historically considered important in wildlife studies, and the technology needed to identify microparasitic pathogens was poorly developed in the early period covered by the study. For example, many of the extinctions they discussed occurred before viruses were discovered; therefore, the IUCN Red List data underestimate extinction due to infectious disease.

Host-pathogen biology research is difficult to do, especially when microparasites such as protozoans, viruses, bacteria, or fungi (e.g., chytrids) are involved, because the organisms are difficult to detect and identify. Molecular biology tools, especially genetic sequencing techniques for identifying pathogens, are helping overcome these limitations and accelerating the pace of discovery. Many of these tools for amphibian research were developed only in the last 5 years.

Finally, pathogen-driven extinction requires special conditions. Models of host-pathogen dynamics for one host and one pathogen, show that, if pathogen transmission is density dependent—that is, pathogen transmission increases directly with host density—then in theory host extinction is unlikely. For example, a human influenza epidemic driven by density-dependent transmission may kill many people, but even one as severe as the pandemic of 1918 is not expected to lead to human population extinction.[2] Here is why.

Flu virus transmission is facilitated in populations in dense situations—such as among passengers on a crowded train or plane or among observers at a parade. Under these conditions, the pathogen infects new susceptible hosts. Eventually, however, transmission stops when the susceptible host density drops below a critical point because individuals die or become resistant and the likelihood of susceptible individuals' contacting infected hosts is reduced. At this threshold, the pathogen goes extinct locally before the host population does, because host density is too low to sustain pathogen transmission. Local extinction of the host may occur if the pathogen reduces the host population to a size at which "demographic stochasticity" leads to extinction. For example, only a few individuals may survive after an epidemic, only to be all killed by a flood; or they may all be of one sex in a sexually reproducing species. Under these conditions, infectious disease made the population prone to extinction but is not technically the cause of the extinction.

Of special interest for understanding how Bd might cause amphibian extinctions are realistic modifications to the basic theory that allow a pathogen to remain at high frequency even at low host densities. Under such conditions, the host species could go extinct before the pathogen does. Here are three such cases.

1. *A pathogen has a transmission rate independent of the density of infected individuals in a pool of potential contacts.* Sexually transmitted diseases are a good example of transmission that can occur with just a few hosts, because, even at low host densities, the infection frequency may remain high as infected hosts continue to seek new partners. Pathogen transmission continues despite the low population size, which means that disease transmission is independent of density. In the case of Bd, the pathogen can be transmitted during sex as well as when animals aggregate for breeding or shelter. Vector-borne diseases, such as avian malaria, can also be transmitted at low host densities.[3]
2. *Hosts harbor and transmit pathogens with no negative effects for themselves.* Early in the 20th century, Typhoid Mary (her actual name was Mary Mallon) was a worker in New York City and surroundings who was healthy and asymptomatic but harbored typhoid bacteria and infected people she contacted. She was the first person in the United States to be identified as a healthy carrier of typhoid fever. The "Typhoid Mary mechanism" is an especially effective way to transmit disease if the agent is a multihost pathogen, which is a pathogen that readily infects many species of hosts. The combination of one or a few species harboring a disease with no ill effects while simultaneously infecting other susceptible species can in theory lead to extinction of the susceptible species. Imagine two, three, or four frog species that harbor a disease with no ill effects surrounded by susceptible species. Even though the susceptible species may reach numbers low enough

that the density-dependent model predicts local pathogen elimination, the pathogen persists in the tolerant species and keeps infecting the susceptible species, driving them to or near extinction. Bd is a multihost pathogen that readily infects a wide range of frog and salamander species across the entire vertebrate class Amphibia.[4]

3. *There is an environmental reservoir in which a pathogen survives independent of the host.* Anthrax spores, for example, survive apart from mammalian hosts for long periods in soil and can continually infect susceptible individuals even if the host density is too low to support density-dependent transmission.

For epidemiologists, cases 2 and 3 are both types of disease reservoirs: One might be another amphibian species, and the other is elsewhere in the environment—perhaps water, soil, or even a nonamphibian host.

Infectious diseases in amphibians and avian malaria, which is implicated in the extinction of native Hawaiian birds, offer opportunities to study how pathogens might cause extinction.[5] If Bd is indeed causing extinction, the epidemiological theory just reviewed yields testable predictions regarding the three cases.

For case 2—an amphibian reservoir—we know that Bd affects species differently; some die quickly, and others are apparently unaffected. Recall the variation in susceptibility of amphibian species as the fungus invaded Central American communities. Several species survived, and it is these that are candidates for playing the role of Typhoid Mary—species that are unaffected or less affected themselves but capable of transmitting the pathogen. A number of temperate and tropical amphibian species readily harbor Bd and are therefore biotic reservoir candidates.

Is there evidence for a Bd reservoir apart from other amphibians that would satisfy the case 3 conditions?

As we learned in chapter 5, laboratory experiments have demonstrated that Bd can live for days under sterile conditions. In addition, Lips and colleagues reported finding *B. dendrobatidis* DNA on rocks from which a dead frog was collected and on other rocks randomly chosen without dead frogs.[6] But once again, finding DNA is not proof of viable zoospores that exist apart from amphibians and can cause infection.

One report suggested that Bd can survive outside of laboratory conditions apart from an amphibian host. Matthew Parris and his students at the University of Memphis introduced adult southern leopard frogs (*Rana sphenocephala*) infected with Bd into mesocosms.[7] In each mesocosm, an infected adult was placed in a mesh cage and left for 12 days, and then the cage and frog were removed. Mesh screening covered the mesocosms to exclude insects, which might carry Bd. Parris assumed that while the host was in the cage, viable Bd zoospores could swim through the mesh into the water. Healthy tadpoles were introduced into some mesocosms immediately, others at 3 weeks, and the remainder at 6 weeks. Even after 6 weeks, tadpoles were still becoming infected, indicating that viable zoospores were still in the mesocosm after a number of weeks without an amphibian host.

The work of Kate Mitchell and her team at Imperial College and the Institute of Zoology in London generalized these empirical results. They developed a mathematical model to study the specific effect of a Bd environmental reservoir on host population dynamics, using *Bufo bufo* in the United Kingdom as a source of values for the population variables.[8] Their model indicated that the longer Bd persisted in water, the greater the likelihood that a *B. bufo* population would go extinct.

We now have evidence supporting case 2 and some evidence for case 3; namely, amphibians as Bd reservoirs and a reservoir elsewhere in the environment. Is there evidence for case 1, density-independent selection? We will discuss the evidence in the context of climate change.

Climate Change and Amphibian Disease

Temperature and Bd Emergence

In chapter 6, we learned that Alexander and Eischeid at NOAA tested the relationship between amphibian declines and climate variations in Colorado, Puerto Rico, Costa Rica–Panama, and Queensland.[9] They concluded that "unusual climate, as measured by regional estimates of temperature and precipitation, is unlikely to be the direct cause of amphibian declines, but it may have indirectly contributed to them." They also tested the possibility that wind in these regions might transmit Bd. The evidence for a consistent pattern of winds was not significant: "if wind is involved in the propagation of amphibian declines, it is through a complex set of processes."

Although this large-scale test for a correlation between climate and incidence of amphibian declines showed no relationship, could there be a relationship at a smaller scale? Peter Daszak, at the Consortium for Conservation Medicine in New York City, joined with researchers at the Savannah River Ecology Laboratory in South Carolina and used 26 years of field population records and museum specimens to test whether Bd might be causing declines at the site.[10] They found Bd in only 3 specimens (two bullfrogs, *Rana catesbeiana*, and one *Rana sphenocephala*) out of 137 frogs examined, or about 2 percent of the sample. Among nine species over 26 years, four species declined, including *R. sphenocephala*, but there was no significant association of losses with Bd. Populations declined independent of the pathogen. An increase in the number of years with low rainfall and a consequent shortened length of time during which breeding sites held water seemed to account for the losses. The Savannah River site had a drying trend through the 1990s, but it was unclear whether climate warming was the cause. So, a close look at a local site showed no clear relationship of long-term amphibian losses to Bd or climate change.

In chapter 6, we also reviewed how Alan Pounds and colleagues hypothesized that large-scale warming is responsible for enigmatic amphibian declines.[11] They analyzed declines and extinction reports for some 110 species of tropical frogs in the genus *Atelopus* and proposed the chytrid-thermal-optimum hypothesis to account for the data (Figure 7.1). Their proposed mechanism is as follows. Temperatures at high elevations in the tropics have been shifting toward the Bd growth optimum as increasing

Figure 7.1 Examples of tropical frogs in the genus *Atelopus*. Both species are still to be formally described. Top photograph by Martha Crump; bottom photograph by Dante Fenolio.

cloud cover caused by global warming alters local temperatures. They argued that one warm year can alter local temperatures in ways that create microenvironmental conditions favoring Bd growth, which enhances the likelihood of amphibian declines and extinction. The model assumes that Bd is endemic in these tropical systems, but so far the best evidence indicates that Bd entered, or is entering, these tropical systems and extinction follows.

Several research teams have reexamined, or are reexamining, the hypothesis proposed by Pounds' team. Ross Alford and colleagues at James Cook University used morphological analyses to show how two species, the waterfall treefrog, *Litoria nannotis* and the green-eyed treefrog, *Litoria genimaculata,* in northern Queensland showed signs of stress based on unusual morphological development 2 years before chytrid-related declines were observed. They concluded that "multi-year warm periods may be more important in amphibian declines than single warm years."[12] The team felt that the patterns from the Australian Old World tropics agreed with the general climate-linked epidemic hypothesis but argued that multiple warm years may be needed to facilitate Bd's emergence as a pathogen.[13] The argument also assumes that Bd is endemic in the populations, but no samples are available to support the assumption.

In a reply to the Australian analysis, Pounds and colleagues agreed with the conclusion that multiyear warm periods are probably more important than single warm years.[14] William Laurance from the Smithsonian Tropical Research Institute found little direct support for the single-warm-year hypothesis.[15] He used 17 sites with at least 20 years of weather records near the same general area used by Alford's team. He found support for a modification of the hypothesis in which frog declines were likely to occur after three consecutive years of unusually warm weather.

A team from the Università di Perugia claimed support for Pounds' theory, but in fact they also supported a modification of the hypothesis. Led by Ines Di Rosa, they argued that their findings for the pool frog (*Rana lessonae*) in Italy's Trasimeno Lake agreed with the climate-linked-epidemic hypothesis by virtue of the fact that Bd was present in the lake since at least 1999, but chytridiomycosis only emerged during the European heat wave of 2003.[16] They argued: "By causing stress to the animals, environmental change may raise their susceptibility to various pathogens, many of which are unknown."[17]

The Australian and Italian data are suggesting that warming could have stressed animals in ways that predisposed them to Bd infections. Neither team is providing evidence for enhanced Bd growth in newly favorable microenvironments. Earth is warming, but the data so far are not supporting the mechanistic details of the chytrid-thermal-optimum hypothesis.

Douglas Woodhams from Vanderbilt University led a team whose research begins to explain why results are varying among groups.[18] They used experiments and models to show how Bd grows well at temperatures varying from 7° to 25°C. At low temperatures (7°–10°C) individual reproductive structures (the zoosporangium) produce greater numbers of zoospores than at high temperatures, while at high temperatures (17°–25°C) zoospores settle faster and grow more quickly into the infectious zoosporangium stage. And at least in cultures a sudden temperature drop triggers zoospore release. Bd, therefore, has evolved trade offs in its life history that allows infections to develop across a range of temperatures. The results suggest that once Bd enters a system, a series of cool or warm years will interact with host immune function, ecology, behavior, and life history to trigger an epidemic.

Climate Change and Bd Emergence

Climate change effects will be complex, with significant variation among regions and species. Here is an example that illustrates the point and also hints at how global warming may facilitate the spread of infectious disease via density-independent selection without a need for changing microclimates and improved Bd growth. Rather, as expected for case 1, climate change affects behavior in ways that enhance disease transmission through movement of individuals and formation of aggregations that increase the risk of pathogen transmission. Woodhams and colleagues called this "risky host behavior."[19]

Jaime Bosch and teammates at the Museo Nacional de Ciencias Naturales in Madrid argued that they had support for the chytrid-thermal-optimum hypothesis, but they also entertained other options.[20] Since 1997, midwife toads (*Alytes obstetricans*), European toads (*Bufo bufo*), and fire salamanders (*Salamandra salamandra*) have declined in the 770-hectare Peñalara Natural Park, an area lying between 1800 and 2200 m in central Spain with 242 ponds and 10 amphibian species. In a pattern reminiscent of Puerto Rico and Costa Rica, the team described an average increase in air temperature at the park's weather station of about 1°C since 1976; hotter and sunnier days are more frequent. They made two other important observations.

First, the Iberian green frog (*Rana perezi*) was historically restricted to low elevations but in recent years entered the high-elevation park, and the species has doubled the number of ponds it occupied 20 years ago. Iberian green frogs harbor Bd, but "the species susceptibility to the disease appears to be very low as no dead individuals or population declines for this species have been recorded. It may be that *R. perezi* or other thus far unidentified vectors are contributing to the spread of the pathogen."[21] Think of *R. perezi* as Typhoid Mary entering the system.

Second, aggregation behavior occurs in the community. Newly metamorphosed amphibians remain close to ponds, searching for cool damp places, and at least midwife toads and fire salamanders—the species most susceptible to Bd—"share terrestrial habitats and both the species breed in permanent ponds and have overwintering larvae."[22] These are ideal behaviors for transmitting pathogens even at low densities. The prospect of increasing average temperatures in the future leads us to expect that such behaviors will increase. The warming will also likely increase stress on some amphibian species, which may increase their susceptibility to disease.

The team concluded, "Our data support the chytrid-thermal-optimum hypothesis and the climate change as a driver of chytridiomycosis in Peñalara, by defining an increase in a climatic envelope that is broadly similar to that described by Pounds et al. (2006) and showing that this climatic envelope is associated with observed increases in chytridiomycosis."[23] Although mean park temperature may be rising and Bd may even have enhanced growth opportunities as suggested by Woodhams' group, there are no data regarding Bd's history in the park. Furthermore, the chytrid-thermal-optimum hypothesis is not needed to explain what is happening. In fact, the authors conceded that "we do not rule out the introduction of *B. dendrobatidis* in initiating the Peñalara epidemic."[24] A parsimonious explanation is that increasing temperatures allowed *R. perezi* to expand its range in elevation just as the Andean frogs have. And, just as in the Andes, the Iberian green frog brought Bd as a pathogen, which now is infecting susceptible species. This "Typhoid Mary species," coupled with the fact that some species aggregate in common refuges, facilitates Bd spread and increases the likelihood of extinction.

This scenario suggests that midwife toads, European toads, and fire salamanders are at great risk of local extinction. An analysis of where Bd can live showed that all of the Iberian Peninsula is suitable Bd habitat, so recent climate warming is unlikely to have increased the climatic envelope at which Bd can persist.[25] More likely, the increasing temperature allowed a new species with Bd to invade Peñalara, and the disease spread was facilitated by the behavior of amphibians in the area—the "risky host behavior" described by the Woodhams team. Warming may also have stressed the animals and made them more susceptible to a novel pathogen.

The origin of *Batrachochytrium dendrobatidis*

At Duke University, Erica Morehouse, Rytas Vilgalys, Timothy James, and others analyzed DNA sequences of 35 Bd strains collected from 20 frog species and 1 salamander species from Africa, Australia, and North and Central

America.[26] Despite the small sample size, the team was able to draw several conclusions.

Bd appears to have a predominantly clonal population structure, and so, not surprisingly, the strains were closely related genetically. Bd genotypes vary little among hosts and therefore show minimal co-evolution of host and pathogen. These are hallmarks of a generalist pathogen infecting diverse amphibian species.

The strains also had few mutations. The low genetic variation suggests that Bd coalesced as a distinct evolutionary group comparatively recently and then dispersed geographically. Unfortunately, the team had no independent way to estimate Bd's time of divergence from an ancestral species and subsequent dispersal.

The only significant geographical structuring was slight genetic differentiation between one North American strain and the rest of the world. Notably, similarity of the Panamanian and Australian genotypes "argues for recent dispersal or introduction of the pathogen into these two areas of amphibian population declines."[27] Overall, the team concluded, "These data do not support the hypothesis that *Batrachochytrium dendrobatidis* has emerged from a pre-existing relationship between the fungus and amphibians via recent climatic change or other abiotic factors, and are more consistent with the recent introduction of the pathogen into naïve populations."[28] These conclusions must be tempered by the fact that their analysis was based on a survey of only 10 genetic loci, and despite the intercontinental samples, only 5 of these loci varied. The molecular method used, which is typical for this type of analysis, sampled only part of the Bd genome; sampling more of the genome or another region could change the results.

Indeed, Daszak and colleagues reported genetic distances of about 10 percent among strains from Australia, North America, Central America, and South America.[29] These distances exceed those determined by the Duke team's multilocus sequence typing, in which the alleles of seven gene segments were sequenced and compared among strains to estimate their genetic relatedness. Daszak sampled a different gene, the internally transcribed spacer, and found more variability. Greater variation is consistent with a longer history. Using a phenotypic approach, Richard Retallick and Verma Miera, in JPC's laboratory at Arizona State University, assessed differences in mortality between two Bd strains isolated from two Arizona populations of boreal chorus frogs, *Pseudacris maculata*.[30] Lee Berger and colleagues at James Cook University conducted similar experiments using three Bd strains and the common green tree frog, *Litoria caerulea*, in Australia.[31] Both groups found variation between strains in Bd pathogenicity.

As a rule of thumb, the more loci sampled, the greater the likelihood of finding variability among samples, which increases the likelihood of discriminating among populations. This is just the logic Jess Morgan and colleagues at the University of California in Berkeley followed in a study of mountain yellow-legged frogs.[32] We already discussed how fish introduced over the last 50 years reduced frog population sizes in the Sierra Nevada. Declining populations and local extinctions also correlate with prevailing toxin-bearing winds from California's Central Valley. In 2001, Bd was discovered in these populations as well. Mountain yellow-legged frogs illustrate how multiple forces can impinge on populations. Morgan and colleagues analyzed variable regions of the Bd genome and drew the following conclusions:

- There are few genetic differences between the Sierra Nevada Bd strains and strains from frogs in South America, Australia, Ghana, and South Africa.
- Bd was introduced as a novel pathogen at least twice into two regions of the Sierra Nevada, at locations some 200 km apart. The pathogen then spread and evolved into local endemic strains, based on the fact that no sites shared the same genotype and some sites had several related genotypes.
- There is evidence of genetic recombination, suggesting that sex occurs somewhere in the life cycle. Other chytrids with sexual reproduction also have resistant sporangia (like a seed stage in higher plants), which would have great implications for long distance dispersal and survival in an environmental reservoir apart from amphibian hosts. The Università di Perugia team led by Di Rosa, which was mentioned earlier, reported morphological evidence for what they interpreted as a resistant stage in the Bd life cycle, but other researchers have questioned this conclusion.[33]
- Humans probably facilitated recent Bd spread in the Sierra Nevada.

Genotypic analyses are showing that, even though the genetic variation is small, Bd, like many pathogens, still has strains that vary across large geographical areas and among populations. Ambystoma tigrinum virus (ATV) shows similar large-scale and local variation where endemic strains have co-evolved with each host salamander population. Phenotypic studies of Bd strains are showing how they differentially affect important life history traits such as survival. It is likely that, before long, studies will uncover genotypic variation in virulence among Bd strains that will help account for the diverse effects of Bd within and among species. Erica Rosenblum at the University of Idaho and her associates are using functional genomic approaches to try to uncover just such a suite of genes in Bd where they have already identified a gene family thought to be pathogenic in another fungal pathogen group that causes vertebrate skin infections.[34]

Recent breakthroughs will help research teams working on Bd's evolutionary origins and spread. The U.S. Department of Energy's Joint Genome Institute managed by its Office of Science just completed sequencing the genome of a Bd strain isolated from *Rana muscosa* in California.[35] In 2006–2007, the Broad Institute in Massachusetts completed sequencing of the genome of a Bd strain from a dying lemur leaf frog, *Phyllomedusa lemur,* collected from El Copé, Panama.[36] These complete sequences should reveal even more variable genomic regions to differentiate Bd strains. Such regions will open the way for further tests of hypotheses related to time of divergence of strains, geographical origin, and differences in virulence.

As we saw in chapter 6, there is also evidence that amphibian behavior mitigates or exacerbates Bd's effects. Bd-infected species basking at laboratory temperatures exceeding the pathogen's thermal maximum survived longer than animals without such a refuge.[37] Frogs can also rid themselves of the pathogen if the temperature exceeds Bd's upper thermal limit.[38] In the field, susceptibility to Bd in three Australian frog species varied with basking, aggregation, and microhabitat.[39] The waterfall treefrog, *Litoria nannotis,* spends most of its time in microhabitats that maximize Bd susceptibility. Aggregation is also common, increasing the likelihood of Bd transmission. Green-eyed treefrogs, *Litoria genimaculata,* spend substantial periods in the upper tropical forest canopy, where temperatures are higher and humidity lower, and frog body temperatures can exceed Bd's optimal growth range. *L. genimaculata* shows an intermediate level of susceptibility to Bd. The least susceptible species is Lesueur's frog (=stoney-creek frog), *Litoria lesueuri,* which frequently basks in exposed positions, elevating its body temperature into the range lethal to the fungus in vitro. Rowley and Alford concluded that variation in frog microhabitat use influences the likelihood of Bd survival and reproduction and strongly affects vulnerability to chytridiomycosis. Consistent with density-dependent versus density-independent transmission, they noted, "while traditional models have assumed that contact rate is directly proportional to host population size or density…, the present study confirms that rates of contact between individuals may be almost entirely independent of population size due to host behaviour."[40]

Burrowes and colleagues also reported aggregating behavior for *Eleutherodactylus* in Puerto Rico, as did Bosch and colleagues for midwife toads and fire salamanders in Spain. Congregating facilitates Bd transmission. Both Rowley and Alford and the Burrowes team, reported single-species aggregations, but Bosch and associates reported multispecies aggregations. The latter, especially if at least one species is a Bd reservoir, is a scenario for facilitating rapid pathogen spread.[41]

Bd infects many amphibian species at all elevations in streams and ponds. The most susceptible species have, on average, small ranges and breeding sites in streams at relatively high, cool elevations.[42] Bd also occurs in terrestrial species with direct development and no aquatic life history stage. While a graduate student at the University of Texas, Santiago Ron used New World records for Bd occurrence to describe its environmental envelope—the ranges of major environmental variables such as temperature and humidity that characterize where it lives worldwide.[43] Bd survives under a wide range of rainfall and temperature regimes, but it grows optimally in the laboratory in moist conditions between 17° and 25°C, just the conditions found in Neotropical and Australian tropical forests where most species' extinctions have occurred. Ron noted that "The [Neotropical] regions with the highest suitability for *B. dendrobatidis* include the habitat types that have the world's most diverse amphibian faunas."[44] His analysis showed that in both the New World and the Eastern Hemisphere, Bd is predicted to survive in a wide geographical range that extends beyond where it is known. In these still unsampled areas Bd may already be present, but if it is not, amphibian species in these regions are at risk.

Madagascar is a striking example. Ron's model predicts that Madagascar has extensive environments that will support Bd. Franco Andreone and colleagues at the Museo Regionale di Scienze Naturali di Torino in Italy summarized the situation in Madagascar as of early 2008.[45] An analysis of more than 500 specimens of 80 frog species showed no Bd. Research is under way to test the susceptibility of Malagasy frogs to a South African Bd strain. A wide variety of such tests are needed, and if Malagasy frogs are susceptible, they recommended taking precautions to prevent Bd's introduction to Madagascar. This, of course, assumes that Bd is not already in the island's ecosystems. Andreone and his 24 collaborators argued that Madagascar must be a top priority for amphibian conservation. We agree. The island offers a realistic opportunity to apply lessons learned in the Neotropics and Australia to prevent declines and extinctions rather than reacting after Bd emerges. Their recommendations are warranted, given all that we know now about Bd's negative effects on amphibian biodiversity.

What do we know about Bd's emergence in areas widely separated geographically? When scientists became aware of amphibian declines in the late 1980s, two things elevated the issue above the usual story of extinction: losses in protected areas, and the nearly simultaneous losses in North and Central America as well as in Australia. What could explain emergence in such widely separated places? There are two theoretical arguments, and both have been suggested as explanations for enigmatic amphibian declines.[46]

The novel or spreading pathogen hypothesis posits that a virulent, easily transmitted, unfamiliar microbe enters a naïve population and causes high mortality, even failure of recruitment of successive generations. Bd showed this trait in the Sierra Nevada, Colorado, Central America, and Australia.[47] Pandemics such as severe acute respiratory syndrome (SARS), HIV-AIDS, or influenza fit this hypothesis, and according to Stephen Morse at Columbia University's Mailman School of Public Health, it is the only explanation known for pandemics.[48] In such cases, an understanding of how the pathogen spreads can identify intervention strategies that break the transmission chain.

Ché Weldon and colleagues at South Africa's North-West University used the spreading pathogen hypothesis to argue that Bd evolved in Africa, spread from there (the "out of Africa hypothesis"), and emerged in naïve populations elsewhere.[49] They based their argument on three key findings: (1) the earliest occurrence of Bd is in an African clawed frog (*Xenopus laevis*) specimen in the South African Museum (Capetown) that was collected in 1938 from the Western Cape lowland; (2) by at least 1973, *Xenopus* specimens positive for Bd had been collected from all regions of South Africa, suggesting that the pathogen was widespread; and (3) *X. laevis* shows no clinical signs of chytridiomycosis in the wild and has not experienced die-offs, suggesting that Bd is a pathogen endemic to South Africa. Transportation of animals for pets or biomedicine that are asymptomatic carriers of Bd is a possible commercial route for introducing the pathogen into naïve populations. Weldon and colleagues argued that African clawed frogs seemed an especially likely candidate, because for decades they were shipped worldwide from South Africa to be used in human pregnancy tests.[50]

The alternative emerging endemic hypothesis posits that a microbe is already present in an ecosystem but is inactive, rare, or only mildly pathogenic. Changing environmental conditions facilitate emergence of a lethal strain. Under the new conditions, the microbe becomes virulent and easily transmitted within a population. High mortality and even failure of recruitment of successive generations can result. Appearance of infected individuals in subpopulations does not depend on movement of either the host or the pathogen across the landscape. Synchrony is caused by environmental factors such as rain, temperature, or resource availability common to different populations, which may be geographically separated, triggering pathogen emergence. The triggering variables are often lumped together under the term "environmental stochasticity," and the resulting synchrony is called the "Moran effect," after P.A.P. Moran at Australian National University.[51] The key point is that an environmental variable beyond the scale of individual habitats synchronizes separated locales and explains coincident epidemics. For example, environmental stochasticity in the form of winter precipitation (mainly snowfall, which controls spring runoff and, therefore, seasonal pond size) appears to synchronize ATV dynamics on the Kaibab Plateau in Arizona, a region about the size of Rhode Island.[52]

The emerging endemic hypothesis is complex and requires taking mechanisms known to cause local emergence of endemic pathogens and scaling them up to regional or global levels.[53] According to Morse, this hypothesis is an unconfirmed cause of pandemics at the global scale. The chytrid-thermal-optimum hypothesis is an instance of the emerging endemic hypothesis: Changing temperatures facilitate Bd's growth in microhabitats. Climate warming is the environmental cue common to widely separated populations that is hypothesized to be the stimulus for pathogen emergence. UV-B radiation and global warming are the only environmental cues proposed as accounting for the worldwide population losses seen in amphibians.

The novel or spreading pathogen hypothesis is the best explanation for the enigmatic decline and even extinction of frog populations across widely separated regions. The low genetic variation of Bd is consistent with the widespread introduction of a clonal pathogen. Given that we know the effects of global warming are varying regionally and by species, we agree with Laurance that "it stretches plausibility to argue that the chytrid pathogen is simply an opportunistic, endemic pathogen that has suddenly begun causing catastrophic species declines as a consequence of contemporary global warming."[54] The currently warming climate may predispose some animals to infection, but—and here we agree with Skerratt and colleagues[55]—it is the amphibian chytrid's invasion of naïve populations that is the best explanation for the enigmatic, worldwide declines and extinctions of frogs and salamanders. In chapter 10 we discuss the wider implications of this conclusion.

Seven years after West Nile virus emerged in New York City's songbirds, Shannon LaDeau and colleagues at the Smithsonian Institution's National Zoo and the Consortium for Conservation Medicine reported significant population declines in seven bird species: American crow, blue jay, American robin, eastern bluebird, chickadee, tufted titmouse, and house wren.[56] Carsten Rahbek, from the University of Copenhagen, commented on their article in the journal *Nature* by alluding to scenes described in *Silent Spring*.[57] This time, pesticides were not to blame for declining bird populations, but outbreaks of West Nile virus. The article was entitled, "The Silence of the Robins," and, as in birds, "the silence of the frogs" also can be traced to emerging disease. Rahbek remarked, "The work demonstrates the potentially devastating impact that an invasive pathogen can have on native wildlife, and serves to highlight the potential dangers of the trade in wildlife."[58]

Ironically, the West Nile study was done by researchers at the U.S. National Zoo, the same place where Joyce Longcore and colleagues first isolated the amphibian chytrid fungus from a dying dendrobatid frog. All of this raises the question: What, if anything, is to be done now? The next chapters describe more about the research enterprise itself and the steps taken by researchers and policy makers.

8

New Approaches to Doing Science and Conservation

Late in the 20th century, the historical and modern forces that cause population declines and extinctions converged on amphibian species in ways that accelerated population and species losses in frogs and salamanders. As mentioned in chapter 1, recent analyses by Kim Roelants and colleagues at the Vrije Universiteit Brussel and by Malcolm McCallum at Texas A&M University reached similar conclusions using different methods: The late 20th century extinction rate for amphibians is between 200 and 2700 times greater than the extinction rate for the class Amphibia over the last 350 million years.[1] The fact that amphibians disappeared in protected areas such as nature reserves, which by definition humans "reserve" for protecting species and conserving biodiversity, drew the world's attention to the general importance of explaining these losses, because they signaled that habitat restrictions alone could not protect amphibians, or, probably, other organisms. Cornelia Dean, a reporter for the *New York Times*, wrote: "Conservation organizations that work to preserve biologically rich landscapes are confronting a painful realization: In an era of climate change, many of their efforts may be insufficient or beside the point."[2] At least for amphibians, "conservation as usual" was not working, which raised the general question: What constitutes a nature reserve at the start of the 21st century? In some cases, a reserve of last resort is a zoo, aquarium, or other conservation facility—but more about that possibility in the next chapter.

Besides reshaping our understanding of the challenges of conserving species, trying to explain amphibian losses is also teaching scientists to broaden their vision of what counts as a research program. In the 1990s, research programs in herpetology, population biology, and evolutionary biology were dominated by single investigators and small groups of students. As a result, "science as usual" was an unlikely route to solving a complex problem such as amphibian declines, because some parts of the problem called for interdisciplinary research that might incorporate experts with specializations from molecular biology to global change.[3] And since solutions were likely to lie at the interface of science and society, solving the problem of amphibian declines could also require integrating concepts, theories, and methods from the natural and social sciences.

The first question we address in this chapter emerged as researchers in the late 1990s tried to understand the scientific causes of the declines: Will *science* as usual work? The second question arose later, as researchers realized the scope of the problem, identified the causes of some declines, and sought to stop or reverse the losses: Will *conservation* as usual work?[4]

Will Science as Usual Work?

As discussed in chapter 1, from year to year the number of adult frogs or salamanders in many populations can fluctuate by one or more orders of magnitude. Because of this great variation, some scientists were understandably skeptical about early reports of declines, because the data were often insufficient to separate a long-term signal of increasing or decreasing population numbers from the noise of the annual variation. As mentioned in chapter 2, in 1998, one of us (JPC) and two colleagues at Arizona State University, Elizabeth Davidson and Andrew Storfer, received support for a workshop at the National Science Foundation (NSF) to address the question: Is the threat of extinction increasing for amphibians? The meeting's goal was to assess the evidence of declines and, if warranted, to recommend a strategy for addressing the causes.

Bruce Babbitt, who was then Secretary of the Interior and a member of President Clinton's Cabinet, met with workshop participants and shortly afterward took a big step in addressing the causes by forming a federal Interagency Task Force on Amphibian Declines and Deformities (TADD), which was succeeded by the U.S. Geological Survey's Amphibian Research and Monitoring Initiative (ARMI)—an organization whose main mission is to study amphibian declines and deformities on Federal land. TADD, and its successor ARMI, were to be team efforts to coordinate the research activities of diverse agencies. The important point is that in the late 1990s, as far as the U.S. government was concerned, amphibian losses were not a problem for one agency to solve; rather, teamwork was required to develop answers and provide solutions. It was also clear to workshop participants that losses were occurring in several nations, so a solution required international cooperation—teamwork across national boundaries.

Step 2 involved understanding how multiple causes affected amphibian populations, and this would require interdisciplinary research teams. Such teams are an increasingly common approach to solving environmental problems, but they were not the regular way of conducting research for the communities of herpetologists, population biologists, and evolutionary biologists who played leading roles in discovering the amphibian declines. These communities were dominated by research approaches that relied on investigators usually working alone or in close collaboration with their students. Groups of two to four senior investigators might collaborate occasionally on a project, but it was uncommon to have teams of many, perhaps even dozens, of researchers like those needed to tackle such a farflung problem. The rarity of large research groups meant that the community needed to change its cultural practices with respect to conducting research. How might such changes occur?

Within 6 months after the Washington meeting, NSF funded two more workshops at the San Diego Zoo. Cynthia Carey and Louise Rollins-Smith, whose work we have discussed already, organized the first meeting, which focused on emerging infectious disease. David Gardner from the University of California at Irvine organized the second meeting, on environmental change as a cause of amphibian abnormalities and perhaps declines.

Together, the three workshops made it clear that, beyond habitat destruction and the effects of exotic species, three research areas offered the best hope of understanding the complex causes of declines in protected reserves: toxic chemicals, global climate change, and infectious disease. Research programs might need to incorporate each of these areas, and even others. For example, Kiesecker and collaborators proposed a link between climate change and outbreaks of a pathogenic water mold that kills amphibian eggs and embryos in western U.S. frog populations.[5]

The complex ecological interactions of the last example and others like it suggested that teams could be more effective than single investigators. A research team can tackle more dimensions of a problem than one investigator can, and collaborators often can reach an understanding of a problem faster than one researcher. So, the ways in which scientists conduct research evolved. Collaboration requires more than assembling researchers with diverse skills. Collaborators must trust and respect each other while learning to "speak" the new scientific "languages" of diverse colleagues.

Coincidentally, by the late 1990s, modern communication systems, especially networked computers that support listservers, e-mail, and Web site postings, were becoming readily available and could integrate researchers over the Internet through cyberinfrastructure—the capacity to acquire data, store it, manage it, integrate it, and visualize it. Cyberinfrastructure-supported research environments connect data, computers, and people in ways that can efficiently solve complex problems. The capacity to conduct interdisciplinary, team-based research using virtual organizations has made huge strides in the last decade.[6]

The types of research programs needed to understand the causes and consequences of amphibian decline are clearer now than they were in 1998.

The problem began with a more or less "applied" question focused on one group: Why are amphibians dying in such numbers? But answers eluded researchers for want of a basic understanding of the science in many areas, including host-pathogen biology, climate change, and the effects of environmental toxins. Our lack of understanding in these areas was not restricted to amphibians, but the absence of this basic information meant that in many cases researchers using amphibians as model systems were starting from scratch in trying to pinpoint causes. Researchers faced a problem analogous to that of the great French biologist Louis Pasteur, who was asked to explain, "Why is the wine spoiling?" To answer this "applied" question, Pasteur first had to understand the basic biology of microbes, and in doing so he made seminal contributions to developing the field of microbiology. While at the Woodrow Wilson School for Public and International Affairs, the late Donald Stokes generalized this lesson, arguing that much progress in science happens when basic and applied questions draw upon each other.[7] Similarly, while puzzling about the causes of amphibian declines, investigators had to move beyond just amphibians. In a manner analogous to Pasteur, scientists had to use amphibian model systems to solve basic research questions en route to understanding the causes of amphibian losses.

As smaller, species-specific questions metamorphosed into larger, more general questions that involved life, physical, and social scientists, researchers began collaborating, networking, quickly comparing results, pooling resources, and integrating findings. Common research questions, common methods, and common models facilitated more, and better, research in hopes of forestalling further amphibian losses while providing a general approach for conserving biodiversity in general. A fundamental 21st century challenge for solving complex problems such as amphibian declines will be conducting team research that spans many levels of biological organization while also integrating the physical and social sciences.

Our environment is changing rapidly, and we must pay attention. But some also argue that scientists should only develop the theories and observations that inform such issues and not advocate policies about what should be done to intervene in stopping the losses.[8] In the case of amphibian declines, many researchers have made the transition to advocacy, and in the rest of this chapter and the next we will explore what happened and why. At the heart of the transition is the conviction on the part of many that conservation as usual will not work for amphibians, and new ways of thinking about the problem of amphibian declines and extinction are needed.

Will Conservation as Usual Work?

The path from a basic research discovery to public policy has two especially important points. The first is when data confirm the predictions that convert scientific hypotheses to "facts." For example, for many the Keeling curve and other direct and proxy measures demonstrating a rise in greenhouse gases since the Industrial Revolution are the facts that lead to the conclusion that humans have a significant role in rising global temperatures.[9]

The second point occurs as policy makers decide how to react to the information. The same decision-making processes are not at work at both junctures. For example, as increasing numbers of scientists called for policy changes because they accepted the fact that the atmosphere is warming and humans are a major cause, many policy makers did not respond. The reasons for the differences are complex, but they center on the fact that once the evidence convinces researchers that there is change due to a particular cause, the issue passes through a point after which scientists are only one of many voices that determine how the information is used. Policy making in such a case occurs at the intersection of scientific results, ethics, culture, economics, and other factors that mix diverse results, opinions, and stakeholders. The scientific evidence confirming that something is happening, coupled with the capacity to predict what will happen next, is a powerful combination that changes fundamentally the nature of the discussion. The consequences of inaction are clearer if we know how the variable of interest will change in the future, and this leads to more stakeholders' wanting to shape policy.

Amphibian declines illustrate how basic discoveries have led to a call for changes in public policy. The 1998 NSF conference assembled leading experts in herpetology, ecology, infectious diseases, ecotoxicology, physiology, climate change, and science policy to address the question: Is the threat of extinction increasing for amphibians? At the time, the reality of the threat was debatable. A decade later, the question is: What research programs and complementary public policies are needed to prevent further declines and extinctions in amphibians?

The questions differ because the causes are at least now characterized, although still not understood completely, and we have better estimates of the magnitude of the problem's scope in terms of the species affected. For example, a key point in transforming the study of amphibian declines from focusing on basic research to embracing policy as well was reached when scientists could identify, detect, and predict the onset of chytridiomycosis at a site and that, within a year, some 50 percent of species and 80 percent of individuals would disappear. Losses of amphibian diversity at this scale result in increasingly predictable impacts on local ecosystems. These results, coupled with other data showing that the pathogen can live in many places, meant that an entire vertebrate class was at risk worldwide and supported calls for policy responses.[10] We will outline particular responses in the next chapter.

The two advances mean that explaining amphibian declines is now in transition, from a period dominated primarily by research to one in which research is focused on causes and complemented by policies to forestall further losses. In making this transition, six issues, not necessarily unique to amphibians, are informing policy development.

- Protected ecosystems are a generally accepted way to conserve species for the long-term, but reserve boundaries offer little or no safeguard against toxins, invading pathogens, or climate change. The current inability to halt the spread of chytridiomycosis illustrates that, despite lines on a map, reserve borders are porous to some stresses. How can reserves be designed to respond to such forces?
- Species Survival Plans of The World Conservation Union (IUCN-SSP) mitigate threats to wild populations and usually develop captive assurance populations for eventual reintroductions. But how can we create enough capacity in accredited institutions to maintain the approximately 2500 globally threatened amphibian species? How do we decide which species to save when multiple taxa are at risk? And who decides?[11]
- Laws controlling trade and exploitation of wildlife (e.g., the Convention on International Trade in Endangered Species of Wild Fauna and Flora [CITES]) are designed to prevent overcollection of wild populations.[12] But implementation can be slow, limiting rapid intervention and removal of animals from the wild, an action that may be required to save amphibian populations. These policies need reevaluation.
- Although they are committed to conservation, for understandable reasons many zoos cannot allocate substantial space or personnel to species with little potential for exhibits. Nevertheless, collections-based institutions are the only ones maintaining species on a continuing basis to ensure their survival, because few governmental agencies have made such commitments. Policies are needed to guide the housing and management of large numbers of threatened or endangered, perhaps noncharismatic, species.[13]
- Funding from traditional sources is insufficient to conserve significant numbers of globally threatened species, such as amphibians.
- When a new pathogen becomes established permanently in a wildlife community, reintroduction efforts are useless unless the reintroduced

hosts are selected for decreased susceptibility or increased pathogen tolerance. Research is needed to understand how to expedite this process.

A comprehensive conservation and research strategy is needed to address a challenge of this magnitude. It must be adaptable for rapid implementation across taxa, habitats, and nations and must include basic research, husbandry, ethics, diplomacy, and law. Amphibian biologists have established an alliance of international experts from academia, government, collections-based institutions, the fields of disease and environmental health, and private stakeholders that we will discuss in the next chapter.[14]

The following are policy recommendations specific to amphibians that are under consideration and afford a model for proceeding in any case with global threats to a large number of species, regardless of the causes. Not everyone agrees with all of these recommendations, and this tension is something that we will also discuss in the next chapter.

- Remove representative species from wild populations if declines and extinctions are predicted. The best available science should guide numbers removed, genetic variability, and treatment for disease.[15] Removal should be complemented by efforts to determine whether any species that disappear are actually surviving in small numbers and whether they have tolerance or resistance to infection or are surviving in microenvironments that protect them. Either way, information about surviving individuals could further guide a conservation program. For example, in some cases, surveying a habitat after Bd enters may reveal animals with traits that fostered their survival, and these individuals could be the basis for a reintroduction program. The effects of the initial epidemic provide a way to mitigate future infections.
- Expand a global network of captive facilities beyond the IUCN-SSP model, in which only institutions accredited by the American Zoo and Aquarium Association or the European Association of Zoos and Aquaria may participate and in which at any time only single species are considered. The network should combine traditional collections-based institutions with new partnerships involving the private sector and botanical gardens, and, to the greatest degree possible, it should do so within the ranges of threatened species.
- Promote open communication between governments and institutions to develop and maintain such programs and to overcome logistical, bureaucratic, diplomatic, and financial barriers. Review laws governing amphibian commercial trading and amend them as needed to allow rapid response to imminent population declines by controlled

harvesting of wild animals, and, whenever necessary and possible, control of infectious diseases.

- Create rapid response teams, like those of the U.S. Centers for Disease Control and Prevention, to deploy research and collecting experts to regions where significant declines are reported. Teams will evaluate populations and safeguard threatened species by depositing animals into facilities of the global network.
- Support more basic research on emerging threats, such as wildlife diseases. If reintroductions are to be possible, unique lines of research must be considered: selecting for resistance in wild or captive organisms, inbreeding or cloning of some species, setting priorities for species to rescue, introducing animals outside their historical ranges, and genetically modifying pathogens for possible reintroduction of attenuated strains.
- Reform the traditional model of collections-based institutions' donating staff members and resources to conservation programs, because it is insufficient. Provide dedicated facilities and full-time staff members to implement field and range-country efforts, manage captive colonies, and coordinate a global coalition of captive-care and research partnerships.

There is always the temptation to call for more research before using the available evidence to make the transition to informed policy recommendations. But amphibian declines share two attributes with global warming that argue against a delay in developing and implementing policies. First, global warming models suggest that delaying a response is a risky strategy, because lags in the system mean that any action to mitigate the causes will not result in a response for years, even decades. Likewise, all available evidence indicates that any delay in responding to amphibian losses will ensure more species' extinctions. Second, signs of global warming are subtle, vary in their effect among large regions, and, at least for now, rarely inconvenience the lives of most people. Likewise, the most dramatic losses of amphibians are occurring in remote regions of the Old and New World tropics where few people live presently, and therefore are going largely unnoticed and unappreciated.

Although we focus on amphibians, current predictions of human population growth, climate change, habitat loss, and emergence of infectious diseases lead to the prediction that other major groups will repeat the pattern of declines seen in amphibians.[16] For example, 129 authors of a 2008 report on the world's mammals conclude: "We estimate that one in four species [of the 5487 extant land and marine mammals] is threatened with extinction and that the population [*sic*] of one in two is declining."[17] Programs to preserve habitat, limit harvesting, and control exotic species remain critical to conservation and reintroduction efforts, but they alone cannot conserve amphibians or other taxa in the face of threats such as infectious diseases, toxins, and climate change that readily breach the borders of national parks and reserves. Collaborative, international, and interdisciplinary approaches will be needed. Vaclav Havel, former president of the Czech Republic, wrote in an essay on climate change: "I'm skeptical that a problem as complex as climate change can be solved by any single branch of science."[18] The same skepticism regarding a single discipline approach applies to solving the problem of global amphibian decline.

Traditional conservation programs are designed to address readily identifiable and localized threats of habitat loss and human exploitation, but these strategies are likely to be of limited effectiveness in the face of the complex, synergistic, and subtle global threats of the 21st century. Entire ecosystems, such as coral reefs,[19] or taxonomic classes, such as Amphibia, face large-scale declines and extinctions that require creative biological, legal, practical, and ethical solutions. Groups such as these can serve as models for understanding how to convert basic research into the policies that will form the intervention strategies needed to stop or at least reduce the global loss of biodiversity. In the next chapter we explain how this is happening for amphibians.

9

Science Policy and Reacting to a Challenge

Extinction is inevitable: More than 99 percent of the species that ever lived on Earth are extinct. Until the late 20th century, we relied mainly on paleontologists to uncover the causes, but the accelerating modern loss of biodiversity affords more chances than scientists working with modern species would like to learn about extinction. For amphibians in particular, we are in the midst of an opportunity to study the biology of extinction across an entire vertebrate class, using the sophisticated methods of modern science. But because we are only just beginning to identify the causes of declines and even extinctions, what should we do?

Hamish McCallum from the University of Queensland (now at the University of Tasmania) and Menna Jones at the University of Tasmania study the emerging wildlife disease called Tasmanian devil facial tumor disease (DFTD). They raised a similar question about how to proceed if knowledge about the causes of population declines is imperfect. They see DFTD "as a case study of the wider issue of how to manage an emerging disease threat that poses a serious conservation threat: how should you proceed when you know very little?"[1] They go on to point out how the need to make decisions in the face of uncertainty is a question common to many ecological problems. But deferring decisions while waiting for better information often incurs costs and the added burden that control may be tougher or impossible if a pathogen becomes established. Which actions are appropriate to take despite incomplete information?

Questions about how humans as a species should respond to the decline and extinction of other species take us beyond science to the intersection of science and society. The emergence of the amphibian chytrid in the El Copé amphibian community is a useful case study for exploring the ethical, legal, and social dimensions of species' declines and extinction.

Ethics, Policy, Science, and Conservation Meet in Panama

As we have discussed already, Karen Lips began studying frog behavior in Costa Rica in the 1980s. By the mid-1990s, Lips's work, in association with other research, had revealed a worldwide pattern of declines. In the late 1990s, a group of us began working together on the problem.

Lips' data on amphibian diversity and population sizes, coupled with the fact that the El Copé, Panama, site was an undisturbed tropical forest, afforded a distinctive opportunity to test the possibility that amphibian declines occurred after the amphibian chytrid fungus (Bd) entered a region. El Copé had two crucial elements that were lacking in prior studies: The amphibian community was intact, and years of data indicated that Bd was absent before 2004. For the first time, researchers had a locality in which we could monitor amphibian hosts free of the fungal pathogen and document Bd emergence—which is what happened. The exciting scientific result was that, for the first time, a so-called enigmatic decline was predicted and actually occurred.

As discussed in chapter 6, the team then made another prediction. El Valle de Anton was a logical spot for the fungus to emerge next and cause declines. A plan was made to duplicate the El Copé observations. But was that sufficient? Was it enough only to document the decline of the El Valle fauna? Or, once we could predict the next site to decline, were we obliged to intervene? And if so, how? An option considered was removing at least some species before the pathogen entered the community. But if that were done, which species should be saved, and where would they be placed?[2]

As the El Copé community declined, researchers met at Zoo Atlanta in the U.S. state of Georgia.[3] The mid-December 2004 meeting had two goals:

- Address the logistical, bureaucratic, and financial challenges of a conservation program for El Valle
- Create a conservation plan for threatened Panamanian amphibians that would be a model for similar situations in other countries.

The plan's centerpiece was the decision to monitor El Valle amphibians and, at the first sign of disease, move a diversity of species to ex situ conservation facilities—that is, sites away from El Valle where they would be safe from infection. The plan was consistent with the recommendations of an October 2004 global amphibian summit that we will discuss later. Meetings in 2005 at Conservation International headquarters in Washington, D.C., refined the plan by identifying stakeholders, especially in Panama, who had to be consulted along with individuals to collect the animals and places to house them.

Housing animals meant working with the American Association of Zoos and Aquariums (AZA), especially their Amphibian Taxon Advisory Group (ATAG). AZA is a nonprofit organization for advancing conservation education, science, and recreation at zoos and aquaria. ATAG produces regional collection plans that evaluate species needing captive management, estimate how many taxa can be managed based on space available, and recommend species for AZA conservation programs.[4] The first step in mitigating Bd's effects is preventing its entry into an ecosystem, as has been argued for Madagascar.[5] Removing species from native habitats is among the last steps for saving a species. Notably, almost to the day of the Zoo Atlanta meeting, an independent group met in Melbourne, Australia, to discuss ex situ conservation strategies for Australian frogs.[6] For some amphibian species, removal to protected habitats with the hope of restoring them to their communities is the best step available, because eradication of the chytrid fungus is currently impossible once it enters a region.[7]

In spring 2006, researchers at El Valle found the first dead frogs infected with Bd. Joseph Mendelson from Zoo Atlanta and Ronald Gagliardo from the Atlanta Botanical Garden worked with Central American and U.S. colleagues to move frog species to ex situ facilities in the United States and to begin construction of the first ex situ facility in Panama (Figure 9.1). The story caught the public's attention and was reported widely in print and electronic media.

> ATLANTA, June 5 — *Of all the things airport security screeners have discovered as they rifle through travelers' luggage, the suitcases full of frogs were a first.*
>
> *In a race to save amphibians threatened by an encroaching, lethal fungus, two conservationists from Atlanta recently packed their carry-ons with frogs rescued from a Central American rain forest...*
>
> *The frogs ... were perhaps the last of their kind, collected from a pristine national park that fills the bowl of El Valle, an inactive volcano in Panama.*
>
> (Brenda Goodman, "To Stem Widespread Extinction, Scientists Airlift Frogs in Carry-On Bags," *New York Times*, June 6, 2006)

Figure 9.1 Ronald Gagliardo (*left*) and Joseph Mendelson III packing a box with frogs in containers. Photograph by Joseph Mendelson III.

Five months later, Panama's frogs were a front page story:

EL VALLE DE ANTON, Panama— *The guests in Rooms 28 and 29 at the Hotel Campestre here in this lush volcano-crater town get the full spa treatment.*

Daily cleansing rinses. Exotic lunches. Even 24-hour room service.

It would all be so lovely, a real dream, if they could only go outside every once in a while. But they can't. Not ever. One step outside, or in their case one hop, and they'd be goners.

Thus is the lot of Panama's — and perhaps the world's — most unusual hotel VIPs, the darling little Panamanian golden frogs of El Valle de Anton. The frogs, considered so lucky in Panama that their images appear on lottery tickets, are in big trouble. They're on the run from a vicious fungus that has already wiped out as many as 120 species of amphibians in Central America.

The Hotel Campestre might be their last hope. If the golden frogs make it, this crumbling backpackers' hangout could very well provide a revolutionary new model for handling one of the world's most endangered species.

(Manuel Roig-Franzia, "Panama Hotel Is Imperiled Frogs' Lifeboat," *Washington Post*, October 26, 2006)

But if frogs moved from El Valle cannot return to their forest habitats—at least for the foreseeable future—are they in a hotel or a jail?[8]

Conservation Biology and Ecological Ethics

The events in Panama raise questions about values and research ethics that increasingly confront ecologists and biodiversity managers.[9]

- What is the scientific community's role and responsibility given the imminent extinction of species? What is the conservation community's role and responsibility? Do they differ?
- Does the zoo community have a role and responsibility to house threatened species?
- Do administrators and policy makers have a responsibility to act? For example, should the government close and isolate forest reserves for fear that people may transmit a pathogen, even though in the case of Bd we have limited evidence (only from California) supporting that possibility? (Figure 9.2)[10]
- Should help be sought from the pet trade community to develop protocols for housing and breeding animals? Should that help be sought only if amphibians can be reclaimed and released back into their original habitats?[11]
- Should scientists working on threatened populations alert researchers and the public to their findings, even if it draws the attention of commercial collectors who might take advantage of the situation (as we learned in chapter 4), further depleting populations and, in the case of amphibians, perhaps spreading infectious disease?
- How should public policy issues that involve an ethical component be decided? What standard rules or principles should apply?

Ecologists and biodiversity managers often encounter complex ethical situations in research and professional activities. Questions can arise regarding duties and obligations to research animals, species, and ecological systems, as well as to the scientific profession and the public interest. There is currently no field within professional or practical ethics that addresses the unique moral concerns of practicing ecologists and managers. Ben Minteer at Arizona State University has collaborated with one of us (JPC) to argue for a new field of research and education, called "ecological ethics," to address these situations.[12]

Ultimately, it is the complexity of cases in ecological ethics and the lack of a clear way forward that makes them intriguing intellectually

Figure 9.2 A Project Golden Frog warning poster announcing measures to forestall dispersal of the amphibian chytrid fungus.

but difficult to solve, a tension we can illustrate using amphibians: What is the ethically appropriate response if it is determined that the diseases now killing amphibians are "natural" in origin? Is conservation intervention warranted, and if so, why? And would it matter if the emergence of Bd were "human associated?" How old must a disease's emergence be before it can be considered a "natural" element of an ecosystem? Are such questions moot?—and the real issue is, When should humans act as planetary managers? Specifically, if an argument can be made for conserving a species, regardless of the reason it is jeopardized, is conservation intervention always warranted? These questions are complex and lack simple answers. The following vignette illustrates how their complexity is causing even the amphibian research community to differ in their opinions on the best tactics for conserving species.

In July 2006, 50 scientists published a paper in the journal *Science* endorsing an Amphibian Conservation Action Plan that was developed at an international meeting in the autumn of 2005.[13] The authors (including the authors of this book) argued that, in light of the spread of chytridiomycosis, "Ex situ [conservation] programs may be the only option to avoid extinction for many species ... while research progresses on disease control, abatement, and evolution of resistance." In a Letter to *Science* published in December 2006, Alan Pounds and four other amphibian researchers questioned the use of ex situ conservation as proposed in the July paper.[14] They argued that the conservation facilities were "costly and of uncertain design." Further, the number of species housed would be limited, and for now few if any would be released back to their original habitats. Finally, they conceded that, "Of course, some captive breeding is worthwhile, especially for research and education, but its efficacy in preserving nature should not be oversold."

The authors of the *Science* paper did not suggest ex situ conservation as the only solution; rather, it was one of several possibilities with the additional advantage of buying time in hope of finding a solution that would allow reintroduction. Pounds and colleagues make the larger point that the "answer" to amphibian declines is halting environmental deterioration. All would agree, but differences in the willingness to use ex situ conservation highlight how opinions can differ regarding the best tactics, even within a group sharing the goal of stemming amphibian losses.

Institutionalizing Responses to Global Amphibian Declines

In chapter 2, we saw how the research and conservation communities responded to the first news about amphibian declines by creating initiatives and new institutions, such as the Declining Amphibian Populations Task Force (DAPTF), the Amphibian Research and Monitoring Initiative (ARMI), and Partners in Amphibian and Reptile Conservation (PARC). In 2001, U.S. and Latin American researchers created the Research and Analysis Network for Neotropical Amphibians to promote an understanding of amphibian population declines through collaborative research in the Neotropics. The members of these organizations are largely amphibian researchers. From the first realization that declines were occurring, the communities of amphibian researchers, conservationists, and managers wanted to influence national and international policy makers, with the aim of educating the general public, developing funding, establishing population monitoring programs, and adopting policies to forestall declines and extinctions. But how was this to happen?

Australia took an early lead in moving from research to policy and in doing so provided a model that others can follow. In 2000, investigators from Australia, New Zealand, and the United States met in Cairns, Queensland, to discuss what was known about amphibian declines.[15] The meeting's organizers saw the gathering as step 1 in a process whose goal was having "infection of amphibians with chytrid fungus resulting in chytridiomycosis" listed as a key threatening process under Australia's Environmental Protection and Biodiversity Conservation (EPBC) Act of 1999.[16] Under EPBC, a process is "a key threatening process" if "the process threatens or may threaten the survival, abundance or evolutionary development of a native species or ecological community." Such a listing would have the practical consequences of causing the federal government to make research funding available and adopt regulations to monitor amphibian diseases.

In July 2002, chytridiomycosis was listed, and the Australian Government Minister for the Environment and Heritage decided that a Threat Abatement Plan (TAP) "was a feasible, effective and efficient way to abate the infection process, and directed a nationally coordinated threat abatement plan be prepared to guide management of the impact of the amphibian chytrid fungus on Australian amphibians."[17] A TAP with the following two goals was adopted in 2006:

- Prevent amphibian populations or regions that are currently chytridiomycosis free from becoming infected by blocking further spread of the amphibian chytrid within Australia
- Decrease the impact of infection with the amphibian chytrid fungus on populations that are currently infected.

The Plan's goals are pursued through five objectives: prevent spread, promote recovery, control infection, share information, and coordinate management. Australia's reaction to the amphibian chytrid is an object lesson for conservation biologists in how to move from realizing that species are at risk to responding with a government-sponsored plan about 6 years later.

While the TAP helps to protect Australia's amphibians, there are other efforts to extend such protections worldwide. In 2006, Daszak, Cunningham, and Speare visited the World Organization for Animal Health (Office International des Epizooties [OIE]) in Paris to recommend that OIE deal with disease threats through the amphibian food trade. OIE's mission is protecting livestock from diseases spread by international trade, so the objective was to adopt policies that eliminate amphibian pathogens from the trade in frogs for food. OIE listing is a powerful way to help prevent the spread of amphibian diseases through trade by screening for infectious pathogens.

In June 2008, OIE member countries unanimously approved the listing of both chytridiomycosis and ranaviral disease as notifiable diseases.[18] These amphibian diseases are in the OIE standards that will allow for health certification requirements to be applied to the international trade in live amphibians and their products, as well as making it obligatory to notify detection of either disease to OIE and for that information to be made public. These actions put in place a World Trade Organization–related set of checks and balances for trade that will help prevent future introductions of both diseases.[19] The fact that there is now an international listing of these diseases 10 years after the first paper describing chytridiomycosis reflects the high-profile work on amphibian declines that made such a significant policy change happen relatively quickly.

The research community is continuing to push for more research, conservation of threatened species, and favorable policy changes through even larger institutional change. Recall that DAPTF was the first community effort to institutionalize a response to amphibian declines, and its evolution exemplifies how diverse communities—research, state, federal, and conservation agencies, the public and private sectors—united to understand and begin to manage amphibian declines. As the causes of declines became better understood, a more concerted global response—one that exceeded even what DAPTF was accomplishing—was needed. The enigmatic declines made the need especially acute, because it was clear, as we saw in the last chapter, that the time-honored practice of conserving species by establishing reserves was not working.

On September 17, 2005, more than 60 delegates to an Amphibian Conservation Summit organized by Conservation International gathered at the Cosmos Club in Washington, D.C. By the afternoon of September 19, the delegates produced a declaration built around an Amphibian Conservation Action Plan (ACAP).[20] The plan's Declaration noted, "The Amphibian Conservation Summit was called out of the conviction that it is morally irresponsible to document amphibian declines and extinctions without also designing and promoting a response to this global crisis. To this end, the Amphibian Conservation Summit has designed the Amphibian Conservation Action Plan (ACAP), and commends it to governments, corporations, civil society and the scientific community for urgent and immediate adoption and implementation." Scientific understanding of the causes had reached a point where prediction about the likelihood of declines was possible, and along with that came a clear understanding of the consequences of inaction. At the transition point where prediction was possible, the science intersected with policy and ethics. Subsequent to the summit, it was announced that one institution would be created to implement the ACAP. The details of that transition are informative for understanding institutional change in response to research discoveries and policy needs.

In the June 30, 2006, issue of *Froglog*, the last newsletter prepared under the auspices of the grassroots DAPTF, it was announced that DAPTF would be integrated into a new organization.[21] Before the summit, the IUCN Species Survival Commission (IUCN/SSC) focused on promoting amphibian conservation through the Global Amphibian Specialist Group (GASG), on decline-related research through the DAPTF, and on monitoring and assessments through an informal network of researchers contributing data to the Global Amphibian Assessment (GAA).[22] These programs were individually successful, but, as Mendelson and colleagues argued in *Science*, the extent of the current impact on amphibians required a coordinated, unified approach to amphibian conservation, research, and assessment that exceeded the scope of the individual organizations.[23] Recognizing this, the GASG, DAPTF, and GAA merged into the Amphibian Specialist Group (ASG), creating one organization to implement a global strategy for amphibian conservation. ASG's mission is to conserve biological diversity by stimulating, developing, and executing practical programs to study, save, restore, and manage amphibians and their habitats around the world.[24]

The ASG will take IUCN's Specialist Group model to the next level of effectiveness by establishing a Secretariat that coordinates regional working groups and a global web of stakeholders to leverage their intellectual, institutional, and financial capacity. When completely in place the Secretariat will have two Co-Chairs; an Executive Officer; and Conservation, Research, Assessment, Development, and Communications Divisions each headed by a director. The Research Division most closely reflects DAPTF's mission: It integrates a global network of national and regional working groups to coordinate research into the causes of global amphibian declines and disseminate results. The Conservation Division complements research with activities directed at saving populations and species. The development of ex situ conservation facilities for Panamanian amphibians is a good example of a project in this area. The Assessment Division provides a baseline for research and conservation activities by using the GAA to evaluate regularly the status of every amphibian species. Regional working group heads initiate projects by coordinating with the three division directors. ASG continues *Froglog*, a bimonthly newsletter founded by DAPTF.

Placing DAPTF within the framework of ASG and under the umbrella of the Species Survival Commission's Specialist Groups expands DAPTF's mission by taking advantage of the international connections and intellectual capital of IUCN's global network. Practically speaking, it also affords a broad, diverse platform for raising funds for amphibian research, conservation, and assessment. ASG, through connections with IUCN and Conservation International, will be able to take advantage of a valuable infrastructure for support. ASG is collaborating with organizations such as the World Association of Zoos and Aquariums (WAZA) to create a federation of biodiversity institutions, with a mission to assess the extent and causes of amphibian declines globally and to promote the means by which declines can be halted or reversed. A notable accomplishment of this partnership has been the creation of Amphibian Ark, an alliance of conservation organizations whose mission is to maintain in captivity selected species that otherwise would go extinct, until they can be restored to the wild.[25] WAZA also declared 2008 Year of the Frog and developed a campaign with an ambitious set of goals to engage the public in amphibian conservation (see chapter 2).

But institutions, such as ASG, that will play a global role must be sensitive to the diverse needs of developed and developing countries.[26] For example,

- How will the role and extent of captive breeding as a conservation tool vary among countries, and what are the implications of the variation?
- How will research and conservation priorities be set for each country?
- Will assignment of a species to IUCN categories affect resource allocation? And if so, how will those decisions be made?
- Will there be a budget for implementing the Action Plan? If so, where will it originate, and how will it be allocated?
- What is the role of conservation education and public outreach in changing public policy?

The ACAP called for four responses to amphibian declines: understanding the causes of declines and extinctions; documenting amphibian diversity and how it is changing; developing and implementing long-term conservation programs; and delivering emergency responses to crises. ASG was designed to implement each of these tasks.

Even the most remote places on earth are no longer completely isolated from human actions. DAPTF researchers and supporters realized this and fostered programs that helped bring the problem of global amphibian declines to the world's attention. DAPTF's vision will continue through the ASG. The observation in the 1980s and 1990s that amphibians were declining led researchers to begin searching for causes. We now recognize commercial exploitation, the introduction of exotic species, and land use change as ongoing causes of amphibian declines characteristic of pressures by humans on many organisms. These three causes predated the concerns raised by amphibian declines in the late 20th century, but they quickly became part of the argument for why amphibians deserved attention. Each is important, and in the final analysis each can be addressed with the right combination of resources and political will.

But such tactics are not necessarily sufficient for enigmatic declines—the ones without a ready explanation. These are the ones that require scientists to examine the theories that might account for the disappearances and initiate research projects. After almost 2 decades of research, we are in a better position to explain the declines and react to them. Indeed, the institutional and policy developments discussed in this chapter illustrate the diversity of responses. They also illustrate how, as research questions were answered, they not only generated more questions about the scientific details of declines but raised issues concerning the ethical, legal, and social implications of the declines.

As we observed at the start of the chapter, failing to act even though information is incomplete can incur substantial costs. It may be difficult or impossible to control or reverse the effects of emerging diseases, toxins, and climate change once they begin. But all of this occurs at the interface of science and society, raising the larger question: When does a researcher stop being "a scientist" and start being an "advocate"? And is that separation of activities ever completely possible?[27]

10

Leaping between Mysteries

Modern science attempts to bring clarity and context to issues that change rapidly, and at times unexpectedly. The decline and extinction of amphibian populations made people wonder: What species were involved? Where? When? And most of all, why? Major magazines and newspapers covered the story—on page one. In an effort to place what was happening in context, scientists reported what they knew and guessed at why. Reviews summarized what was known, but mostly what was not. After nearly 2 decades of research, we are in a better position to state what we know and what we do not know. The story of amphibian declines now has a context; its dimensions are clearer, and the problem is better defined.

Some amphibian population declines and extinctions can be explained as the effects of introduced exotic species, commercial use, or land use change acting alone or in combination. We understand these causes increasingly well for two reasons. First, their role in changing numbers of individuals and species in a region has acted for hundreds of years on a wide range of amphibians, other animals, and plants. Second, each acts in a direct or relatively direct way. Imagine a marsh with a population of frogs that call each year when breeding. Fill the wetland for a housing development or agriculture, and the calling will stop. Or, imagine a fire that destroys the only population of a plant or animal species. The action is direct, the species is gone, and the cause is clear.

Late 20th century amphibian population declines are not isolated events. For centuries, humans have drained wetlands and felled forests, to the detriment of their inhabitants. Aquatic habitats, prime amphibian locations, are the most threatened ecosystems in the world.[1] Just like the American buffalo, California red-legged frogs were hunted to low numbers in the late 1800s and early 1900s for commercial reasons. We have known for a long time that an exotic species can enter an ecosystem and displace natives. The late 20th century reports of amphibian declines stimulated researchers to look at old issues with a fresh perspective.

For these three causes, amphibian declines fit a well-known pattern of biodiversity losses. Commercial use can cause amphibian populations to decline and perhaps even go extinct locally. There is no evidence of global extinction caused by commerce, but overexploitation is clearly endangering some species, notably giant salamanders.[2] Another subject needing further study is the role that commercial trade plays in moving exotic amphibians bearing pathogens into contact with native populations. Are pathogens moving from exotic species to natives and thereby placing local species at risk?

Amphibian introductions to new habitats have results predictably similar to those observed in many other exotic species: Establishment and range expansion of an introduced frog or salamander species typically comes at the expense of natives. Or, if there are no native species, as was the case for the coquí frog in Hawaii, exotic species multiply quickly to fill habitats, with potentially negative effects on native species' interactions and ecosystem processes.[3] Exotic amphibians such as bullfrogs can drive native species extinct locally, but there is no evidence that introduced amphibians have driven any frog or salamander species extinct globally.

Land use change has more complex consequences. In some cases, change creates new habitats, such as artificial water impoundments, that facilitate range expansion; this has been the case for salamanders in the arid western United States.[4] But often, as exemplified by Sri Lanka, habitat change leads to the decline or extinction of species.[5] In a similar manner, reduction of Brazil's Amazonian forest due to logging, urbanization, and conversion to agricultural lands suggests that species were lost, and are being lost, in these areas.[6]

Land use change that results in habitat destruction is the leading cause of amphibian declines and extinctions. This is a straightforward, important fact. If a habitat is lost, more often than not species typical of the habitat are also lost. The more specialized a species is for living in a particular habitat, the more likely it is that extinction follows habitat loss. Species are being lost to land use change in many taxa, and amphibian losses are not unique as far as this mechanism is concerned. An exception to this generalization is the loss of freshwater habitats. Carmen Revenga and colleagues from the Nature Conservancy concluded that "freshwater systems and their dependent species are faring worse than forest, grassland, and coastal ecosystems." In particular, "the rate of decline in the conservation status of freshwater amphibians is worse than that of terrestrial species."[7]

In the late 20th century, national parks and reserves were considered sanctuaries for many species. Reports in the 1980s and 1990s, however, demonstrated that at least some sanctuaries were no longer safe havens. Animals were dying or were suddenly missing within reserve boundaries for no obvious reason. Researchers quickly settled on three suspected causes—toxins, climate change, and infectious disease—because each might act in subtle, complex ways that are difficult to detect. If proven as causes, these could provide a context for understanding why populations declined, even to extinction, in protected areas. Any one of the three, however, would be difficult to mitigate.

Many laboratory and mesocosm studies show that a wide range of toxins negatively affect diverse amphibian species by killing them outright, slowing development or causing abnormal development, increasing predation risk, or killing the animals' prey. For all of these reasons, toxins might affect amphibians negatively. Two sets of studies help bridge the gap between laboratory results and declining amphibian populations in the field. First, Sierra Nevada frog populations have declined downwind of California's Central Valley, where pesticide use is high. Toxins are transported by wind, as evidenced by pesticide residues in Sierra Nevada frogs far from the Central Valley.[8] Second, use of the herbicide atrazine is common in the central United States, and between 10 and 90 percent of male leopard frogs sampled were found to be castrated and feminized, presumably as a result of the herbicide, at seven of eight locations in the region.[9] Similar results from agricultural regions were reported for cane toads in south Florida and green frogs and leopard frogs in southern Ontario.[10]

Combined laboratory and field data are part of a chain of causation needed to connect negative effects of toxins in the laboratory with declines in the field. We need more research that fills the gap between laboratory experiments and field populations, especially in relation to how the level and kind of toxin affects amphibian population sizes.[11] To date, no data support a conclusion that toxins have caused extinctions, but it would be a small step to extinction from data showing that toxins greatly reduce population sizes.

The data on toxins are worrisome. Frogs and humans are both vertebrates. At the very least, we should wonder whether the same toxins can affect us negatively, and how the loss of amphibians might affect ecosystems on which humans depend. Teams of researchers led by Checo Colón-Gaud and Matt Whiles at Southern Illinois University and Scott Connelly at the University of Georgia are beginning to show how the loss of amphibians is changing stream structure and function in Panama: algae and inorganic sediments increased, while the kinds, numbers, and body sizes of invertebrates such as aquatic insects changed.[12] In the few studies available, the food web dynamics and energy flow in these streams are being altered by the loss of amphibian larvae. Beyond direct effects on amphibians, research also reveals how toxins negatively influence other animals and plants in the same communities. Prudence dictates at least that more information should be gathered to understand what is happening, and even more conservatively, that use of the toxins should cease until we have more facts in hand.

The results for environmental contaminants raise the general question that applies to toxins, climate change, and pathogens: What data are needed, and when do we need them, before we act? Or, to paraphrase a recent European Environment Agency report, how can we avoid always drawing lessons about harmful chemicals, changing climate regimes, or emerging infectious diseases that are too late from warnings that we had years or decades earlier?[13]

Earth's environment has changed since its origin, and it will keep changing. But the rate of change is notably different in the 21st century. For example, atmospheric carbon dioxide and temperature increases exceed rates not experienced on Earth for tens of hundreds of thousands of years. Human actions are the cause, according to the most recent United Nations Intergovernmental Panel on Climate Change report.[14] Global warming has affected and will affect amphibians into the future. Throughout the 21st century, the ranges of amphibian species will respond to global warming. A melting glacier has a direct effect when amphibian habitats are created. Effects are indirect when a community loses or gains microbial, protozoan, fungal, plant, or animal species that positively or negatively affect native frogs, salamanders, or caecilians. Research is still needed to clarify the role of climate change in amphibian population declines or extinctions.

Among the hypothesized causes of enigmatic amphibian population declines and extinction, pathogens show a clear causal link between declines and extinction. That claim alone is noteworthy, because infectious disease is not among the usual causes of extinction cited by biologists.[15] The fact that there is increasing evidence that pathogens are driving populations of at least some modern frogs, and perhaps salamanders, to extinction makes amphibians worthy of special study, not only for reasons related to conservation but also because it challenges our understanding of what is theoretically possible.

Paleontologists study bygone worlds. In their trained minds and skilled hands, remnants of past ages—bones, shells, a shadow on a rock—become a living and breathing species imagined as part of a community from a lost era. Paleontologists hypothesize that cataclysms often cause extinctions, including volcanic eruptions, rapid climate change, meteor impacts, continental movements, and shifting sea levels. These are the great physical mechanisms of extinction, and they yield hypotheses that are testable by logical deductions and correlations; paleontologists cannot run experiments in quite the same way that researchers studying living organisms do.[16]

Paleontologists also posit biological causes of extinction, such as competition or predation. Scientists studying living populations and species know, however, how hard it is to demonstrate that such mechanisms cause extinction—which serves as fair warning to any paleontologist proposing a biotic mechanism for a species long gone. Nonetheless, explaining the late Pleistocene megafaunal extinction is one instance in which paleontologists excitedly debate both physical and biological causes.[17]

At the end of the Pleistocene, about 11,000 years ago, many large North American mammals such as mastodons and mammoths went extinct. Rapid climate change that included global warming, predation by humans who migrated from Asia about that time, and emerging infectious diseases are three suggested causes of those extinctions. The third hypothesis is relevant to us for understanding amphibian extinctions.

In 1997, Ross MacPhee, of the American Museum of Natural History, and Preston Marx, who at the time was at the Aaron Diamond AIDS Research Center, posed the "Pleistocene hyperdisease hypothesis."[18] The hypothesis is not given a lot of credence by other paleontologists. Regardless of its applicability to large Pleistocene mammals, however, the hypothesis anticipates several features of how disease affects modern amphibians.

The hyperdisease hypothesis has four conditions. The pathogen must have the following characteristics:

- It has a reservoir species—perhaps humans or companion animals
- It infects any and all age classes while killing quickly, thereby slowing the evolution of immunity
- It has high virulence
- It infects multiple species

MacPhee thought that the last criterion was the hardest to fulfill: "in order for this (hyperdisease) notion to work, you have to have multiple, simultaneous, epizootics in a large number of species. They have to be sharing the disease-provoking organism in a way that is unprecedented. You have to have a disease that crosses species lines with incredible ease."[19]

An epidemiologist reflecting on the hyperdisease hypothesis would immediately see the conditions of density-independent selection within a host-pathogen system, which we discussed in chapter 7. These are the conditions under which a pathogen could drive a host to extinction. The chytrid fungus, *Batrachochytrium dendrobatidis* (Bd), meets all four criteria: There are multiple reservoir species, it infects all age classes while killing metamorphosed animals quickly, it has high virulence, and it infects multiple species. In meeting these conditions, the emergence of amphibian chytrid has a great deal to teach us about extinction and the novelty of what is happening in amphibians today.

In his book *Extinction: Bad Genes or Bad Luck*, David Raup, a paleontologist at the University of Chicago, makes several key points that provide an important framework for understanding extinction and for placing modern amphibian declines in the larger context of what we know about the causes of extinctions in general.

Species require a minimum population size to survive. Across a wide range of species, this number is in the tens to hundreds of reproductive, adult individuals.[20] Below that size, unusual events, such as one bad breeding year, can push a species close to extinction.

Small populations lack genetic diversity. In such cases, bad genes can become dominant and increase the extinction risk. Small populations are also more susceptible to dysfunctional behavior, such as males failing to find females in bisexual species. Even normally advantageous behavior, such as frogs aggregating during periods of low moisture, becomes dysfunctional if one or a few individuals harbor an infectious disease. Finally, small populations are more at risk from accidents—forces extrinsic to the system that kill all or most population members, such as a hurricane that floods an island with an endemic lizard population and drowns all individuals. Many tropical and subtropical amphibians have small ranges. Small population sizes are therefore typical, which means that they are inherently more susceptible to extinction than a species that is distributed across a very large area, such as the Canadian north. Late 20th and early 21st century amphibian population declines and extinctions are typically occurring in species with comparatively small ranges—thousands of hectares as opposed to thousands of square kilometers.[21]

Widespread species are harder to drive extinct, for reasons opposite to those that put small populations at risk.[22] Declines in species with large ranges move them into the "risk of extinction" category. Yellow-legged frogs are notable in this last regard, because they are now gone from 90 percent of their habitats in a range that formerly extended from southern to northern California in the Sierra Nevada.[23] These frogs are highly susceptible to Bd infection and, as we have discussed, a variety of other pressures. Such pressures may predispose a species to extinction.

Widespread species are resilient to many factors. This means that a severe physical or biological stress is unlikely to affect the entire species, and the species will recover because there are individuals and populations that survive. Raup argued, however, that such species may be made susceptible by a "first strike," typically an external stressor that makes the species extinction-prone by reducing the number of populations and/or the size of populations. Exotic fish species may be a possible first strike in yellow-legged frogs. The trout were a 20th century exotic or introduced stress, a selection pressure that these frogs had not previously experienced, in a Sierra Nevada system that probably had been fishless since the Pleistocene. Pesticides from California's Central Valley may be playing a role yet to be fully understood in making the frogs susceptible to Bd infection. But even without toxins, we know that Bd kills healthy frogs in the laboratory, so the introduction of Bd into a system with an already reduced number of populations due to trout predation would increase the species' extinction risk.

Simultaneous extinguishing of many species requires a stressor that cuts across taxa and ecological niches, which is why paleontologists look to cataclysmic events to explain the really big extinctions in life's history, such as the demise of the dinosaurs. Species are not adapted to cataclysmic events, because such events are too rare. And aside from such cataclysms, even extreme stresses, such as a forest fire that destroys tens of thousands of hectares, rarely affect more than a few ecosystem types.

Raup did not see most normal stresses as potential extinction mechanisms, with one caveat: "A possible exception is epidemic disease."[24] He continued, "Rapid devastation of species over wide areas by disease are [sic] well documented, including attacks on the human species by various plague organisms. However, cases of widespread species actually pushed to complete, global extinction by disease are vanishingly rare—even though it is easy to point to cases, like that of the American chestnut, where species have been pushed *toward* [italics his] extinction by disease." And later, "The virtual lack of extinctions by disease may stem merely from the fact that human civilization has had only a few thousand years to observe. So, epidemic disease remains a 'normal' stress with the unproven potential of killing widespread species."[25]

Species infected with Bd show three outcomes. Some species tolerate this chytrid fungus and may even act as carriers infecting other species. Second, a species such as the Eungella torrent frog (*Taudactylus eungellensis*) in Australia appears to have undergone a selective sweep in which susceptible individuals were killed but resistant individuals survived. Population sizes have rebounded from the lowest values but have not yet reached pre-epidemic levels.[26] Finally, Bd appears to drive some species to local extinction or at least to take them below their minimum viable population size, which exposes them to other negative factors.[27]

Bd can be a major stress in species with either large or small ranges. Widespread species may be placed at greater risk, but species with small ranges are pushed close to or over the brink of extinction. And it is just these tropical and subtropical species with small ranges that are experiencing population extinctions. So far, the mechanism of infectious disease seems to apply to species distributed over small areas. But Bd also affects widespread species such as yellow-legged frogs, raising the possibility that amphibians will be a general example of how pathogens play a role in the extinction of widely and narrowly distributed species.

As amphibian declines were first reported in the late 1980s, some biologists suggested that frogs and salamanders were acting as "canaries in coal mines," in the sense that their declines were harbingers of poor environmental quality. As we already discussed, that is true only in that, after the animals are gone, we recognize in hindsight that something about the environment had changed for the worse as far as the frogs and salamanders were concerned. Amphibians are not a very good "canary in the coal mine" to serve as a warning that something untoward may be happening in terms of current environmental quality: By the time amphibians respond, environmental quality has already deteriorated.

Nonetheless, amphibians are a warning and a glimpse of possible future events in the following sense: It is possible for an infectious disease to emerge and place many species within a class of organisms at risk of extinction. There are more than 6000 amphibian species, and Bd infects all that have been tested. Evidence for that sort of widespread biological risk due to infectious disease is notable. To be clear, the message is not that we should worry about risk of extinction from infectious disease for humans. That probability is very low, because, consistent with theory, we are a populous, genetically diverse, widespread species with many biological and cultural traits or technological mechanisms that would forestall extinction due to an infectious disease.

Amphibians are a cautionary tale alerting us to the possibility that there are conditions under which a pathogen can emerge and rapidly spread, causing declines and even extinctions in species with just the right characteristics. What if the next group were closely related plants such as cereal grains or pine trees, one whose demise would not threaten humans with extinction but could certainly affect the quality of our lives?

Modern amphibian extinctions bring losses that paleontologists previously relegated to cataclysmic events into the realm of a modern ecological process. Take for example a community ecologist wanting to model and predict ecosystem processes in a montane, Neotropical ecosystem. The researcher would have to account for the fact that within the foreseeable future a significant fraction of the grazers (tadpoles) and predators (metamorphosed amphibians) could be gone from the aquatic and terrestrial components of the ecosystem. A formerly remote, ancient process—extinction—must now be included within modern ecosystem dynamics.[28]

The studies we have reviewed reflect real advances in our understanding of the causes of amphibian population declines and extinctions over the last 2 decades. The enigmatic or so-called mysterious declines now have explanations, or at least the hypotheses are framed much more tightly than was the case in the 1990s. But our summary reveals large questions still needing answers. Here are just a few.

- What are the characteristics of a pathogen that can put a large group of organisms, such as the class Amphibia, at risk? Has such an infectious disease occurred before, and are amphibians a modern example of how such a pathogen would cause the demise of many species? Can we predict the next infectious disease to emerge that might cause extinction? What can such infectious diseases teach us about fundamental biological and evolutionary mechanisms?
- What are the ecological and evolutionary consequences of losing amphibian species? Research to date has focused largely on defining the dimensions of the problem and what might be done to conserve species. Less attention has been given to the consequences of the losses for ecosystem stability, resilience, the delivery of goods and services to humans, economics, and public health.
- How will climate change affect Earth's biodiversity in general and amphibians in particular?
- How should we respond to the growing awareness that some amphibian species are declining even to extinction? It is straightforward to imagine closing a coal-fired power plant responsible for air pollution. But how do we create safe havens for amphibians in a time of climate change, toxic exposures, and emerging infectious diseases that are not controlled by reserve boundaries? Since the classic idea of a reserve is ineffective, what role should ex situ holding facilities play in the long-term preservation of Earth's biodiversity?
- What are the long-term effects of toxic chemicals on Earth's ecosystems? How can we improve our capacity to reconcile the short-term gains accrued from using chemicals to improve human quality of life with the likely long-term negative effects that use of such chemicals can have on the ecosystems on which all life, including human life, depends?
- What do we need to know to react sooner to the next case of population decline and extinction in a major group of organisms? Widespread amphibian losses were first evident in the late 1980s to early 1990s.

> It took some 15 years to understand the problem's dimensions well enough to mount the response outlined at the Amphibian Conservation Summit. The 2001 European Environment Agency Report, *Late Lessons from Early Warnings*, makes it clear that the need for decision making under uncertainty is not unique to amphibians and that the question of how to make wise environmental decisions in the face of incomplete information deserves the attention of the world's best minds.[29]

The story of amphibian population declines and extinction began as a mystery. As scientists we have improved our capacity to explain that mystery. We do not yet know all the details of how the six forces of population decline and extinction noted earlier are affecting amphibians, but we have the larger dimensions of the story in place. With this knowledge, we can more effectively set priorities for addressing each of the causes. Nonetheless, scientific discovery often raises more questions than we had at the start. That is the nature of the enterprise: Mystery provokes questions that are answered while raising more questions.

Early in 2009, as this book was going to press, two new reports indicate how quickly new information is arising. Sean Rovito, at the University of California, Berkeley, and his colleagues reported major declines within plethodontid assemblages at high-elevation and upper-cloud forest sites in Mexico and Guatemala. In the San Marcos region of Guatemala, terrestrial salamander species that do not use arboreal bromeliads as refuges declined more than did species that use terrestrial microhabitats but also rely at least in part on arboreal bromeliads. A research team led by Matthew Fisher at Imperial College reported evidence that the amphibian chytrid showed evidence for genotypic differentiation among isolates along with some evidence for differential virulence of isolates in *Bufo bufo*. This research extends our discussion of variation among isolates (p. 168) and confirms a prediction we made (p. 169) that variation in virulence among strains was likely to be discovered.[30]

We began this project to summarize our understanding of why amphibian populations declined, some to extinction, late in the 20th and early in the 21st century. We know enough now to place this problem in a larger context of historical and present-day biodiversity losses. The knowledge we have raises several final questions: How will humans react to an increased awareness that Earth's biodiversity is diminishing? What are these losses telling us about our place on the planet, our role in the biosphere? What is our role in conserving biodiversity as we become custodians of a planet that has clear limitations? And how can we pass to future generations the wisdom needed to make sound environmental decisions? The answers to these questions will tell us much about ourselves, and science will take us only part of the way along that journey.

One mystery yields another.

Notes

Chapter 1

1. Noss and Cooperrider, *Saving Nature's Legacy,* 5.
2. Wilson, *The Diversity of Life,* 132.
3. Hammond, "The current magnitude of biodiversity," 113–138.
4. Numbers of amphibian species from Frost, *Amphibian Species of the World.* See also *AmphibiaWeb* for a running total of amphibian species described.
5. Myers, "Threatened biotas," 187–208.
6. Duellman, "Global distribution of amphibians," 1–30.
7. Ehrlich and Ehrlich, *Extinction: The Causes and Consequences of the Disappearance of Species;* Myers, "Mass extinctions," 175–185; Raven, "The politics of preserving biodiversity," 769–774; Soulé, "Conservation: Tactics for a constant crisis," 744–750; Wilson, *The Diversity of Life,* 424 pp.; Thomas et al., "Comparative losses of British butterflies, birds, and plants," 1879–1881.
8. Wilson, *The Diversity of Life,* 280.
9. At the first DIVERITAS Open Science Conference, held November 9–12, 2005, in Oaxaca, Mexico, leading scientists from 60 countries convened to discuss the global loss of biodiversity. Peter Raven, conservation biologist and Director of the Missouri Botanical Garden, suggested that at the current rate, by the end of the 21st century as many as two thirds of all species could be headed for extinction.
10. Stevens, "The latitudinal gradient," 240–256.
11. Wilson, *The Diversity of Life,* 280.
12. Pimenta et al., "Comment on 'Status and Trends,'" pointed out additional problems with using threatened species lists to evaluate conservation status. Their comments were in reference to a paper published in 2004 by Stuart et al., "Status and trends of amphibian declines," 1783–1786, which concluded that 1856 species of amphibians are threatened worldwide. Pimenta and his colleagues argued

that the conclusions of the Stuart et al. paper, based on the Global Amphibian Assessment (GAA), are questionable. They pointed out that the GAA results for Brazilian amphibians differ from what specialists have determined. Specialists have categorized 24 Brazilian species as threatened, whereas Stuart et al. designated 110 species as threatened. Why the difference? The GAA used the same threshold for distribution range for all animal groups. But the minimal area needed by a tiny leaf-litter frog is NOT the same as that for a tiger! Besides, we don't know the distribution of many species of amphibians. Another problem with the Stuart et al. analysis is the way in which "rapidly declining species" were determined. Pimenta et al. pointed out that if threatened species lists are to be used, they must be updated continually to allow for meaningful comparisons over time and space. Unfortunately, for many species no such data are available. How can we evaluate population trends for species for which there are no data? Pimenta et al. ended their paper by stating, "The use of the criteria as adopted by the GAA is a straightjacket that has artificially forced a great number of healthy species/populations into threatened categories. A proper assessment should take into account the ecological specificities of naturally endemic species that occupy a more restricted area, so as to mirror more realistically what probably is or is not a threatened species."

13. Meegaskumbura et al., "Sri Lanka: An amphibian hot spot," 379.

14. Pennisi, "100 Frogs-a-leaping for biodiversity," 339, 341.

15. Groom et al., *Principles of Conservation Biology,* 173–251.

16. Groom et al., *Principles of Conservation Biology,* 293–331.

17. Groom et al., *Principles of Conservation Biology,* 253–291.

18. Groom et al., *Principles of Conservation Biology,* 188–197.

19. Groom et al., *Principles of Conservation Biology,* 333–374.

20. Groom et al., *Principles of Conservation Biology,* 67–68.

21. Noss and Cooperrider, *Saving Nature's Legacy,* 17–23; Hunter, *Fundamentals of Conservation Biology,* 41–65; Primack, *Essentials of Conservation Biology,* 87–155; Groom et al., *Principles of Conservation Biology,* 111–135.

22. Perrings, "The economic value of biodiversity," 823–914.

23. Ehrlich and Wilson, "Biodiversity studies: Science and policy," 758–762.

24. Lawton and Brown, "Redundancy in ecosystems," 255–270.

25. For example, see Walker et al., "Stability of semi-arid savanna grazing systems," 473–498; Holling, "Cross-scale morphology," 447–502; Perrings, "The economic value of biodiversity," 823–914.

26. Lawton and Brown, "Redundancy in ecosystems," 255–270.

27. Baskin, "Ecologists dare to ask," 202–203.

28. Wilson, *Biophilia.* 1.

29. Groom et al., *Principles of Conservation Biology,* 119–124.

30. As quoted in Noss and Cooperrider, *Saving Nature's Legacy,* 23.

31. Daszak et al., "Emerging infectious diseases and amphibian," 735.

32. Wells, "Frogs missing?" as quoted in Noss and Cooperrider, *Saving Nature's Legacy,* 327.

33. For example, see Tyler et al., "How frogs and humans interact," 1–18; Beebee, *Ecology and Conservation of Amphibians,* 144–145.

34. Teixeira et al., *The World Market for Frog Legs,* 1–44.

35. Stebbins and Cohen, *A Natural History of Amphibians,* 4–6.

36. Pough et al., *Herpetology,* 15.

37. Bruce and Christiansen, "The food and food habits of Blanchard's cricket frog," 63–74.
38. Dickerson, *The Frog Book*, 234.
39. Oliver, *The Natural History of North American Amphibians and Reptiles*, 29.
40. For example, Conant, *A Field Guide to Reptiles and Amphibians;* Stebbins, *A Field Guide to Western Reptiles and Amphibians.*
41. Gibbs et al., "The live frog is almost dead," 1027–1034.
42. Kagarise Sherman, and Morton, "Population declines of Yosemite toads," 186–198.
43. Davidson et al., "Declines of the California red-legged frog," 464–479.
44. Carey, "Hypothesis concerning the causes of the disappearance of boreal toads," 355–362.
45. Corn and Fogleman, "Extinction of montane populations," 147–152.
46. Corn et al., "Acid precipitation studies in Colorado and Wyoming," 56 pp.
47. Blair, "Note on the herpetology of the Elk Mountains, Colorado," 239–240.
48. Harte and Hoffman, "Possible effects of acidic deposition," 149–158.
49. Lannoo, "Amphibian conservation and wetland management," 330–339.
50. Hay, "Blanchard's cricket frogs in Wisconsin," 79–82.
51. Corser, "Decline of disjunct green salamander," 119–126.
52. Highton, "Declines of eastern North American woodland salamanders," 34–46.
53. Daigle, "Distribution and abundance of the chorus frog," 73–77.
54. Joglar and Burrowes, "Declining amphibian populations in Puerto Rico," 371–380.
55. La Marca et al., "Catastrophic population declines and extinctions," 190–201.
56. Bustamante et al., "Cambios en la diversidad," 180–189.
57. Vaira, "Report of a breeding aggregation extirpation," 3.
58. Banks et al., "Conservation of the natterjack toad," 111–118.
59. Bosch et al., "Evidence of a chytrid fungus infection," 331–337.
60. Beebee, *Ecology and Conservation of Amphibians*, 174–176.
61. Beebee, *Ecology and Conservation of Amphibians*, 176–179.
62. Rastegar-Pouyani, "Ecology and conservation of the genus *Neurergus*," 1–2.
63. Fellers et al., "Status of amphibians at the Zoige Wetlands," 1.
64. "Amphibians," *Grzimek's Animal Life Encyclopedia*, 256.
65. Krajick, "The lost world of the Kihansi toad," 1230–1232; Ché Weldon, personal communication to J. Collins November 2006.
66. Melvani, "Frog tea?," 2.
67. Alcala, "The Negros cave frog is critically endangered," 1.
68. Dolino et al., "Populations of two species of forest frogs," 1–2.
69. Tyler, "Declining amphibian populations," 43–50.
70. Bell, "Frog declines in New Zealand," 3.
71. Drost and Fellers, "Collapse of a regional frog fauna," 414–425.
72. Vredenburg et al., "Concordant molecular and phenotypic data," 361–374.
73. Pounds et al., "Tests of null models," 1307–1322.
74. Heyer et al., "Decimations, extinctions, and colonizations," 230–235.
75. Weygoldt, "Changes in the composition," 249–255.
76. Richards et al., "Declines in populations," 66–77.

77. Laurance et al., "Epidemic disease and the catastrophic decline," 406–413; Schloegel et al., "The decline of the sharp-snouted day frog," 35–40.

78. Gower and Wilkinson, "Conservation biology of caecilian," 45–55.

79. Marvalee Wake, personal communication to M. Crump, August 2008.

80. Wake, "Reproduction, growth," 244–256.

81. See Nussbaum and Pfrender, "Revision of the African," 1–32; Hofer, "A short note," 1; Gower and Wilkinson, "Conservation biology of caecilian," 45–55.

82. Marvalee Wake, personal communication to M. Crump, August 2008.

83. Wake and Vredenburg, "Are we in the midst," 11471.

84. For example, see Pechmann et al., "Declining amphibian populations: The problem," 892–895; Semlitsch et al., "Structure and dynamics of an amphibian community," 217–248.

85. Alford and Richards, "Global amphibian declines," 133–165; Marsh, "Fluctuations in amphibian populations," 327–335.

86. Green, "The ecology of extinction," 331–343.

87. Blaustein, "Chicken Little or Nero's fiddle?," 94.

88. Travis, "Calibrating our expectations," 104–108.

89. See Pechmann, "Natural population fluctuations and human influences," 85–93, for a critical examination of four null models that have been used to analyze and interpret declining amphibian populations.

90. Pounds et al., "Tests of null models," 1307–1322.

91. Pechmann et al. "Declining amphibian populations," 892–895.

92. Houlahan et al., "Quantitative evidence for global amphibian declines," 752–755; Alford et al., "Global amphibian population declines," 449–450.

93. Stuart et al., "Status and trends of amphibian declines," 1783–1786. But see note 12 for critique by Pimenta et al. of the conclusions of the Stuart et al. paper.

94. The percentages for amphibians in these two assessments (32.5 and 30 percent) are slightly different because different total numbers of amphibians were used.

95. Gibbons et al., "The global decline of reptiles," 653–666.

96. Thomas et al., "Comparative losses of British butterflies, birds, and plants," 1879–1881.

97. McCallum, "Amphibian decline," 483–491; quote from p. 487.

98. Roelants et al., "Global patterns of diversification," 891.

99. For example, see Blaustein and Wake, "Declining amphibian populations," 203–204; Blaustein and Wake, "The puzzle of declining amphibian populations," 56–61; Blaustein, "Chicken Little or Nero's fiddle?" 85–97; Vitt et al., "Amphibians as harbingers of decay," 418.

100. Phillips, *Tracking the Vanishing Frogs*, 211.

101. Pechmann and Wilbur, "Putting declining amphibian populations in perspective," 65–84.

102. Noss, "Indicators for monitoring biodiversity," 355–364; Karr, "Ecological integrity and ecological health," 97–109; McGeoch, "The selection, testing and application of terrestrial insects," 181–201; Poiani et al., "Biodiversity conservation at multiple scales," 133–146; Feinsinger, *Designing Field Studies for Biodiversity Conservation*, 113–130.

103. Odum, "Great ideas in ecology for the 1990s," 542–545.

104. McGeoch, "The selection, testing and application of terrestrial insects," 181–201; Hilty and Merenlender, "Faunal indicator taxa selection," 185–197.

105. Beebee and Griffiths, "The amphibian decline crisis," 278.

106. Feinsinger, *Designing Field Studies for Biodiversity Conservation,* 113–130.
107. Phillips, *Tracking the Vanishing Frogs,* 211.
108. "Take action to aid amphibians," www.conservation.org/FMG/Articles/Pages/10160701.aspx.

Chapter 2

1. Adler, *Herpetology: Current Research,* 4.
2. Crump et al., "Apparent decline of the golden toad," 413–420; Pounds and Crump, "Amphibian declines and climate disturbance," 72–85.
3. Phillips, *Tracking the Vanishing Frogs,* 22–27.
4. Wake and Morowitz, "Declining amphibian populations," 33–42.
5. Phillips, *Tracking the Vanishing Frogs,* 30.
6. See chapter 1; Phillips, *Tracking the Vanishing Frogs,* 210–211; Beebee and Griffiths, "The amphibian decline crisis," 271–285.
7. Phillips, *Tracking the Vanishing Frogs,* 207–208.
8. Heyer and Murphy, "Declining Amphibian Populations Task Force," 17–21.
9. Heyer et al., *Measuring and Monitoring,* xvii. The manual begins with an introduction and overview of amphibian diversity and natural history. Subsequent chapters are as follows: (1) Essentials of Standardization and Quantification, (2) Research Design for Quantitative Amphibian Studies, (3) Keys to a Successful Project: Associated Data and Planning, (4) Standard Techniques for Inventory and Monitoring, (5) Supplemental Approaches to Studying Amphibian Biodiversity, (6) Estimating Population Size, (7) Analysis of Amphibian Biodiversity Data, and (8) Conclusions and Recommendations.
10. *Froglog.* See issue #42 (December 2000) for a detailed list of fields in the database.
11. Bishop, *Declines in Canadian Amphibian Populations;* Green, *Amphibians in Decline;* Lannoo, *Status and Conservation of Midwestern Amphibians.*
12. *Froglog.* Issue #74 (April 2006).
13. For example, see Cunningham et al., "Pathological and microbiological findings," 1539–1557. Also see Laurance et al., "Epidemic disease," 406–413.
14. Collins et al., "Meeting the challenge," 23–27.
15. Raloff, "Common pesticide clobbers amphibians," 150.
16. Dodd, *"Monitoring Amphibians,"* 1; Corn et al., *"Amphibian Research and Monitoring Initiative,"* 1.
17. For more information on ARMI, visit their Web site at http://armi.usgs.gov.
18. IUCN, *The IUCN Red List of Threatened Species;* Global Amphibian Assessment Web site, http://www.globalamphibians.org.
19. Weir and Mossman, "North American Amphibian Monitoring Program," 307–313.
20. Souder, *A Plague of Frogs;* Souder, "Of men and deformed frogs," 344–347.
21. Murphy, "A Thousand Friends of Frogs," 341–342.
22. Murphy, "A Thousand Friends of Frogs," 341–342.
23. Hamline University Center for Global Environmental Education, A Thousand Friends of Frogs. http://hamline.edu/frogs/
24. Gibbons, "Embracing human diversity in conservation," 15.
25. Partners in Amphibian and Reptile Conservation, Web site. http://www.parcplace.org.

26. Snider and Arbaugh, "The National Amphibian Conservation Center," 339–340. For more information on NACC, visit their Web site at http://www.detroitzoo.org/Attracions/Amphibiville/NACC/.

27. Pavajeau et al., "Amphibian Ark," 24–29; the Amphibian Ark, Web site is available at http://amphibianark.org/; see also Amphibian Ark 2008 YOTF Campaign at http://www.2008yearofthefrog.org.

28. Dodd and Seigel, "Relocation," 336–350; Dodd, "Population manipulations," 265–270.

29. Lewis et al., "Possible extinction of the Wyoming toad," 166–168.

30. Taylor et al., "Causes of mortality of the Wyoming toad," 49–57.

31. Odum and Corn, "*Bufo baxteri,*" 390–392.

32. Krajick, "The lost world of the Kihansi toad," 1230–1232; Lee et al., "Captive-breeding programme," 241–253.

33. Ché Weldon, personal communication to J. Collins, November 2006.

34. *Froglog,* issue #53 (October 2002), 3–4.

35. See chapter 1 for a discussion on the IUCN Red List of Threatened Species.

36. *Froglog,* the bimonthly newsletter of the former DAPTF, is now the bimonthly newsletter of the ASG. It is free and available on the ASG Web site, http;//www.amphibians.org.

Chapter 3

1. Pechmann et al., "Declining amphibian populations: The problem," 892–895.

2. Pounds et al., "Test of null models," 1307–1322; Alford and Richards, "Global amphibian declines," 133–165; Houlahan et al., "Quantitative evidence," 752–755; Alford et al., "Global amphibian population declines," 449–450; Storfer, "Amphibian declines: Future directions," 151–163.

3. Phillips, *Tracking the Vanishing Frogs,* 30.

4. Heyer et al., *Measuring and Monitoring Biological Diversity.*

5. For analyses and discussion of characteristics of declining species, see Williams and Hero, "Rainforest frogs," 597–602; Crump, "Conservation of amphibians," 53–69; Lips et al., "Ecological traits," 1078–1088; Stuart et al., "Status and trends," 1783–1786; Crump, "Why are some species in decline," 7–9; Lips and Donnelly, "Lessons from the tropics," 198–205; Murray and Hose, "Life-history and ecological correlates," 564–571.

6. Young et al., "Population declines and priorities," 1213–1223; Stuart et al., "Status and trends," 1783–1786; Crump, "Conservation of amphibians," 53–69; Lips and Donnelly, "Lessons from the tropics," 198–205.

7. Brown and Kodric-Brown, "Turnover rates in insular biogeography," 445–449.

8. Lips et al., "Ecological traits," 1078–1088.

9. Cooper et al., "Macroecology and extinction risk," 211–221; see also Bielby et al., "Predicting susceptibility, 82–90.

10. Wyman, "What's happening to the amphibians," 350–352; Hoffman and Blows, "Evolutionary genetics and climate change," 165–178.

11. Pounds et al., "Test of null models," 1307–1322.

12. Lips, "Decline of a tropical montane amphibian fauna," 106–117.

13. Lips et al., "Ecological traits," 1078–1088.

14. Laurance et al., "Epidemic disease," 406–413; Williams and Hero, "Rainforest frogs," 597–602; McDonald and Alford, "A review of declining frogs," 14–22.

15. Stuart et al., "Status and trends," 1783–1786.

16. Salthe, "Reproductive modes," 467–490; Salthe and Duellman, "Quantitative constraints," 229–249; Crump, *Reproductive Strategies for a Tropical Anuran Community,* 1–68; Duellman and Trueb, *Biology of Amphibians,* 21–28.

17. Heyer et al., "Decimations, extinctions, and colonizations," 230–235.

18. Weygoldt, "Changes in the composition," 249–255.

19. Crump, "Conservation of amphibians," 53–69; data used for the analysis were from the literature through 2000.

20. Williams and Hero, "Rainforest frogs," 597–602; McDonald and Alford, "A review of declining frogs," 14–22.

21. Murray and Hose, "Life-history and ecological correlates," 564–571.

22. Corey and Waite, "Phylogenetic autocorrelation," 614–629.

23. Williams and Hero, "Rainforest frogs," 597–602; McDonald and Alford, "A review of declining frogs," 14–22.

24. Williams and Hero, "Rainforest frogs," 597–602; McDonald and Alford, "A review of declining frogs," 14–22; Lips et al., "Ecological traits," 1078–1088.

25. Crump, "Why are some species in decline," 7–9.

26. Collins and Storfer, "Global amphibian declines: Sorting," 89–98.

27. For selected references concerning the effects of introduced species and amphibians, see Gamrandt and Kats, "Effect of introduced crayfish," 1155–1162; Komak and Crossland, "An assessment of the introduced mosquitofish," 185–189; Kats and Ferrer, "Alien predators and amphibian declines," 99–110; Kiesecker, "Invasive species as a global problem," 113–126; Vredenberg, "Reversing introduced species effects," 7646–7650; Knapp, "Effects of nonnative fish," 265–279; Meshaka, "Exotic species," 271–274.

28. For selected references concerning overexploitation of amphibians, see Gibbs et al., "The live frog is almost dead," 1027–1034; Patel, "French may eat Indonesia out of frogs," 7; Gorzula, "The trade in dendrobatid frogs," 116–123; Jensen and Camp, "Human exploitation of amphibians," 199–213; Wilson, "Commercial trade," 146–148.

29. For selected references on land use change and amphibians, see Ash and Bruce, "Impacts of timber harvesting," 300–301; Delis et al., "Decline of some west-central Florida anuran populations," 1579–1595; Lehtinen et al., "Consequences of habitat loss," 1–12; Dodd and Smith, "Habitat destruction and alteration," 94–112; Marsh and Beckman, "Effects of forest roads," 1882–1891.

30. For selected references concerning pollutants and amphibians, see Berrill et al., "Effects of pesticides," 233–245; Sparling et al., "Pesticides and amphibian population declines," 1591–1595; Boone and Bridges, "Effects of pesticides on amphibian populations," 152–167; Davidson, "Declining downwind," 1892–1902; Bridges and Semlitsch, "Variation in pesticide tolerance," 93–95; Relyea et al., "Pesticides and amphibians," 1125–1134.

31. For selected references concerning climate change and amphibians, see Pounds and Crump, "Amphibian declines and climate disturbance," 72–85; Beebee, "Amphibian breeding and climate," 219–220; Ovaska, "Vulnerability of amphibians in Canada," 206–225; Donnelly and Crump, "Potential effects of climate change," 541–561; Pounds et al., "Biological response to climate change," 611–615; Blaustein

et al., "Amphibian breeding and climate change," 1804–1809; Carey and Alexander, "Climate change and amphibian declines," 111–121; Blaustein and Belden, "Ultraviolet radiation," 87–88; Daszak et al., "Amphibian population declines at Savannah River," 3232–3237; Pounds et al., "Widespread amphibian extinctions," 161–167.

32. For selected references concerning diseases and amphibians see Berger et al., "Chytridiomycosis causes amphibian mortality," 9031–9036; Berger et al., "Chytrid fungi and amphibian declines," 21–31; Carey et al., "Pathogens, infectious disease, and immune defenses," 127–136; Collins et al., "Ecology and evolution of infectious disease," 137–151; Daszak et al., "Infectious disease and amphibian population declines," 141–150; Briggs et al., "Investigating the population-level effects of chytridiomycosis," 3149–3159; Rachowicz et al., "Emerging infectious disease," 1671–1683.

33. Halliday, "Diverse phenomena," 3–6.

34. Mendelson et al., "Confronting amphibian declines," 48.

Chapter 4

1. Definition from U.S. Geological Survey Web site at http://www.usgs.gov/science/science.php?term=602; see U.S. Executive Order 13122 for details (http://www.invasivespeciesinfo.gov/laws/execorder.shtml).

2. Vences et al., "Phylogeography of *Ptychadena mascareniensis*," 593–601.

3. Meshaka, *The Cuban Treefrog in Florida.*

4. Specific regions summarized in Tables 4.1 and 4.2 are covered by the following sections in Duellman, *Patterns of Distribution of Amphibians: A Global Perspective:* Australia, including Tasmania (Tyler, 541–563); New Guinea, including the archipelagos and islands in the Australo-Papuan region (Tyler, 541–563); Europe, delimited in the east by the Ural Mountains and Ural River and including the northern slopes of the Greater Caucasus Mountains (Borkin, 329–420); the West Indies (Hedges, 211–254); and Nearctic, including Canada, the continental United States, and parts of northern Mexico (Duellman and Sweet, 31–109). Also for Canada, see Weller and Green, "Checklist and current status" 309–328; for Europe, Arnold and Ovenden, *Reptiles and Amphibians: Britain and Europe;* for U.S. salamanders, Petranka, *Salamanders of the United States and Canada;* for introduced U.S. and Canadian amphibians, Smith and Kohler, "A survey of herpetological introductions," 1–24. For freshwater fishes, see Mayden, *Systematics, Historical Ecology, and North American Freshwater Fishes* and Fuller et al., *Nonindigenous Fishes Introduced into Inland Waters in the United States;* for birds, see Sibley, *The Sibley Guide to Birds.*

5. See note 4.

6. Global Invasive Species Database, *Bufo marinus,* http://www.invasivespecies.net/database/species/ecology.asp?si=1138&fr=1&sts=sss. This database is managed by the Invasive Species Specialist Group of the IUCN Species Survival Commission.

7. Nonindigenous Aquatic Species, http://nas.er.usgs.gov/.

8. Bryan, "Frogs in Hawaii," 61–64.

9. The reasons behind the arrival of *Bufo boreas halophilus, Bufo gargarizans gargarizans, Rana clamitans melanota, Rana nigromaculata nigromaculata,* and *Hyla aurea* are unclear. For information on *E. coqui* and selected other invasive species, see The

Earlham College Web site (Introduced Species in Hawaii, http://www.earlham.edu/~biol/hawaii/amphibians.htm), and for an overview of *E. coqui*, see the Global Invasive Species Database.

10. Beard et al., "Amphibian effects on its prey and ecosystem," 607–617.

11. Kraus et al. "*Eleutherodactylus* frog introductions to Hawaii," 21–25; Kraus and Campbell, "Human-mediated escalation," 327–332.

12. Proyecto Coquí, http://coqui.uprrp.edu/eng/species.html.

13. Raloff, "Hawaii's hated frogs," 11. See also, Gorman, "A frog brings cacophony," D2.

14. Meshaka, *The Cuban Treefrog in Florida.*

15. Kats and Ferrer, "Alien predators and amphibian declines," 99–100.

16. Rahel, "Homogenization of freshwater faunas," 291.

17. Vredenburg et al., "Concordant molecular and phenotypic data," 361–374.

18. Our summary from here largely follows David Bradford's review in Carey et al., "Biotic factors in amphibian population declines," 153–208. Additional details are from the following: Knapp et al., "Resistance and resilience of alpine lake fauna," 401–421; Knapp, "Effects of nonnative fish," 265–279; Vredenberg, "Reversing introduced species effects," 7646–7650; Knapp et al., "Removal of nonnative fish," 11–20. See also Hayes and Jennings, "Decline of ranid frog species in western North America," 490–509, for a general discussion of the role of fish in amphibian declines in California.

19. For example, see Gillespie, "The role of introduced trout," 187–198.

20. Knapp et al., "The introduction of nonnative fish," 277.

21. Jensen and Camp, "Human exploitation of amphibians," 199–213; Tyler et al., "How frogs and humans interact," 1–18.

22. Cost varies with availability and source, but one example is here: http://www.amazonreptilecenter.com/popup.php3?id=2565.

23. Tyler, "Distribution patterns," 554.

24. Global Amphibian Assessment, http://www.globalamphibians.org/servlet/GAA?searchName=Ceratobatrachus+guentheri.

25. Jensen and Camp, "Human exploitation of amphibians," 202.

26. Go to the Global Amphibian Assessment Web page, http://www.globalamphibians.org/servlet/GAA, and search for *Mantella aurantiaca.*

27. Simon Stuart, personal communication via e-mail to JPC, June, 2006. See also Andreone et al., "Species review of amphibian extinction risks," 1790–1802, for more details on the pet trade and Malagasy frogs.

28. Wistar, *The Virginian,* 162, 164.

29. Jennings and Hayes, "Pre-1900 overharvest of the California red-legged frog," 94–103.

30. Helfrich et al., "Commercial frog farming."

31. Lannoo et al., "An altered amphibian assemblage," 311–319.

32. Borkin, "Distribution of amphibians," 395.

33. Emmons, "Problems of an amphibian supply house," 91.

34. Nace, "The amphibian facility," 767–775.

35. McKay, *An Evaluation of Captive Breeding and Sustainable Use of the Mexican Axolotl (Ambystoma mexicanum).*

36. Xie et al., "Conservation needs of amphibians in China," 265–276.

37. Teixeira et al., *The world market for frog legs,* 1–44.

38. Schlaepfer et al., "Challenges in evaluating the impact," 256–264.

39. The nine frog families were Pipidae, Ranidae, Discoglossidae, Rhacophoridae, Hylidae, Bufonidae, Hyperolidae, Pelobatidae, and Lectodactylidae; the two salamander families were Plethodontidae and Salamandridae.

40. Schlaepfer et al., "Challenges in evaluating the impact," 263.

41. Courchamp and his research team at Universite Paris-Sud provided an exception to this generalization ("Rarity value and species extinction," 2405–2410). They showed that, in theory, the human predisposition to place an exaggerated value on rare species may provoke greater exploitation. The result is that such species become rarer and even more desirable, ultimately following an irreversible path to extinction.

42. For details on Laotian salamanders sold as pets in Japan, see the article by Kenneth Chang in *The New York Times,* available at http://www.nytimes.com/2006/04/25/science/25find.html?_r=1&oref=slogin, or the Japanese report at http://www.caudata.org/forum/messages/8/62694.html?1146003064, which is supported by Caudata.org.

43. For information on the bait trade see Collins, "Distribution, habitats, and life history variation," 666–675; Collins et al., "Conserving genetically-distinctive populations," 45–53; and Jensen and Camp, "Human exploitation of amphibians," 199–213.

44. Inflation calculator at http://www.westegg.com/inflation/.

45. Riley et al., "Hybridization between a rare," 1263–1275; Storfer et al., "Evidence for introgression," 78–96.

46. Fitzpatrick and Shaffer, "Hybrid vigor," 1579–5798.

47. The summary of pathogen movement with amphibians and the discussion of pathogen pollution are from Picco and Collins, "Amphibian commerce;" which contains all supporting citations.

48. Cunningham et al., "Pathogen pollution," S78–S83.

49. Jancovich et al., "Evidence for emergence," 213–224; Storfer et al., "Phylogenetic concordance analysis," 1–9.

50. Fisher and Garner, "The relationship between the emergence," 2–9.

51. Barbassa, "Habitat or homes?" The history of the jumping frog story is from eNotes.com, http://www.enotes.com/celebrated-jumping/. For Calaveras County history, see http://www.gocalaveras.com/.

52. Meyer and Turner, "Human population growth," 39–61.

53. Vitousek, "Beyond global warming," 1861–1876.

54. The following discussion follows Collins and Halliday, "Forecasting changes in amphibian biodiversity," 309–314, which has additional supporting citations.

55. Alford and Richards, "Global amphibian declines," 133–165.

56. Green, "The ecology of extinction," 33–43.

57. Collins and Wilbur, *Breeding Habits and Habitats,* 1–34.

58. Skelly et al., "Long-term distributional dynamics," 232–337.

59. Hecnar, "Amphibian pond communities," 1–15; Hecnar and M'Closkey, "Regional dynamics," 2091–2097; Hecnar and M'Closkey, "Species richness patterns," 763–772.

60. Hecnar, "Amphibian pond communities," 13.

61. Hecnar and M'Closkey, "Species richness patterns," 770.

62. Achard et al., "Determination of deforestation rates," 999–1002.

63. Ficetola and DeBernardi, "Amphibians in a human-dominated landscape," 219–230.

64. Ficetola and DeBernardi, "Amphibians in a human-dominated landscape," 227–228.

65. Collins and Halliday, "Forecasting changes in amphibian biodiversity," 309–314.

66. Lannoo, *Amphibian Declines: The Conservation Status of United States Species.*

67. Bradford, "Factors implicated in amphibian population declines," 915–925.

68. Stuart et al., "Status and trends of amphibian declines," 1783–1786.

69. For U.S. habitat losses, see Dodd and Smith, "Habitat destruction and alteration," 94–112; for losses worldwide, see Hanski, *The Shrinking World,* or Groombridge and Jenkins, *World Atlas of Biodiversity.*

70. Dodd and Smith, "Habitat destruction and alteration," 94.

71. Gallant et al., "Global rates of habitat loss," 967–979.

72. Meegaskumbura et al., "Sri Lanka," 379.

73. Stuart et al., "Status and trends of amphibian declines," 1783–1786.

Chapter 5

1. Introductory material modified from Collins, "Where have all the frogs gone?," 44–49.

2. From Souder, *A Plague of Frogs: The Horrifying True Story.*

3. Carson, *Silent Spring.*

4. Souder, *A Plague of Frogs: The Horrifying True Story.*

5. Sparling and Fellers, "Comparative toxicity of chlorpyrifos," 535–539, reported that the breakdown products of the three most commonly used organophosphorus pesticides in California's Central Valley are more toxic to amphibians than their parent compounds.

6. Burkhart et al., "Chemical stressors," 111–128.

7. Relyea, "A cocktail of contaminants," 363–376.

8. Several overviews of toxins and amphibian declines are available. Two places to start are Linder et al., *Amphibian Decline,* and Semlitsch, *Amphibian Conservation.*

9. Boone and Bridges, "A hierarchical approach," 265–270.

10. ECOTOX, http://www.epa.gov/ecotox.

11. Boone and Bridges, "Effects of pesticides on amphibian populations," 152–167; Bridges and Semlitsch, "Variation in pesticide tolerance," 93–95; Bridges and Semlitsch, "Genetic variation," 7–13.

12. Sparling, "A review of the role of contaminants," 1099–1128.

13. Relyea and Hoverman, "Assessing the ecology in ecotoxicology," 1157–1171.

14. Burkhart et al., "Chemical stressors," 111–128.

15. For a discussion of compensatory deaths, see Vucetich et al., "Influence of harvest," 259–270.

16. Sparling, "A review of the role of contaminants," 1102.

17.For a review of using mesocosms in ecotoxicology see Boone and James, "Aquatic and terrestrial," 231–257.

18. Boone et al., "Effects of an insecticide," 685–691.

19. Vertucci and Corn, "Evaluation of episodic acidification," 449–457; Sparling, "A review of the role of contaminants," 1099–1128; Rowe and Freda, "Effects of acidification," 545–571.

20. Rouse et al., "Nitrogen pollution," 799–803.

21. Hecnar, "Acute and chronic toxicity," 2131–2137; Vitousek et al., "Human alterations of the global nitrogen cycle," 737–750; Moffat, "Global nitrogen overload problem," 988–989; Sparling, "A review of the role of contaminants," 1099–1128; Hamer et al., "Amphibian decline and fertilizers," 299–305.

22. Relyea and Hoverman, "Assessing the ecology in ecotoxicology," 1157–1171.

23. Relyea and Dieks, "An unforeseen chain of events," 1728–1742.

24. Relyea and Hoverman, "Assessing the ecology in ecotoxicology," 1167.

25. Davidson et al., "Declines of the California red-legged frog," 464–479; Davidson et al., "Spatial tests of the pesticide drift," 1588–1601; Davidson, "Declining downwind," 1892–1902.

26. Daly and Wania, "Organic contaminants in mountains," 385–398.

27. Hageman et al., "Atmospheric deposition," 3174–3180. See also Kolpin et al., "Pharmaceuticals, hormones," for a discussion of contaminants in protected areas.

28. Davidson and Knapp, "Multiple stressors and amphibian declines," 587–597.

29. Lannoo has thoroughly reviewed abnormal frog development and amphibian declines in *Malformed Frogs.*

30. FrogWeb, http://www.frogweb.gov/portal/server.pt. This Web site incorporates the North American Center for Amphibian Deformities, which was previously part of the North American Reporting Center for Amphibian Malformations (NARCAM).

31. Ouellet, "Amphibian deformities," 617–661.

32. Lannoo, *Malformed Frogs.*

33. Sessions, "What is causing deformed amphibians?," 168–186.

34. Ouellet, "Amphibian deformities," 617–661; Burkhart et al., *Amphibian Decline,* 111–128.

35. Carey et al., "Biotic factors in amphibian population declines," 153–208.

36. Sessions, "What is causing deformed amphibians?," 168–186.

37. Sessions and Ruth, "Explanation for naturally occurring," 38–47.

38. Johnson et al., "The effect of trematode infection," 802–804.

39. Johnson et al., "Limb deformities," 1724–1737.

40. Personal communication with JPC, August 2007.

41. Blaustein and Johnson, "Explaining frog deformities," 60–65; Kiesecker, "Synergism between trematode infection and pesticide exposure," 9900–9904; Leutwyler, "Parasites or pollution?"

42. Kiesecker, "Synergism between trematode infection and pesticide exposure," 9900–9904.

43. Johnson, et al., "Aquatic eutrophication," 15781–15786.

44. Rohr et al., "Agrochemicals increase," 1235–1239. See also Rohr et al., "Understanding the net effects," 1743–1753.

45. Lannoo, *Malformed Frogs.*

46. Skelly et al., "*Ribeiroia* infection," 156–163.

47. Lannoo, *Malformed Frogs.*

48. Matson et al., "Agricultural intensification," 504–509; Tilman et al., "Agricultural sustainability," 671–677; Tilman et al., "Forecasting agriculturally driven," 281–284; Foley et al., "Global consequences of land use," 570–574.

49. Hayes et al., "Hermaphroditic, demasculinized frogs," 5476–5480; Hayes, "Feminization of male frogs," 895–896; Hayes et al., "Characterization of

atrazine-induced gonadal malformations," 1–53; Hayes, "Endocrine disruption in amphibians," 573–593.

50. Ouellet, "Amphibian deformities," 617–661.

51. Hayes et al., "Hermaphroditic, demasculinized frogs," 5476–5480.

52. Krimsky, *Hormonal Chaos.*

53. Hayes, "Characterization of atrazine-induced gonadal malformations," 5.

54. McCoy et al. "Agriculture alters gonadal form and function," 1526–1532.

55. McDaniel, et al. "Potential endocrine disruption," 230–242.

56. Hayes et al., "Hermaphroditic, demasculinized frogs," 5479.

57. Plowright et al., "Causal inference in disease ecology," 420–429.

58. See the analysis by Skelly, "Experimental venue," 2097–2101, which demonstrates the general difficulty of translating mesocosm experiments into estimates of changes in natural populations.

59. Poulin, "Global warming and temperature-mediated increases," 143–151.

60. Corn, "Climate change and amphibians," 59.

61. IPCC, http://www.ipcc.ch/. For details see the IPCC report, especially Frequently Asked Questions.

62. Parmesan, "Ecological and evolutionary responses," 637–639.

63. McCallum, "Amphibian decline or extinction?" 483–491; Roelants et al., "Global patterns of diversification," 887–892.

64. For details see the IPCC report, especially Frequently Asked Questions.

65. For examples see Avise et al., "Speciation durations," 1707–1712; Green et al., "Postglacial range fluctuation," 374–390; Demastes, et al., "Phylogeography of the blue-spotted salamander," 149–161; Hedges et al., "Ancestry of unisexual salamanders," 708–710; Vieites et al., "Rapid Diversification and dispersal," 19903–19907; Alsos et al., "Frequent long-distance plant colonization," 1606–1609.

66. Vieites et al., "Rapid diversification and dispersal," 19903–19907.

67. Thuiller, "Climate change and the ecologist," 550–552. Williams et al., "Projected distributions," 5738–5742, report 6 km per decade.

68. Smol and Douglas, "Crossing the final ecological threshold," 12395.

69. McMenamin et al., "Climate change and wetland desiccation," 16988–16993.

70. For a discussion of this point, see Payne and Finnegan, "The effect of geographic range on extinction risk," 10506–10511.

71. Seimon et al., "Upward range extension of Andean anurans," 288–299.

72. Andean wetlands and cities' water supplies threatened, A5; see also Wynn's article at http://www.ecoearth.info/shared/reader/welcome.aspx?linkid=80438&keybold=Rainforest%20Birds%20Habitat.

73. Thuiller, "Climate change and the ecologist," 550–552.

74. Gibbs and Breisch, "Climate warming and calling phenology," 1175–1178.

75. Beebee, "Amphibian breeding and climate," 219–220.

76. Blaustein et al., "Amphibian breeding and climate change," 1804–1809.

77. Whitfield, "Amphibian and reptile declines," 8352–8356. See also the companion commentary by Wake, "Climate change implicated," 8201.

78. For example, Sigel et al., "Avian community response," attributed declines in a number of La Selva bird species from 1960 to 1999 to reduced insect availability.

79. Carey et al., "Amphibian declines and environmental change," 903–913.

80. Carey et al., "Amphibian declines and environmental change," 903.

81. Blaustein and colleagues have published a number of papers related to UV-B's effects. Two recent summary papers from the group are Bancroft et al., "Effects of UVB radiation," 332–345, and Bancroft et al., "A meta-analysis."

82. Broomhall et al., "Comparative effects of ambient ultraviolet-B radiation," 420–427.

83. Blaustein, "UV-B radiation."

84. Corn, "Effects of ultraviolet radiation," 18–26.

85. Hossack et al., "Distribution of boreal toad populations," 98–107.

86. Palen et al., "Optical characteristics," 2951–2957.

87. Blaustein et al., "Variable breeding phenology," 1747–1754.

88. Davidson et al., "Spatial tests of the pesticide drift," 1588–1601.

89. Davidson et al., "Declines of the California red-legged frog," 464–479.

90. Araújo et al., "Climate warming," 1712–1728.

91. Araújo et al., "Climate warming," 1725.

92. Bernardo and Spotila, "Physiological constraints," 135–139.

93. Raxworthy et al., "Extinction vulnerability," 1703–1720.

94. Raxworthy et al., "Extinction vulnerability," 1715.

95. Seimon et al., "Upward range extension of Andean anurans," 288–299.

96. Storfer et al., "Phylogenetic concordance analysis," 1–9.

97. LaDeau et al., "West Nile virus emergence," 710–713.

98. Kiesecker and Blaustein, "Synergism between UV-B radiation," 11049–11052.

99. Kiesecker et al., "Complex causes," 681–684.

100. Johnson and Wellehan, "Amphibian virology," 53–65.

101. Chinchar, V. G. 2002. "Ranaviruses (family Iridoviridae)," 447–470.

102. Cullen and Owens, "Experimental challenge," 83–92.

103. Jancovich et al., "Transmission of the *Ambystoma tigrinum* virus," 159–163.

104. Carey et al., "Biotic factors in amphibian population declines," 153–208.

105. Brunner et al., "Intraspecific reservoirs," 560–566; Brunner et al., "Dose and host characteristics," 399–406.

106. Carey et al., "Biotic factors in amphibian population declines," 153–208.

107. Robinson et al., "The development of a cane toad biological control," 1–35.

108. Pfennig et al., "Pathogens as a factor," 161–166; Bolker et al., "Disease as a selective force,"105–128.

109. Jancovich et al., "Transmission of the *Ambystoma tigrinum* virus," 159–163.

110. Brunner et al., "Transmission dynamics of the amphibian ranavirus," 87–95; Greer et al., "Testing a key assumption of host-pathogen theory," 1667–1673.

111. Blackwell et al., "Fungi: Eumycota."

112. Burrowes et al., "Potential causes," 141–154.

113. Whitfield, "Amphibian and reptile declines," 8352–8356.

114. Burrowes et al., "Potential causes," 141–154.

115. Daszak and Cunningham, "Extinction," 279.

116. Amphibians: Schloegel et al." The decline of the sharp-snouted day frog," 35–40; Berger, et al., "Chytridiomycosis causes amphibian mortality," 9031–9036; Birds: Smith et al., "Evidence for the role," 1349–1357; Mammals: MacPhee and Marx, "The 40,000-year plague," 169–217.

117. The following discussion follows James et al., "A molecular phylogeny," 860–871. See also Gleason et al., "The ecology of chytrids."

118. Ron, "Distribution of the amphibian pathogen *Batrachochytrium dendrobatidis*," 209–221. Ron reports Bd on all continents except Asia. See Watanabe, "Deadly frog fungus spreads to Japan," for a report from Asia.
119. Voyles et al., "Electrolyte depletion," 113–118.
120. Woodhams et al., "Resistance to chytridiomycosis," 409–417.
121. Harris et al., "Amphibian pathogen," 53–56; Lauer et al., "Common cutaneous bacteria," 630–640.
122. Johnson and Speare, "Survival of *Batrachochytrium dendrobatidis*," 922–925.
123. Lips et al., "Emerging infectious disease," 3165–3170.
124. Kirshtein et al., "Quantitative PCR detection," 11–15; Walker et al., "Environmental detection," 105–112.

Chapter 6

1. Richards et al., "Declines in populations," 66–77.
2. Trenerry et al., "Further evidence," 150–153.
3. Hero, "Where are Queensland's missing frogs?," 8–13.
4. Laurance et al., "Epidemic disease," 406–413.
5. Two sets of alternative interpretations appeared in one issue of *Conservation Biology* the following year: Hero and Gillespie, "Epidemic disease and amphibian declines," 1023–1025 and Alford and Richards, "Lack of evidence," 1026–1029. Laurance and his colleagues followed with a rebuttal: Laurance et al., "In defense of the epidemic disease hypothesis," 1030–1034.
6. Cunningham, "Disease and Pathology Working Group report," 3–4.
7. Nichols, "Tracking down," 101–104.
8. Berger et al., "Chytridiomycosis causes," 9031–9036.
9. Longcore et al., "*Batrachochytrium dendrobatidis*," 219–227.
10. Anderson, "A great leap forward," 4.
11. Halliday, "A declining amphibian conundrum," 418.
12. McDonald and Alford, "A review of declining frogs," 14–22.
13. Woodhams et al., "Emerging disease of amphibians cured," 6–7.
14. Woodhams and Alford, "Ecology of chytridiomycosis," 144–459.
15. Woodhams and Alford, "Ecology of chytridiomycosis," 144–459.
16. Woodhams et al., ""Population trends," 531–540.
17. Retallick et al., "Endemic infection," 1965–1971.
18. Schloegel et al., "The decline of the sharp-snouted day frog," 35–40.
19. Skerratt et al., "Spread of chytridiomycosis," 125.
20. Rowley and Alford, "Behaviour of Australian rainforest stream frogs," 1–9.
21. Hammerson, "Field surveys of amphibians."
22. Corn et al., "Acid precipitation studies."
23. Corn and Vertucci, "Descriptive risk assessment," 361–369.
24. Carey, "Hypothesis concerning the causes," 355–362; quote is from p. 360.
25. Blaustein et al., "UV repair and resistance," 1791–1795.
26. Corn, "Effects of ultraviolet radiation," 18–26.
27. Carey et al., "Amphibian declines: an immunological perspective," 459–472.
28. Muths et al., "Evidence for disease-related amphibian decline," 357–365.
29. Alexander and Eischeid, "Climate variability in regions," 930–942.
30. Carey et al., "Amphibian declines and environmental change," 903–913.
31. Carey et al., "Experimental exposures," 5–21.

32. Johnson and Speare, "Survival of *Batrachochytrium*," 922–925.
33. Carey et al., "Factors limiting the recovery," 222–240.
34. Carey et al., "Factors limiting the recovery," 222–240.
35. Carey et al., "Factors limiting the recovery," 222–240.
36. Cynthia Carey, personal communication to M. Crump, January 2008.
37. Crump, *In Search of the Golden Frog*, 149–165.
38. Crump, *In Search of the Golden Frog*, 149–165.
39. Crump et al., "Apparent decline," 413–420.
40. Pounds and Crump, "Amphibian declines and climate disturbance," 72–85.
41. Lips, "Decline of a tropical montane," 106–117.
42. Lips, "Decline of a tropical montane," 106–117.
43. Lips, "Decline of a tropical montane," 106.
44. Laurance et al., "Epidemic disease," 406–413.
45. Lips, "Mass mortality and population declines," 117–125.
46. Berger et al., "Chytridiomycosis causes," 9031–9036.
47. Lips, "Mass mortality and population declines," 117–125.
48. Lips et al., "Chytridiomycosis in wild frogs," 215–218.
49. Puschendorf, "*Atelopus varius* (harlequin frog)," 355.
50. Lips et al., "Ecological traits predicting," 1078–1088.
51. Pounds et al., "Tests of null models," 1307–1322.
52. Pounds et al., "Biological response to climate change," 611–615.
53. Pounds et al., "Biological response to climate change," 611–615.
54. Pounds et al., "Biological response to climate change," 614.
55. Berger et al., "Chytridiomycosis causes amphibian mortality," 9031–9036.
56. Pounds and Puschendorf, "Clouded futures," 107–109.
57. Woodhams et al., "Emerging disease of amphibians cured," 65–67.
58. Pounds et al., "Widespread amphibian extinctions," 161–167; see La Marca et al., "Catastrophic population declines," 190–201, for an overview of population declines in the genus *Atelopus*.
59. Pounds et al., "Widespread amphibian extinctions," 165.
60. Lips et al., "Riding the wave."
61. Puschendorf et al., "The amphibian chytrid fungus," 136–142.
62. Lips et al., "Riding the wave."
63. Lips et al., "Emerging infectious disease and the loss," 3165–3170.
64. Goodman, "To stem widespread extinction."

Chapter 7

1. Smith et al., "Evidence for the role of infectious disease," 1349–1357.
2. Barry, *The Great Influenza*.
3. Rachowitz and Briggs, "Quantifing the disease," 711–721, estimated the transmission rate of Bd among tadpoles of the mountain yellow-legged frog. They found some support for a density independent transmission function, but they did not have sufficient statistical power to distinguish between a density independent form of transmission and other similar forms of transmission. Some evidence for vertical transmission of Bd from adult to eggs is reported for *Eleutherodactylus coqui* by Burrowes et al., "Potential fitness cost," 51–57.
4. Corey and Waite, "Phylogenetic autocorrelation," 614–629. These authors report a concentration of endangered species and those showing enigmatic decline

in the superfamily Hyloidea (=Hyloides in Frost et al., The Amphibian Tree of Life, 1–370). This group includes several New World temperate and tropical families like toads, hylids, eleutherodactylids, and dendrobatids in which Bd has had a particularly negative effect. The concentration also reflects the fact that these families occur in what are to date, outside of Australia, the world's regions where Bd is most studied. Even with this caveat, Corey and Waite report that Bd is already found in 24 of 51 families of frogs and salamanders recognized by Frost and colleagues.

For a discussion of multihost pathogens see Woolhouse et al., "Population biology," 1109–1112.

5. De Castro and Bolker, "Mechanisms of disease-induced extinction," 117–126.
6. Lips et al., "Emerging infectious disease," 3165–3170.
7. Matthew Parris, personal communication to JPC, October 16, 2007. The results are in preparation for publication.
8. Mitchell et al., "Persistence of the emerging pathogen," 329–334.
9. Alexander and Eischeid, "Climate variability," 930–942. Both quotes are from p. 930.
10. Daszak et al., "Amphibian population declines at Savannah River," 3232–3237.
11. Pounds et al., "Widespread amphibian extinctions," 161–167.
12. Alford et al., "Global warming and amphibian losses," E4.
13. The "climate-linked epidemic hypothesis" posits a general link between climate change and emerging infectious disease in amphibians (see Pounds and Crump, "Amphibian declines and climate disturbance," 72–85). The "chytrid-thermal-optimum hypothesis" posits a specific mechanism linking climate change "in which daytime cooling (local or microscale) and night time warming accelerate disease [Bd] development" (Pounds et al., "Widespread amphibian extinctions," 165). They originally argued that temperature change during a single year was sufficient to initiate conditions favoring Bd emergence and spread but later revised this conclusion.
14. Pounds et al., "Global warming and amphibian losses," E5–E6.
15. Laurance, "Global warming and amphibian extinctions," 1–9.
16. Di Rosa et al., "The proximate cause," E4–E5.
17. Di Rosa et al., "The proximate cause," E4.
18. Woodhams et al., "Life-history trade-offs," 1627–1639.
19. Woodhams et al., "Life-history trade-offs," 1637.
20. Bosch et al., "Climate change and outbreaks," 253–260.
21. Bosch et al., "Climate change and outbreaks," 259.
22. Bosch et al., "Climate change and outbreaks," 259.
23. Bosch et al., "Climate change and outbreaks," 259.
24. Bosch et al., "Climate change and outbreaks," 259.
25. Ron, "Distribution of the amphibian pathogen *Batrachochytrium dendrobatidis*," 209–221.
26. Morehouse et al., "Multilocus sequence typing," 395–403.
27. Morehouse et al., "Multilocus sequence typing," 401.
28. Morehouse et al., "Multilocus sequence typing," 400.
29. Daszak et al., "Infectious disease and amphibian population declines," 141–150.
30. Retallick and Miera, "Strain differences in the amphibian chytrid," 201–207.
31. Berger et al., "Virulence of the amphibian chytrid fungus," 47–50.

32. Morgan et al., "Population genetics," 13845–13850.

33. One of us (JPC) discussed this result with Joyce Longcore, a leading expert in Bd biology. The Di Rosa team identified the Bd resistant stage using molecular primers that differed from those employed by most other investigators studying the amphibian chytrid. The possibility that Bd has a resistant stage is important enough that the results should be corroborated using standard techniques. Another expert commented to one of us (JPC) that the published figure illustrating the putative Bd resistant stage looks like a bacterial spore, and if samples were collected from animals also infected with Bd, that could account for any molecular positives. In general, replication is needed to confirm that Bd has a resistant stage.

34. Rosenblum et al., "Global gene expression." 17034–17039.

35. Joint Genome Institute, "Why sequence a frog-killing fungus?"

36. Broad Institute, *Batrachochytrium dendrobatidis* Database.

37. Woodhams et al., "Emerging disease of amphibians cured," 65–67.

38. Retallick and Miera, "Strain differences in the amphibian chytrid," 201–207.

39. Rowley and Alford, "Behaviour of Australian rainforest stream frogs," 1–9.

40. Rowley and Alford, "Behaviour of Australian rainforest stream frogs," 5. A similar conclusion is reached by Woodhams et al., "Life-history trade-offs," 1627–1639.

41. Burrowes et al., "Potential causes," 141–154; Bosch et al., "Climate change and outbreaks," 253–260; Rowley and Alford, "Behaviour of Australian rainforest stream frogs," 1–9.

42. See Lips et al., "Ecological traits," 1078–1088, for data from Central America; Kriger and Hero, "The chytrid fungus *Batrachochytrium dendrobatidis*," 781–788, for southeast Queensland data; Drew et al., "Analysis of climatic and geographic factors," 245–250, for 56 sites across Australia; and Buckley and Jetz, "Linking global turnover," 17836–17841. For amphibians worldwide see: Cooper et al. "Macroecology and extinction risk," 1–11; Bielby et al., "Predicting susceptibility," 82–90.

43. Ron, "Distribution of the amphibian pathogen," 209–221.

44. Ron, "Distribution of the amphibian pathogen," 215.

45. Andreone et al., "The challenge of conserving," e118.

46. This summary is drawn from Berger et al., "Chytridiomycosis causes amphibian mortality," 9031–9036; Rachowicz et al., "The novel and endemic pathogen hypotheses," 1441–1448; Skerratt et al., "Spread of chytridiomycosis," 125–134; Laurance, "Global warming and amphibian extinctions," 1–9.

47. Studies in Australia support the emergence of Bd in naïve populations, although there is debate regarding the timing of emergence and whether the pattern is wavelike, as in Central America. For a discussion, see Laurence et al., "Catastrophic declines of Australian rain forest frogs," 1–9; Alford and Richards, "Lack of evidence for epidemic disease," 1026–1029; Laurence et al., "In defence of the epidemic disease hypothesis," 1030–1034.

48. Morse, "Factors in the emergence," 7–15.

49. Weldon et al., "Origin of the amphibian chytrid fungus," 2100–2105.

50. Recent, unpublished analyses are not supporting the "Out of Africa" hypothesis for Bd's emergence and spread (Matthew Fisher and colleagues, personal communication to JPC, November, 2007).

51. Moran, "The statistical analysis," 291–298.

52. Greer, *The Impact of Disease on the Host-Population Dynamics in a Salamander-Virus System.*

53. At the local and regional scales, ATV can be both a novel and an endemic pathogen, as shown by Storfer et al. ("Phylogenetic concordance analysis," 1075–1083).
54. Laurance, "Global warming and amphibian extinctions," 7.
55. Skerratt et al., "Spread of chytridiomycosis," 125–134.
56. LaDeau et al., "West Nile virus emergence," 710–713.
57. Rahbek, "Disease ecology," 652–653.
58. Rahbek, "Disease ecology," 652.

Chapter 8

1. Roelants et al., "Global patterns of diversification," 104:887–892; McCallum, "Amphibian decline or extinction?" 483–491. See also Wake and Vredenberg, "Are we in the midst," 11466–11473.
2. Dean, "The preservation predicament," D1.
3. For a general discussion of this point in herpetology, see Wake, "Eye of newt," 1–11, and for biology in general, see Collins, "May you live in interesting times," 75–83, and Wake, "Integrative biology," 349–353.
4. The section on changes in how scientists conduct research is based largely on Collins et al., "Meeting the challenge of amphibian declines," 23–27. The argument that amphibian losses cannot be addressed with "conservation as usual" is drawn from conversations, drafts, and ultimately publication of a paper by ourselves and our colleagues: Mendelson III et al., "Confronting amphibian declines and extinctions," 48 + 13 pp. online supplemental material.
5. Kiesecker et al., "Complex causes," 681–683. See also Blaustein and Kiesecker, "Complexity in conservation," 597–608.
6. Atkins et al., *Revolutionizing Science and Engineering Through Cyberinfrastructure;* Ellisman, "Cyberinfrastructure and the future of collaborative work."
7. Stokes, *Pasteur's Quadrant.*
8. Marris, "Should conservation biologists push policies?" 442:13.
9. Oreskes, "Beyond the ivory tower," 1686; Oreskes, "The long consensus on climate change," A15; see also the Intergovernmental Panel on Climate Change (IPCC) reports at http://www.ipcc.ch/.
10. Mendelson et al., "Confronting amphibian declines," 48 + 13 pp. online supplemental material.
11. For a sample of the discussion surrounding this issue, see Hutchins and Wiese, "Beyond genetic and demographic management," 285–292; Forest et al., "Preserving the evolutionary potential," 757–760; Mooers, "The diversity of diversity," 717–718; Rodrigues et al., "Integrating phylogenetic diversity," 101–119.
12. Schlaepfer et al., "Challenges in evaluating the impact," 256–263.
13. Rabb, "The evolution of zoos," 237–246.
14. Two global, multidisciplinary collaborations with innovative conservation approaches are models for this plan: the Turtle Survival Alliance (http://www.turtlesurvival.org) and the global coral reef research (http://www.GEFCoral.org).
15. See, for example, Russello and Amato, "On the horns of a dilemma," 2405–2406.
16. Vitousek et al., "Human domination of Earth's ecosystems," 494–499; Levitus et al., "Anthropogenic warming," 267–270; Food and Agriculture Organization of the United Nations, *Global Forest Resources Assessment 2000,* www.fao.org/forestry/

site/fra2000report/en; Daszak et al., "Emerging infectious diseases of wildlife," 443–449.

17. Schipper et al., "The status of the world's," 228.
18. Havel, "Our moral footprint," A31.
19. Bellwood et al., "Confronting the coral reef crisis," 827–833.

Chapter 9

1. McCallum and Jones. "To lose both would look like carelessness." For additional background, as noted by McCallum and Jones, see Burgman, *Risks and Decisions for Conservation and Environmental Management;* Sakai et al., "The population biology of invasive species," 305–332; Owen and Pemberton, *Tasmanian Devil: A Unique and Threatened Animal.*

2. For a discussion of issues that need to be considered see Loyola et al., "Hung out to dry," e2120.

3. The Zoo Atlanta meeting was held December 17–19, 2004, and included James Collins, Peter Daszak, Ronald Gagliardo, Roberto Ibáñez, Dwight Lawson, Karen Lips, Joseph Mendelson, George Rabb, Kevin Wright, and Kevin Zippel. Conservation International meetings regularly included Donald Church, Collins, Claude Gascon, Lips, Mendelson, Robin Moore, Rabb, Simon Stuart, and Zippel.

4. Association of Zoos and Aquariums, http://www.aza.org/.

5. Andreone et al., "The challenge of conserving," e118.

6. Amphibian Research Centre, http://frogs.org.au/arc/conference.html.

7. For a database on amphibian reintroductions, see Declining Amphibian Populations Task Force, http://www.open.ac.uk/daptf/about/abou7.htm.

8. Marris, "Bagged and boxed," 394–395; Morell, "Into the wild," 742–743.

9. Ak, "The moral element," 397, raises these questions in a review of Jonathan Baron's *Against bioethics.* Minteer and Collins, "Ecological ethics," 1803–1812, raise similar questions focusing on environmental issues.

10. The U.S. Agency for International Development (USAID) sponsored the production of 400 posters (13×17-inch PVC sheets) and donated them to the Autoridad Nacional del Ambiente (ANAM), Panama's National Authority for the Environment, in October 2006. A small insert in a local Panamanian newspaper was planned to announce the event. The impact on personnel and visitors' behavior is unknown. Roberto Ibáñez, a research scientist affiliated with the Smithsonian Tropical Research Institute in Panama, reported to JPC that in the past another version was posted in a few places without much impact. Once frogs disappeared from El Copé, however, ANAM's personnel became eager about the poster. At present, the poster is intended only for Panama, not Costa Rica.

11. One Web site argued, "Freshwater biodiversity is highly threatened today—a fact that should be on the mind of every serious aquarist. The natural habitats of tropical freshwater fish are increasingly threatened by human activities, and while at times the hobby has been been [sic] at odds with conservation, the role of aquariasts in preserving species is growing in importance." And, in language that would work if "amphibians" were substituted for "fish": "Aquariasts are helping to maintain species (such as Cherry barbs and certain Killifish) that are essentially extinct in the wild. By keeping these species and populations viable, the fish-keeping community is protecting against extinction. When and if reintroduction to natural habitats becomes possible, it will be in part thanks to aquariasts. Time has effectively run out

for many species. Aquariasts can do their part in preventing the further extinction of some freshwater fish." (Mongabay.com, http://fish.mongabay.com/aquariast_role_conservation.htm.

12. Minteer and Collins, "Why we need," 332–337; Minteer and Collins, "Ecological ethics," 1803–1812; Minteer and Collins, 483–501, "From environmental to ecological ethics."

13. Mendelson, "Biodiversity: Confronting amphibian declines," 48.

14. Pounds et al., "Letters." For a response, see Mendelson, "Letters," 154–542.

15. National Heritage Trust (Australia) conference: Getting the jump! on amphibian disease.

16. Commonwealth of Australia, *Threat Abatement Plan.*

17. Commonwealth of Australia, *Threat Abatement Plan.*

18. The summary in this paragraph is from an e-mail message sent by Peter Daszak to JPC, June 9, 2008. For a discussion of protocols for managing the spread of Bd see Young et al., "Amphibian chytridiomycosis," 1–11.

19. The consequences of listing are as follows: "Essentially, it requires all 172 member countries of OIE to immediately (within 24 hrs) notify them of detection for the first time in the country or a part of the country not previously found affected. Following that, follow-up reports must be submitted at approximately monthly intervals until the matter is resolved or it is accepted as endemic. All countries are required to submit routine reports at six monthly intervals (used to be annual) on the occurrence of all the listed diseases in their territories and all this information is made public on the OIE website in the World Animal Health Information (WAHID) database. The listing also legitimises [sic] countries applying health certification requirements for imports subject to some conditions. The guidelines on this are in the specific disease chapters that will go into the Aquatic Code (very soon on the web version and later in the year in the printed version)." (From an e-mail sent by Barry Hill, Vice-chair of the OIE Aquatics Commission, to JPC and others, June 14, 2008.)

20. A copy of the Declaration and Action Plan is provided in the Supplemental Material for Mendelson, "Confronting amphibian declines," 48, and at the Global Amphibian Assessment Web site, http://www.globalamphibians.org/future_steps.htm.

21. Collins, "Transitioning," 1–3. Our summary is taken from the article. The ASG Web site can be accessed at http://www.amphibians.org.

22. Parts of the text follow Moore et al., The Amphibian Specialist Group: Past, present and future. In: *Threatened Amphibians of the World.*

23. Mendelson, "Confronting amphibian declines," 48.

24. After the Amphibian Action Plan Summit in 2005, Holly Dublin, Chair of IUCN's Species Survival Commission, asked Claude Gascon, Executive Vice-President at Conservation International, and one of us (JPC) if we would co-chair the ASG. We agreed, and since then we have worked on developing ASG activities and merging DAPTF and GASG.

25. Amphibian Ark, http://amphibianark.org/; Amphibian Ark 2008 Year of the Frog (YOTF) Campaign, http://www.2008yearofthefrog.org.

26. Modified from a presentation by A. Angulo in 2006, entitled, "Policy issues related to amphibian declines: A Latin American perspective," at a symposium sponsored by the Research and Analysis Network for Neotropical Amphibians, Annual Meeting of the American Society of Ichthyologists and Herpetologists, New Orleans, Louisiana, USA.

27. The complexity of this question was raised at the 2006 meeting of the Society for Conservation Biology—see Marris, "Should conservation biologists push policies?," 13.

Chapter 10

1. Revenga et al., "Prospects for monitoring," 397–413.
2. See Nickerson and Briggler, "Harvesting as a factor," 207–216, for hellbenders in the United States; for *Andrias* in China, see Xie et al., "Conservation needs," 265–276.
3. Beard et al., "Amphibian effects," 607–617. See Collins, "Distribution, habitats, and life history," 666–675 for the United States; for Australia, see Tyler et al., "How frogs and humans interact," 1–18.
4. See Collins, "Distribution, habitats, and life history," 666–675 for the United States; for Australia, see Tyler et al., "How frogs and humans interact," 1–18.
5. Meegaskumbura et al., "Sri Lanka: An amphibian hot spot," 379.
6. Gascon et al., "Matrix habitat and species persistence," 223–229; Groombridge and Jenkins, *World Atlas of Biodiversity;* Gallant et al., "Global rates of habitat loss," 967–979.
7. Revenga et al., "Prospects for monitoring," 400.
8. Davidson, "Declining downward," 1892–1902; Sparling et al., "Pesticides and amphibian population declines," 1591–1595.
9. Hayes et al., "Hermaphroditic, demasculinized frogs," 5476–5480; Hayes et al., "Feminization of male frogs," 895–896.
10. McCoy et al., "Agriculture alters gonadal form and function," 1526–1532; McDaniel, et al. "Potential endocrine disruption," 230–242.
11. Research by Rohr and colleagues begins to close this gap: Rohr et al., "Agrochemicals increase," 1235–1239; Rohr et al., "Understanding the net effects," 1743–1753.
12. Ranvestel, et al., "Neotropical tadpoles," 274–285; Whiles et al., "The effects of amphibian population declines," 27–34; Colón-Gaud et al., "Allochthonous litter inputs," 301–312; Connelly et al., "Changes in stream," 1262–1276; Colón-Gaud et al., "Assessing ecological responses," 331–343.
13. See the European Environment Agency Report by Harremoës et al., *Late Lessons from Early Warnings,* especially Chapter 12 on precautionary principles applied to toxins.
14. Intergovernmental Panel on Climate Change, http://www.ipcc.ch/.
15. Schloegel et al., "The decline of the sharp-snouted day frog," 35–40.
16. Raup, *Extinction,* 181.
17. Martin and Klein, "*Quarternary Extinctions.*"
18. MacPhee and Marx, "The 40,000-year plague," 169–217. Lyons and colleagues ("Was a 'hyperdisease' responsible," 859–868) at the University of New Mexico and the Field Museum of Natural History analyzed the spread of West Nile virus in birds. Using this model system, they concluded that it offered no support for disease as a cause of the late Pleistocene megafaunal extinction.
19. MacPhee, "Lightning strikes twice."
20. Raup, *Extinction,* 126.
21. For amphibians see: Buckley and Jetz, "Linking global turnover,"17836–17841; Cooper et al., "Macroecology and extinction risk," 211–221. Purvis et al.

("Predicting extinction risk," 1947–1952) at Imperial College analyzed data for contemporary carnivores and primates and concluded that, among other traits, small geographical range size is significantly related to high extinction risk in declining species.

22. See Johnson, "Species extinction," 272–274, for a discussion of the general relationship expected between range size, abundance, and likelihood of extinction.

23. Vredenburg et al., "Concordant molecular and phenotypic data," 361–374.

24. Raup, *Extinction*, 183.

25. Raup, *Extinction*, 184.

26. Retallick et al., "Endemic infection of the amphibian chytrid fungus," e351.

27. The research of Hamish McCallum and colleagues on the Tasmanian devil affords another look at how evolutionary and ecological processes are closely tied (See McCallum et al., "Distribution and impacts," 318–325; Jones et al., "Life-history change," 10023–10027; McCallum, "Tasmanian devil," 631–637). Tasmanian devil facial tumour disease (DFTD) emerged in 1996. DFTD is an infectious cancer probably spread among adults by biting in the mating season—the cancer cells are the infective agent—and animals die within about 6 months of infection. The disease persists at low adult densities suggesting transmission is density independent, which we learned can place a species at risk of extinction. Two things are notable about the remaining devil populations. First, historically, females produced multiple litters in a lifetime. They now produce an average of only one as natural selection has favored adults that breed earlier and earlier. A life history character has been altered by a pathogen within a decade. Second, models suggest that extinction of the Tasmanian devil in the wild could occur in 20 to 25 years. Just as for tropical amphibians, the loss of this predator would have consequences for the entire ecosystem.

28. In contrast, ranaviruses appear to be within the range of typical stresses for populations harboring these pathogens, which means that populations may be low for a while, but they rebound in the course of host-pathogen dynamics.

29. Harremoës et al., *Late Lessons from Early Warnings.*

30. Rovito, S. M., G. Parra-Olea, C. R. Vásquez-Alamazán, T. J. Pappenfuss, and D. B. Wake. 2009. Dramatic declines in neotropical salamander populations are an important part of the global amphibian crisis. *Proceedings of the National Academy of Sciences (USA)* 106:3231–3236; Fisher, M. C., J. Bosch, Z. Yin, et al. 2009. Proteomic and phenotypic profiling of the amphibian pathogen *Batrachochytrium dendrobatidis* shows that genotype is linked to virulence. *Molecular Ecology* 18:415–429.

References

Achard, F., H. D. Eva, H-J. Stibig, et al. 2002. Determination of deforestation rates of the world's humid tropical forests. *Science* 297:999–1002.

Adler, K., ed. 1992. *Herpetology: Current Research on the Biology of Amphibians and Reptiles.* St. Louis: Society for the Study of Amphibians and Reptiles.

Ak, P. 2006. The moral element. *Nature* 443:397.

Alcala, A. C. 2000. The Negros cave frog is critically endangered. *Froglog* (39):1.

Alexander, M. A., and J. K. Eischeid. 2001. Climate variability in regions of amphibian declines. *Conservation Biology* 15:930–942.

Alford, R. A., and S. J. Richards. 1997. Lack of evidence for epidemic disease as an agent in the catastrophic decline of Australian rain forest frogs. *Conservation Biology* 11:1026–1029.

Alford, R. A., and S. J. Richards. 1999. Global amphibian declines: A problem in applied ecology. *Annual Review of Ecology and Systematics* 30:133–165.

Alford, R. A., K. S. Bradfield, and S. J. Richards. 2007. Global warming and amphibian losses. *Nature* 447, E3-E4.

Alford, R. A., P. M. Dixon, and J. H. K. Pechmann. 2001. Global amphibian population declines. *Nature* 414:449–450.

Alsos, I. G., P. B. Eidesen, D. Ehrlich, I. Skrede, K. Westergaard, G. H. Jacobsen, J. Y. Landvik, P. Taberlet, and C. Brochmann. 2007. Frequent long-distance plant colonization in the changing Arctic. *Science* 316:1606–1609.

Amphibian Ark. 2008. Web site. http://amphibianark.org/.

Amphibian Ark 2008 Year of the Frog (YOTF) Campaign. 2008. http://www.2008yearofthefrog.org.

Amphibian Research and Monitoring Initiative (ARMI). 2008. Web site. http://armi.usgs.gov.

Amphibian Research Centre Husbandry and Hygiene Conference. 2004. Captivity, Reintroduction and Disease Control Technologies for Amphibians. http://frogs.rg.au/arc/conference.html.

Amphibian Specialist Group. 2008. Web site. http://www.amphibians.org.

Amphibians. 2003. In *Grzimek's Animal Life Encyclopedia,* 2nd edition, volume 6, edited by M. Hutchins, W. E. Duellman, and N. Schlager. Farmington Hills, Michigan: Gale Group.

AmphibiaWeb: Information on Amphibian Biology and Conservation. 2008. Electronic database. Biodiversity Sciences Technology Group (BSCIT), Berkeley Natural History Museums, Berkeley, California. http://amphibiaweb.org/.

Anderson, I. 1998. A great leap forward. *New Scientist* 158(2140):4–5.

Andreone, F., J. E. Cadle, N. Cox, F. Glaw, R. A. Nussbaum, C. J. Raxworthy, S. N. Stuart, D. Vallan. 2005. Species review of extinction risk in Madagascar: Conclusions from the Global Amphibian Assessment. *Conservation Biology* 19:1790–1802.

Andreone, F., A. I. Carpenter, N. Cox, et al. 2008. The challenge of conserving amphibian megadiversity in Madagascar. *PLoS Biology* 6(5):e118.

Araújo, M. B., W. Thuiller, and R. G. Pearson. 2006. Climate warming and the decline of amphibians and reptiles in Europe. *Journal of Biogeography* 33:1712–1728.

Arnold, N., and D. Ovenden. 2002. *Reptiles and Amphibians: Britain and Europe.* London: HarperCollins.

Ash, A. N., and R. C. Bruce. 1994. Impacts of timber harvesting on salamanders. *Conservation Biology* 8:300–301.

Association of Zoos and Aquariums. 2008. Web site. http://www.aza.org.

Atkins, D. E., K. K. Droegemeier, S. I. Feldman, et al.2003. *Revolutionizing Science and Engineering Through Cyberinfrastructure: Report of the National Science Foundation Blue-Ribbon Advisory Committee on Cyberinfrastructure.* pp. 1–84. http://www.communitytechnology.org/nsf_ci_report/.

Avise, J. C., D. Walker, and G. C. Johns. 1998. Speciation durations and Pleistocene effects on vertebrate phylogeography. *Proceedings Royal Society London B: Biological Sciences* 265:1707–1712.

Bancroft, B. A., N. J. Baker, and A. R. Blaustein. 2007. Effects of UVB radiation on marine and freshwater organisms: A synthesis through meta-analysis. *Ecology Letters* 10:332–345.

Bancroft, B. A., N. J. Baker, and A. R. Blaustein. 2008. A meta-analysis of the effects of ultraviolet B radiation and other stressors on survival in amphibians. *Conservation Biology* 22:987–996.

Banks, B., T. J. C. Beebee, and A. S. Cooke. 1994. Conservation of the natterjack toad *Bufo calamita* in Britain over the period 1970–1990 in relation to site protection and other factors. *Biological Conservation* 67:111–118.

Barbassa, J. 2006. Habitat or homes? Frog jumps into debate. Species made famous by Mark Twain is being squeezed out. *Associated Press, MSNBC News.* http://www.msnbc.msn.com/id/12515045/from/ET/.

Barry, J. M. 2004. *The Great Influenza: The Epic Story of the Deadliest Plague in History.* London: Penguin Books.

Baskin, Y. 1994. Ecologists dare to ask: How much does diversity matter? *Science* 264:202–203.

Beard, K. H., A. K. Eschtruth, K. A. Vogt, D. J. Vogt, and F. N. Scatena. 2003. Amphibian effects on its prey and ecosystem: Evidence from two spatial scales. *Journal of Tropical Ecology* 19:607–617.

Beebee, T. J. C. 1995. Amphibian breeding and climate. *Nature* 374:219–220.

Beebee, T. J. C. 1996. *Ecology and Conservation of Amphibians.* London: Chapman & Hall.

Beebee, T. J. C., and R. A. Griffiths. 2005. The amphibian decline crisis: A watershed for conservation biology? *Biological Conservation* 125:271–285.

Bell, B. D. 1999. Frog declines in New Zealand. Abstract. *Froglog* (35):3.

Bellwood, D. R., T. P. Hughes, C. Folke, and M. Nyström. 2004. Confronting the coral reef crisis. *Nature* 429:827–833.

Berger, L., G. Marantelli, L. F. Skerratt, and R. Speare. 2005 Virulence of the amphibian chytrid fungus *Batrachochytrium dendrobatidis* varies with the strain. *Diseases of Aquatic Organisms* 68:47–50.

Berger, L., R. Speare, P. Daszak, et al. 1998. Chytridiomycosis causes amphibian mortality associated with population declines in the rain forests of Australia and Central America. *Proceedings of the National Academy of Sciences (USA)* 95:9031–9036.

Berger, L., R. Speare, and A. D. Hyatt. 2000. Chytrid fungi and amphibian declines: Overview, implications and future directions. In *Declines and Disappearances of Australian Frogs,* edited by A. Campbell. Canberra: Environment Australia. pp. 21–31.

Bernardo, J., and J. Spotila. 2006. Physiological constraints on organismal response to global warming: Mechanistic insights from clinally varying populations and implications for assessing endangerment. *Biology Letters* 2:135–139.

Berrill, M., S. Bertram, and B. Pauli. 1997. Effects of pesticides on amphibian embryos and larvae. *Herpetological Conservation* 1:233–245.

Bielby, J., N. Cooper, A. A. Cunningham, T. W. J. Garner, and A. Purvis. 2008. Predicting susceptibility to future declines in the world's frogs. *Conservation Letters* 1:82–90.

Bishop, C. A. 1992. *Declines in Canadian Amphibian Populations: Designing a National Monitoring Strategy.* Proceedings of a Workshop Sponsored by the Canadian Wildlife Service (Ontario Region) and the Metropolitan Toronto Zoo, October 5–6, 1991. Ottawa: Canadian Wildlife Service.

Blackwell, M., R. Vilgalys, T. Y. James, and J. W. Taylor. 2007, July 13. Fungi: Eumycota. Mushrooms, sac fungi, yeast, molds, rusts, smuts, etc. *The Tree of Life Web Project.* http://tolweb.org/Fungi/2377/2007.07.13.

Blair, A. P. 1951. Note on the herpetology of the Elk Mountains, Colorado. *Copeia* 1951:239–240.

Blaustein, A. R. 1994. Chicken Little or Nero's fiddle? A perspective on declining amphibian populations. *Herpetologica* 50:85–97.

Blaustein, A. R. 2004. UV-B radiation. Posted March 23, 2004, on AmphibiaWeb, Berkeley, California. http://www.amphibiaweb.org/declines/UV-B.html.

Blaustein, A. R., and L. K. Belden. 2005. Ultraviolet radiation. In *Amphibian Declines: The Conservation Status of United States Species,* edited by M. Lannoo. Berkeley: University of California Press. pp. 87–88.

Blaustein, A. R., and P. J. Johnson. 2003. Explaining frog deformities. *Scientific American* 288:60–65.

Blaustein, A. R., and J. M. Kiesecker. 2002. Complexity in conservation: Lessons from the global decline of amphibian populations. *Ecology Letters* 5:597–608.

Blaustein, A. R., and D. B. Wake. 1990. Declining amphibian populations: A global phenomenon? *Trends in Ecology and Evolution* 5:203–204.

Blaustein, A. R., and D. B. Wake. 1995. The puzzle of declining amphibian populations. *Scientific American* 272:56–61.

Blaustein, A. R., L. K. Belden, D. H. Olson, D. L. Green, T. L. Root, and J. M. Kiesecker. 2001. Amphibian breeding and climate change. *Conservation Biology* 15:1804–1809.

Blaustein, A. R., P. D. Hoffman, D. G. Hokit, J. M. Kiesecker, S. C. Walls, and J. B. Hays. 1994. UV repair and resistance to solar UV-B in amphibian eggs: A link to population declines? *Proceedings of the National Academy of Sciences (USA)* 91:1791–1795.

Blaustein, A. R., B. Han, B. Fasy, et al. 2004. Variable breeding phenology affects the exposure of amphibian embryos to ultraviolet radiation and optical characteristics of natural waters protect amphibians from UV-B in the U.S. Pacific Northwest: A comment. *Ecology* 85:1747–1754.

Bolker, B. M., F. de Castro, A. Storfer, S. Mech, E. Harvey, and J. P. Collins. 2008. Disease as a selective force precluding widespread cannibalism: A case study of an iridovirus of tiger salamanders, *Ambystoma tigrinum. Evolutionary Ecology Research* 10:105–128.

Boone, M., and C. Bridges. 2003. A hierarchical approach in studying the effects of an insecticide on amphibians. In Linder, G. L., S. K. Krest, and D. W. Sparling, eds. *Amphibian Decline: An Integrated Analysis of Multiple Stressor Effects.* Pensacola, Florida: Society of Environmental Toxicology and Chemistry (SETAC). pp. 265–270.

Boone, M. D., and C. M. Bridges 2003. Effects of pesticides on amphibian populations. In *Amphibian Conservation,* edited by R. D. Semlitsch. Washington, D. C.: Smithsonian Books. pp. 152–167.

Boone, M. D., R. D. Semlitsch, J. F. Fairchild, and B. B. Rothermel. 2004. Effects of an insecticide on amphibians in large-scale experimental ponds. *Ecological Applications* 14:685–691.

Boone, M. D., and S. M. James. 2005. Aquatic and terrestrial mesocosms in amphibian ecotoxicology. *Applied Herpetology* 2:231–257.

Borkin, L. J. 1999. Distribution of amphibians in North Africa, Europe, Western Asia, and the former Soviet Union. In *Patterns of Distribution of Amphibians: A Global Perspective,* edited by W. E. Duellman. Baltimore: Johns Hopkins University Press. pp. 329–420.

Bosch, J., L. M. Carrascal, L. Durán, S. Walker, and M. C. Fisher. 2006. Climate change and outbreaks of amphibian chytridiomycosis in a montane area of Central Spain: Is there a link? *Proceedings of the Royal Society B* 274:253–260.

Bosch, J., I. Martínez-Solano, and M. García-París. 2001. Evidence of a chytrid fungus infection involved in the decline of the common midwife toad (*Alytes obstetricans*) in protected areas of central Spain. *Biological Conservation* 97:331–337.

Bradford, D. F. 2005. Factors implicated in amphibian population declines. In *Amphibian Declines: The Conservation Status of United States Species,* edited by M. Lannoo. Berkeley: University of California Press. pp. 915–925.

Bridges, C. M., and R. D. Semlitsch. 2001. Genetic variation in insecticide tolerance in a population of southern leopard frogs (*Rana sphenocephala*): Implications for amphibian conservation. *Copeia* (1):7–13.

Bridges, C. M., and R. D. Semlitsch. 2005. Variation in pesticide tolerance. In *Amphibian Declines: The Conservation Status of United States Species,* edited by M. Lannoo. Berkeley: University of California Press. pp. 93–95.

Briggs, C. J., V. T. Vredenburg, R. A. Knapp, and L. J. Rachowicz. 2005. Investigating the population-level effects of chytridiomycosis: An emerging infectious disease of amphibians. *Ecology* 86:3149–3159.

Broad Institute. 2008. Batrachochytrium dendrobatidis Database. http://www.broad.mit.edu/annotation/genome/batrachochytrium_dendrobatidis.

Broomhall, S. D., W. S. Osborne, and R. B Cunningham. 2000. Comparative effects of ambient ultraviolet-B radiation on two sympatric species of Australian frogs. *Conservation Biology* 14:420–427.

Brown, J. H., and A. Kodric-Brown. 1977. Turnover rates in insular biogeography: Effect of immigration on extinction. *Ecology* 58:445–449.

Bruce, K. J., and J. L. Christiansen. 1976. The food and food habits of Blanchard's cricket frog, *Acris crepitans blanchardi* (Amphibia, Anura, Hylidae), in Iowa. *Journal of Herpetology* 10:63–74.

Brunner, J. L., K. Richards, and J. P. Collins. 2005. Dose and host characteristics influence virulence of ranavirus infections. *Oecologia* 144:399–406.

Brunner, J. L., D. M. Schock, and J. P. Collins. 2007. Transmission dynamics of the amphibian ranavirus Ambystoma tigrinum virus. *Diseases of Aquatic Organisms:* 77:87–95.

Brunner, J. L., D. M. Schock, E. W. Davidson, and J. P. Collins. 2004. Intraspecific reservoirs: Complex life history and the persistence of a lethal ranavirus. *Ecology:* 85:560–566.

Bryan, E. H., Jr. 1932. Frogs in Hawaii. *Mid-Pacific Magazine* 43:61–64.

Buckley, L. B., and W. Jetz. 2008. Linking global turnover of species and environments. *Proceedings of the National Academy of Sciences (USA)* 105:17836–17841.

Burgman, M. A. 2005. *Risks and Decisions for Conservation and Environmental Management.* New York: Cambridge University Press.

Burkhart, J. G., J. R. Bidwell, D. J. Fort, and S. R. Sheffield. 2003. Chemical stressors. In *Amphibian Decline: An Integrated Analysis of Multiple Stressor Effects,* edited by G. L. Linder, S. K. Krest, and D. W. Sparling. Pensacola, Florida: Society of Environmental Toxicology and Chemistry (SETAC). pp. 111–128.

Burrowes, P. A., R. L. Joglar, and D. E. Green. 2004. Potential causes for amphibian declines in Puerto Rico. *Herpetologica* 60:141–154.

Burrowes, P. A., A. V. Longo, and C. A. Rodríguez. 2008. Potential fitness cost of *Batrachochytrium dendrobatidis* in *Eleutherodactylus coqui*, and comments on environment-related risk of infection. *Herpetotropicos* 4:51–57.

Bustamante, M. R., S. R. Ron, and L. A. Coloma. 2005. Cambios en la diversidad en siete comunidades de anuros en los Andes de Ecuador. *Biotropica* 37:180–189.

Carey, C. 1993. Hypothesis concerning the causes of the disappearance of boreal toads from the mountains of Colorado. *Conservation Biology* 7:355–362.

Carey, C., and M. A. Alexander. 2003. Climate change and amphibian declines: Is there a link? *Diversity and Distributions* 9:111–121.

Carey, C., D. F. Bradford, J. L. Brunner, et al. 2003. Biotic factors in amphibian population declines. In *Amphibian Decline: An Integrated Analysis of Multiple Stressor Effects,* edited by G. L. Linder, S. K. Krest, and D. W. Sparling. Pensacola, Florida: Society of Environmental Toxicology and Chemistry (SETAC). pp. 153–208.

Carey, C., J. E. Bruzgul, L. J. Livo, et al. 2006. Experimental exposures of boreal toads (*Bufo boreas*) to a pathogenic chytrid fungus (*Batrachochytrium dendrobatidis*). *EcoHealth* 3:5–21.

Carey, C., N. Cohen, and L. Rollins-Smith. 1999. Amphibian declines: An immunological perspective. *Developmental and Comparative Immunology* 23:459–472.

Carey, C., P. S. Corn, M. S. Jones, L. J. Livo, E. Muths, and C. W. Loeffler. 2005. Factors limiting the recovery of boreal toads (*Bufo b. boreas*). In *Amphibian Declines: The Conservation Status of United States Species,* edited by M. Lannoo. Berkeley: University of California Press. pp. 222–240.

Carey, C., W. R. Heyer, J. Wilkinson, et al. 2001. Amphibian declines and environmental change: Use of remote-sensing data to identify environmental correlates. *Conservation Biology* 15:903–913.

Carey, C., A. P. Pessier, and A. D. Peace. 2003. Pathogens, infectious disease, and immune defenses. In *Amphibian Conservation,* edited by R. D. Semlitsch. Washington, D. C.: Smithsonian Books. pp. 127–136.

Carson, R. 1962. *Silent Spring.* New York: Houghton-Mifflin.

Caudata.org Newt and Salamander Forum. 2006. Pet dealers target newfound newt. *JPN Press* (*Yomiuri Shimbun,* Tokyo), April 24, 2006. http://www.caudata.org/forum/messages/8/62694.html?1146003064.

Chang, K. 2006. Scientists discover a Salamander; now profiteers have found it too. *The New York Times,* April 25, 2006. Published online at http://www.nytimes.com/2006/04/25/science/25find.html?_r=1&oref=slogin.

Chinchar, V. G. 2002. Ranaviruses (family Iridoviridae): Emerging cold-blooded killers. *Archives of Virology* 147:447–470.

Collins, J. P. 1981. Distribution, habitats, and life history variation in the tiger salamander, *Ambystoma tigrinum,* in east-central and southeast Arizona. *Copeia* (3):666–675.

Collins, J. P. 2002. May you live in interesting times: Using multidisciplinary and interdisciplinary programs to cope with change in the life sciences. *BioScience* 52:75–83.

Collins, J. P. 2004. Where have all the frogs gone? *Natural History* 113:44–49.

Collins, J. P. 2006. Transitioning the Declining Amphibian Populations Task Force. *Froglog* (75):1–3.

Collins, J. P., and T. Halliday. 2005. Forecasting changes in amphibian biodiversity: Aiming at a moving target. *Philosophical Transactions B, Royal Society, London* 360:309–314.

Collins, J. P., and A. Storfer. 2003. Global amphibian declines: Sorting the hypotheses. *Diversity and Distributions* 9:89–98.

Collins, J. P., and H. M. Wilbur. 1979. *Breeding Habits and Habitats of the Amphibians of the E. S. George Reserve, Michigan, with Notes on the Local Distribution of Fishes.* Occasional Papers of the Museum of Zoology, University of Michigan 686:1–34.

Collins, J. P., J. L. Brunner, V. Miera, M. J. Parris, D. M. Schock, and A. Storfer. 2003. Ecology and evolution of infectious disease. In *Amphibian Conservation,* edited by R. D. Semlitsch. Washington, D. C.: Smithsonian Books. pp. 137–151.

Collins, J. P., N. Cohen, E. W. Davidson, J. E. Longcore, and A. Storfer. 2005. Meeting the challenge of amphibian declines with an interdisciplinary research program. In *Amphibian Declines: The Conservation Status of United States Species,* edited by M. Lannoo. Berkeley: University of California Press. pp. 23–27.

Collins, J. P., T. R. Jones, and H. J. Berna. 1988. Conserving genetically-distinctive populations: The case of the Huachuca tiger salamander (*Ambystoma tigrinum stebbinsi* Lowe). In *Management of Amphibians, Reptiles, and Small Mammals in North America,* edited by R. C. Szaro, K. C. Severson, and D. R. Patton. USDA

Forest Service GTR-RM-166. Fort Collins, Colorado: Rocky Mountain Forest and Range Experiment Station. pp. 45–53.

Colón-Gaud, C., S. Peterson, M. R. Whiles, S. S. Kilham, K. R. Lips, and C. M. Pringle. 2008. Allochthonous litter inputs, organic matter standing stocks, and organic seston dynamics in upland Panamanian streams: potential effects of larval amphibians on organic matter dynamics. *Hydrobiologia* 603:301–312.

Colón-Gaud, C., M. R. Whiles, S. S. Kilham, K. R. Lips, C. M. Pringle, S. Connelly, and S. D. Peterson. 2009. Assessing ecological responses to catastrophic amphibian declines: Patterns of macroinvertebrate production and food web structure in upland Panamanian streams. *Limnology and Oceanography* 54: 331–343.

Commonwealth of Australia. 2006. *Threat Abatement Plan: Infection of Amphibians with Chytrid Fungus Resulting in Chytridiomycosis.* Canberra: Department of the Environment and Heritage. http://www.deh.gov.au/biodiversity/threatened/publications/tap/chytrid/pubs/chytrid-report.pdf.

Conant, R. 1958. *A Field Guide to Reptiles and Amphibians of the United States and Canada East of the 100th Meridian.* Boston: Houghton Mifflin.

Connelly, S., C. M. Pringle, R. J. Bixby, R. Brenes, M. R. Whiles, K. R. Lips, S. Kilham, and A. D. Huryn. 2008. Changes in stream primary producer communities resulting from large-scale catastrophic amphibian declines: Can small-scale experiments predict effects of tadpole loss? *Ecosystems* 11:1262–1276.

Cooper, N., J. Bielby, G. H. Thomas, and A. Purvis. 2008. Macroecology and extinction risk correlates of frogs. *Global Ecology and Biogeography* 17:211–221.

Coral Reef: Targeted Research & Capacity Building for Management. 2008. Web site. htt://www.GEFCoral.org.

Corey, S. J., and T. A. Waite. 2008. Phylogenetic autocorrelation of extinction threat in globally imperiled amphibians. *Diversity and Distributions* 14:614–629.

Corn, P. S. 1998. Effects of ultraviolet radiation on boreal toads in Colorado. *Ecological Applications* 8:18–26.

Corn, P. S. 2005. Climate change and amphibians. *Animal Biodiversity and Conservation* 28:59–67.

Corn, P. S., and J. C. Fogleman. 1984. Extinction of montane populations of the northern leopard frog (*Rana pipiens*) in Colorado. *Journal of Herpetology* 18:147–152.

Corn, P. S., and F. A. Vertucci. 1992. Descriptive risk assessment of the effects of acidic deposition on Rocky Mountain amphibians. *Journal of Herpetology* 26:361–369.

Corn, P. S., M. J. Adams, W. A. Battaglin, et al. 2005. *Amphibian Research and Monitoring Initiative: Concepts and Implementation.* USGS Scientific Investigations Report 2005–5015.

Corn, P. S., W. Stolzenburg, and R. B. Bury. 1989. Acid precipitation studies in Colorado and Wyoming: Interim report of surveys of montane amphibians and water chemistry. Report no. 26. *U. S. Fish and Wildlife Service, Research and Development, Biological Report 80* (40.26).

Corser, J. D. 2001. Decline of disjunct green salamander (*Aneides aeneus*) populations in the southern Appalachians. *Biological Conservation* 97:119–126.

Courchamp, F., E. Angulo, P. Rivalan, et al. 2006. Rarity value and species extinction: The anthropogenic Allee Effect. *PLoS Biology* 4:2405–2410.

Crump, M. L. 1974. *Reproductive Strategies in a Tropical Anuran Community.* University of Kansas Museum of Natural History Miscellaneous Publications No. 61., Lawrence: University of Kansas.

Crump, M. L. 2000. *In Search of the Golden Frog.* Chicago: University of Chicago Press.
Crump, M. L. 2003. Conservation of amphibians in the New World tropics. In *Amphibian Conservation,* edited by R. D. Semlitsch. Washington, D. C.: Smithsonian Books. pp. 53–69.
Crump, M. L. 2005. Why are some species in decline but others not? In *Amphibian Declines: The Conservation Status of United States Species,* edited by M. Lannoo. Berkeley: University of California Press. pp. 7–9.
Crump, M. L., F. R. Hensley, and K. L. Clark. 1992. Apparent decline of the golden toad: Underground or extinct? *Copeia* 1992:413–420.
Cullen, B. R., and L. Owens. 2002. Experimental challenge and clinical cases of Bohle iridovirus (BIV) in native Australian anurans. *Diseases of Aquatic Organisms* 49:83–92.
Cunningham, A. A. 1998. Disease and Pathology Working Group report: A breakthrough in the hunt for a cause of amphibian declines. *Froglog* (30):3–4.
Cunningham, A. A., P. Daszak, and J. P. Rodriguez. 2003. Pathogen pollution: Defining a parasitological threat to biodiversity conservation. *Journal of Parasitology* 89(suppl.):S78–S83.
Cunningham, A. A., T. E. S. Langton, P. M. Bennett, et al. 1996. Pathological and microbiological findings from incidents of unusual mortality of the common frog (*Rana temporaria*). *Philosophical Transactions of the Royal Society of London B* 351:1539–1557.
Daigle, C. 1997. Distribution and abundance of the chorus frog *Pseudacris triseriata* in Québec. In *Amphibians in Decline: Canadian Studies of a Global Problem,* edited by D. M. Green. *Herpetological Conservation* 1. St. Louis: Society for the Study of Amphibians and Reptiles. pp. 73–77.
Daly, G. L., and F. Wania. 2005. Organic contaminants in mountains. *Environmental Science and Technology* 39:385–398.
Daszak, P., L. Berger, A. A. Cunningham, A. D. Hyatt, D. E. Green, and R. Speare. 1999. Emerging infectious diseases and amphibian population declines. *Emerging Infectious Diseases* 5:735–748.
Daszak, P., and A. A. Cunningham. 1999. Extinction by infection. *Trends in Ecology and Evolution* 14:279.
Daszak, P., A. A. Cunningham, and A. D. Hyatt. 2000. Emerging infectious diseases of wildlife: Threats to biodiversity and human health. *Science* 287:443–449.
Daszak, P., A. A. Cunningham, and A. D. Hyatt. 2003. Infectious disease and amphibian population declines. *Diversity and Distributions* 9:141–150.
Daszak, P., D. E. Scott, A. M. Kilpatrick, C. Faggioni, J. W. Gibbons, and D. Porter. 2005. Amphibian population declines at Savannah River site are linked to climate, not chytridiomycosis. *Ecology* 86:3232–3237.
Davidson, C. 2004. Declining downwind: Amphibian population declines in California and historical pesticide use. *Ecological Applications* 14:1892–1902.
Davidson, C., and R. A. Knapp. 2007. Multiple stressors and amphibian declines: Dual impacts of pesticides and fish on yellow-legged frogs. *Ecological Applications* 17:587–597.
Davidson, C., H. B. Shaffer, and M. R. Jennings. 2001. Declines of the California red-legged frog: Climate, UV-B, habitat, and pesticides hypotheses. *Ecological Applications* 11:464–479.
Davidson, C., H. B. Shaffer, and M. R. Jennings. 2002. Spatial tests of the pesticide drift, habitat destruction, UV-B and climate change hypotheses for California amphibian declines. *Conservation Biology* 16:1588–1601.

Dean, C. 2008. The preservation predicament. *New York Times*, January 29, 2008: D1, D4.

De Castro, F., and B. Bolker. 2005. Mechanisms of disease-induced extinction. *Ecology Letters* 8:117–126.

Declining Amphibian Populations Task Force (DAPTF). 2008. Amphibian reintroduction database. http://www.open.ac.uk/daptf/about/abou7.htm.

Delis, P. R., H. R. Mushinsky, and E. D. McCoy. 1996. Decline of some west-central Florida anuran populations in response to habitat degradation. *Biodiversity and Conservation* 5:1579–1595.

Demastes, J. W., J. M. Eastman, J. S. East, and C. Spolsky. 2007. Phylogeography of the blue-spotted salamander, *Ambystoma laterale* (Caudata: Ambystomatidae). *American Midland Naturalist* 157:149–161.

Dickerson, M. C. 1906. *The Frog Book.* New York: Doubleday, Page and Company.

Di Rosa, I., F. Simoncelli, A. Fagotti, and R. Pascolini. 2007. The proximate cause of frog declines. *Nature* 447:E4–E5. doi:10.1038/nature05941.

Dodd, C. K., Jr. 2003. *Monitoring Amphibians in Great Smoky Mountains National Park.* USGS Circular 1258. Washington, D.C.: U.S. Department of the Interior, U.S. Geological Survey.

Dodd, C. K., Jr. 2005. Population manipulations. In *Amphibian Declines: The Conservation Status of United States Species,* edited by M. Lannoo. Berkeley: University of California Press. pp. 265–270.

Dodd, C. K., Jr., and R. A. Seigel. 1991. Relocation, repatriation, and translocation of amphibians and reptiles: Are they conservation strategies that work? *Herpetologica* 47:336–350.

Dodd, C. K., Jr., and L. L. Smith. 2003. Habitat destruction and alteration: Historical trends and future prospects for amphibians. In *Amphibian Conservation,* edited by R. D. Semlitsch. Washington, D. C.: Smithsonian Books. pp. 94–112.

Dolino, C. N., E. L. Alcala, and A. C. Alcala. 2003. Populations of two species of forest frogs, genus *Platymantis,* are diminishing on Negros Island, the Philippines. *Froglog* (60):1–2.

Donnelly, M. A., and M. L. Crump. 1998. Potential effects of climate change on two neotropical amphibian assemblages. *Climatic Change* 39:541–561.

Drew, A., E. J. Allen, and L. J. S. Allen. 2006. Analysis of climatic and geographic factors affecting the presence of chytridiomycosis in Australia. *Diseases of Aquatic Organisms* 68:245–250.

Drost, C. A., and G. M. Fellers. 1996. Collapse of a regional frog fauna in the Yosemite area of the California Sierra Nevada, USA. *Conservation Biology* 10:414–425.

Duellman, W. E. 1999. Global distribution of amphibians: Patterns, conservation, and future challenges. In *Patterns of Distribution of Amphibians: A Global Perspective,* edited by W. E. Duellman. Baltimore: Johns Hopkins University Press. pp. 1–30.

Duellman, W. E., ed. 1999. *Patterns of Distribution of Amphibians: A Global Perspective.* Baltimore: Johns Hopkins University Press.

Duellman, W. E., and S. S. Sweet. 1999. Distribution patterns of amphibians in the Nearctic region of North America. In *Patterns of Distribution of Amphibians: A Global Perspective,* edited by W. E. Duellman. Baltimore: Johns Hopkins University Press. pp. 31–109.

Duellman, W. E., and L. Trueb. 1986. *Biology of Amphibians.* New York: McGraw-Hill.

ECOTOX. 2008. ECOTOXicology database, Version 4, Web site. U.S. Environmental Protection Agency. http://www.epa.gov/ecotox.

Ehrlich, P. R., and A. H. Ehrlich. 1981. *Extinction: The Causes and Consequences of the Disappearance of Species.* New York: Random House.

Ehrlich, P. R., and E. O. Wilson. 1991. Biodiversity studies: Science and policy. *Science* 253:758–762.

Ellisman, M. H. 2005. Cyberinfrastructure and the future of collaborative work. *Issues in Science and Technology.* http://www.issues.org/22.1/ellisman.html.

Emmons, M. B. 1973. Problems of an amphibian supply house. *American Zoologist* 13:91–92.

Feinsinger, P. 2001. *Designing Field Studies for Biodiversity Conservation.* Washington, D. C.: Island Press.

Fellers, G. M., W. Yuezhao, and L. Shaoyin. 2003. Status of amphibians at the Zoige Wetlands, Sichuan Province, China. *Froglog* (58):1.

Ficetola, G. F., and F. DeBernardi. 2004. Amphibians in a human-dominated landscape: The community structure is related to habitat features and isolation. *Biological Conservation* 119:219–230.

Fisher, M. C., and T. W. J. Garner. 2007. The relationship between the emergence of *Batrachochytrium dendrobatidis,* the international trade in amphibians and introduced amphibian species. *Fungal Biology Reviews* 21:2–9.

Fitzpatrick, B. M., and H. B. Shaffer. 2007. Hybrid vigor between native and introduced salamanders raises new challenges for conservation. *Proceedings of the National Academy of Sciences (USA)* 104:15793–15798.

Foley, J. A., R. DeFries, G. P. Asner, et al. 2005. Global consequences of land use. *Science* 309:570–574.

Food and Agriculture Organization of the United Nations. 2001. *Global Forest Resources Assessment 2000: Main Report.* FAO Forestry Paper 140. Rome: FAO. http://www.fao.org/forestry/site/fra2000report/en.

Forest, F., R. Grenyer, M. Rouget, et al. 2007. Preserving the evolutionary potential of floras in biodiversity hotspots. *Nature* 445:757–760.

Froglog. Newsletter of the IUCH/SSC Amphibian Specialist Group. Bimonthly. http://www.amphibians.org. Back issues are archived at http://www.open.ac.uk/daptf/froglog/.

FrogWeb. 2008. National Biological Information Infrastructure (NBII): Focus on Amphibians. http://www.frogweb.gov/portal/server.pt.

Frost, D. R. 2008. *Amphibian Species of the World: An Online Reference.* Electronic database. American Museum of Natural History. http://research.amnh.org/herpetology/amphibia/index.html.

Frost, D. R., T. Grant, J. Faivovich, et al. 2006. The Amphibian Tree of Life. *Bulletin of the American Museum of Natural History* 297:1–370.

Fuller, P. L., L. G. Nico, and J. D. Williams. 1999. *Nonindigenous Fishes Introduced into Inland Waters in the United States.* Special Publication 27. Bethesda, Maryland: American Fisheries Society.

Gallant, A. L., R. W. Klaver, G. S. Casper, and M. J. Lannoo. 2007. Global rates of habitat loss and implications for amphibian conservation. *Copeia* 2007:967–979.

Gamrandt, S. C., and L. B. Kats. 1996. Effect of introduced crayfish and mosquitofish on California newts. *Conservation Biology* 10:1155–1162.

Gascon, C., T. E. Lovejoy, R. O. Bierregaard, et al. 1999. Matrix habitat and species persistence in tropical forest remnants. *Biological Conservation* 91:223–229.

Gibbons, W. 2005. Embracing human diversity in conservation. In *Amphibian Declines: The Conservation Status of United States Species,* edited by M. Lannoo. Berkeley: University of California Press. pp. 15–16.

Gibbons, J. W., D. E. Scott, T. J. Ryan, et al. 2000. The global decline of reptiles, déjà vu amphibians. *BioScience* 50:653–666.

Gibbs, E. L., G. W. Nace, and M. B. Emmons. 1971. The live frog is almost dead. *BioScience* 21:1027–1034.

Gibbs, J. P., and A. R. Breisch. 2001. Climate warming and calling phenology of frogs near Ithaca, New York, 1900–1999. *Conservation Biology* 15:1175–1178.

Gillespie, G. R. 2001. The role of introduced trout in the decline of the spotted tree frog (*Litoria spenceri*) in south-eastern Australia. *Biological Conservation* 100:187–198.

Gleason, F. H., M. Kagami, E. Lefevre, and T. Sime-Ngando. 2008. The ecology of chytrids in aquatic ecosystems: roles in food web dynamics. *Fungal Biology Reviews* 22:17–25.

Global Amphibian Assessment. 2008. Web site. http://www.globalamphibians.org.

Global Invasive Species Database. 2008. Electronic database. Invasive Species Specialist Group of the IUCN Species Survival Commission. http://www.invasivespecies.net/database/.

Goodman, B. 2006. To stem widespread extinction, scientists airlift frogs in carry-on bags. *New York Times,* June 6, 2006. Published online at http://www.nytimes.com/2006/06/06/science/06frog.html.

Gorman, J. 2005. A frog brings cacophony to Hawaii's soundscape. *New York Times,* January 25, 2005:D2.

Gorzula, S. 1996. The trade in dendrobatid frogs from 1987 to 1993. *Herpetological Review* 27:116–123.

Gower, D. J., and M. Wilkinson. 2005. Conservation biology of caecilian amphibians. *Conservation Biology* 19:45–55.

Green, D. M., ed. 1997. *Amphibians in Decline: Canadian Studies of a Global Problem.* St. Louis: Society for the Study of Amphibians and Reptiles.

Green, D. M. 2003. The ecology of extinction: Population fluctuation and decline in amphibians. *Biological Conservation* 111:331–343.

Green, D. M., T. F. Sharbel, J. Kearsley, and H. Kaiser. 1996. Postglacial range fluctuation, genetic subdivision and speciation in the western North American spotted frog complex, *Rana pretiosa. Evolution* 50:374–390.

Greer, A. 2007. The impact of disease on the host-population dynamics in a salamander-virus system. Ph.D. dissertation, Arizona State University, Tempe, Arizona.

Greer, A. L., C. J. Briggs, and J. P. Collins. 2008. Testing a key assumption of host-pathogen theory: Density and disease transmission. *Oikos* 117:1667–1673.

Groom, M., G. K. Meffe, and C. R. Carroll. 2006. *Principles of Conservation Biology,* 3rd edition. Sunderland, Massachusetts: Sinauer Associates.

Groombridge, B., and M. D. Jenkins. 2002. *World Atlas of Biodiversity: Earth's Living Resources in the 21st Century.* Berkeley: University of California Press.

Hageman, K. J., S. L. Simonich, D. H. Campbell, G. R. Wilson, and D. H. Landers. 2006. Atmospheric deposition of current-use and historic-use pesticides in snow at national parks in the western United States. *Environmental Science and Technology* 40:3174–3180.

Halliday, T. 1998. A declining amphibian conundrum. *Nature* 394:418–419.

Halliday, T. 2005. Diverse phenomena influencing amphibian population declines. In *Amphibian Declines: The Conservation Status of United States Species,* edited by M. Lannoo. Berkeley: University of California Press. pp. 3–6.

Hamer, A. J., J. A. Makings, S. J. Lane, and M. J. Mahony. 2004. Amphibian decline and fertilizers used on agricultural land in south-eastern Australia. *Agriculture, Ecosystems and Environment* 102:299–305.

Hamline University Center for Global Environmental Education. 2008. A Thousand Friends of Frogs. http://www.hamline.edu/cgee/frogs/.

Hammerson, G. A. 1992. Field surveys of amphibians in the mountains of Colorado. Unpublished report. Denver: Colorado Division of Wildlife.

Hammond, P. 1995. The current magnitude of biodiversity. In *Global Biodiversity Assessment,* edited by V. H. Heywood and R. T. Watson. Cambridge, England: Cambridge University Press. pp. 113–138.

Hanski, I. 2005. *The Shrinking World: Ecological Consequences of Habitat Loss.* Excellence in Ecology 14. Oldendorf/Luhe, Germany: Ecology Institute.

Harremoës, P., D. Gee, M. MacGarvin, et al., eds. 2001. *Late Lessons from Early Warnings: The Precautionary Principle 1896–2000.* Environmental Issue Report No 22. European Environment Agency. http://reports.eea.europa.eu/environmental_issue_report_2001_22/en/.

Harris, R. N., T. Y. James, A. Lauer, M. A. Simon, and A. Patel. 2006. Amphibian pathogen *Batrachochytrium dendrobatidis* is inhibited by the cutaneous bacterial flora of amphibian species. *EcoHealth* 3:53–56.

Harte, J., and E. Hoffman. 1989. Possible effects of acidic deposition on a Rocky Mountain population of the tiger salamander *Ambystoma tigrinum. Conservation Biology* 3:149–158.

Havel, V. 2007. Our moral footprint. *New York Times,* September 27, 2007: A31.

Hay, R. 1998. Blanchard's cricket frogs in Wisconsin: A status report. In *Status and Conservation of Midwestern Amphibians,* edited by M. J. Lannoo. Iowa City: University of Iowa Press. pp. 79–82.

Hayes, M. P., and M. R. Jennings. 1986. Decline of ranid frog species in western North America: Are bullfrogs (*Rana catesbeiana*) responsible? *Journal of Herpetology* 20: 490–509.

Hayes, T. B. 2000. Endocrine disruption in amphibians. In *Ecotoxicology of Amphibians and Reptiles,* edited by D. W. Sparling, G. Linder, and C. A. Bishop. Pensacola, Florida: Society of Environmental Toxicology and Chemistry (SETAC). pp. 573–593.

Hayes, T. B., A. Collins, M. Lee, et al. 2002. Hermaphroditic, demasculinized frogs after exposure to the herbicide atrazine at low ecologically relevant doses. *Proceedings of the National Academy of Sciences (USA)* 99:5476–5480.

Hayes, T., K. Haston, M. Tsui, A. Hoang, C. Haeffele, and A. Vonk. 2002. Feminization of male frogs in the wild. *Nature* 419:895–896.

Hayes, T. B., A. A. Stuart, M. Mendoza, et al. 2006. Characterization of atrazine-induced gonadal malformations in African clawed frogs (*Xenopus laevis*) and comparisons with effects of an androgen antagonist (cyproterone acetate) and exogenous estrogen (17Beta-estradiol): Support for the demasculinization/feminization hypothesis. *Environmental Health Perspectives* 14(S1):1–53. Published online.

Hecnar, S. J. 1995. Acute and chronic toxicity of ammonium nitrate fertilizer to amphibians from southern Ontario. *Environmental Toxicology and Chemistry* 14:2131–2137.

Hecnar, S. J. 1997. Amphibian pond communities in southwestern Ontario. In *Amphibians in Decline: Canadian Studies of a Global Problem,* edited by D. M. Green. *Herpetological Conservation* 1:1–15. St. Louis: Society for the Study of Amphibians and Reptiles.

Hecnar, S. J., and R. T. M'Closkey. 1996. Regional dynamics and the status of amphibians. *Ecology* 77:2091–2097.

Hecnar, S. J., and R. T. M'Closkey. 1998. Species richness patterns of amphibians in southwestern Ontario ponds. *Journal of Biogeography* 25:763–772.

Hedges, S. B. Distribution patterns of amphibians in the West Indies. In *Patterns of Distribution of Amphibians: A Global Perspective,* edited by W. E. Duellman. Baltimore: Johns Hopkins University Press. pp. 211–254.

Hedges, S. B., J. P. Bogart, and L. R. Maxson. 1992. Ancestry of unisexual salamanders (genus *Ambystoma*). *Nature* 356:708–710.

Helfrich, L. A., R. J. Neves, and J. Parkhurst. 2001. *Commercial Frog Farming.* Virginia Cooperative Extension of Virginia Polytechnic and State University Publication Number 420–255. http://www.ext.vt.edu/pubs/fisheries/420-255/420-225.html.

Hero, J-M. 1996. Where are Queensland's missing frogs? *Wildlife Australia* 33:8–13.

Hero, J-M., and G. R. Gillespie. 1997. Epidemic disease and amphibian declines in Australia. *Conservation Biology* 11:1023–1025.

Heyer, W. R., and J. B. Murphy. 2005. Declining Amphibian Populations Task Force. In *Amphibian Declines: The Conservation Status of United States Species,* edited by M. Lannoo. Berkeley: University of California Press. pp. 17–21.

Heyer, W. R., M. A. Donnelly, R. W. McDiarmid, L-A. C. Hayek, and M. S. Foster, eds. 1994. *Measuring and Monitoring Biological Diversity: Standard Methods for Amphibians.* Washington, D.C.: Smithsonian Institution Press.

Heyer, W. R., A. S. Rand, C. A. G. da Cruz, and O. L. Peixoto. 1988. Decimations, extinctions, and colonizations of frog populations in southeast Brazil and their evolutionary implications. *Biotropica* 20:230–235.

Highton, R. 2005. Declines of eastern North American woodland salamanders (*Plethodon*). In *Amphibian Declines: The Conservation Status of United States Species,* edited by M. Lannoo. Berkeley: University of California Press. pp. 34–46.

Hilty, J., and A. Merenlender. 2000. Faunal indicator taxa selection for monitoring ecosystem health. *Biological Conservation* 92:185–197.

Hofer, D. 2000. A short note about the status and abundance of caecilian populations. *Froglog* 42:1.

Hoffman, A. A., and M. W. Blows. 1993. Evolutionary genetics and climate change: Will animals adapt to global warming? In *Biotic Interactions and Global Change,* edited by P. M. Kareiva, J. G. Kingsolver, and R. B. Huey. Sunderland, Massachusetts: Sinauer Associates. pp. 165–178.

Holling, C. S. 1992. Cross-scale morphology, geometry, and dynamics of ecosystems. *Ecological Monographs* 62:447–502.

Hossack, B. R., S. A. Diamond, and P. S. Corn. 2006. Distribution of boreal toad populations in relation to estimated UV-B dose in Glacier National Park, Montana, USA. *Canadian Journal of Zoology* 84:98–107.

Houlahan, J. E., C. S. Findley, B. R. Schmidy, A. H. Meyer, and S. L. Kuzmin. 2000. Quantitative evidence for global amphibian declines. *Nature* 404:752–755.

Hunter, M. L., Jr. 2002. *Fundamentals of Conservation Biology,* 2nd edition. Malden, Massachusetts: Blackwell Science.

Hutchins, M., and R. Wiese. 1991. Beyond genetic and demographic management: The future of the Species Survival Plan and related AAZPA conservation efforts. *Zoo Biology* 10:285–292.

Intergovernmental Panel on Climate Change (IPCC). 2008. IPCC Fourth Assessment Report: Climate Change 2007. Available online at http://www.ipcc.ch.

International Union for the Conservation of Nature and Natural Resources (IUCN). 2008. *The IUCN Red List of Threatened Species.* http://www.iucnredlist.org/.

Introduced Species in Hawaii. 2002. Senior Seminar, Earlham College. Published online. http://www.earlham.edu/~biol/hawaii/amphibians.htm.

James, T. Y., P. M. Letcher, J. E. Longcore, et al. 2006. A molecular phylogeny of the flagellated fungi (Chytridiomycota) and description of a new phylum (Blastocladiomycota). *Mycologia* 98:860–871.

Jancovich, J. K., E. W. Davidson, N. Parameswaran, et al. 2005. Evidence for emergence of an amphibian iridoviral disease because of human-enhanced spread. *Molecular Ecology* 14:213–224.

Jancovich, J. K., E. W. Davidson, A. Seiler, B. L. Jacobs, and J. P. Collins. 2001. Transmission of the Ambystoma tigrinum virus to alternative hosts. *Diseases of Aquatic Organisms* 46:159–163.

Jennings, M. R., and M. P. Hayes. 1985. Pre-1900 overharvest of the California red-legged frog (*Rana aurora draytonii*): The inducement for bullfrog (*Rana catesbeiana*) introduction. *Herpetologica* 41:94–103.

Jensen, J. B., and C. D. Camp. 2003. Human exploitation of amphibians: Direct and indirect impacts. In *Amphibian Conservation,* edited by R. D. Semlitsch. Washington, D. C.: Smithsonian Books. pp. 199–213.

Joglar, R. L., and P. A. Burrowes. 1996. Declining amphibian populations in Puerto Rico. In *Contributions to West Indian Herpetology: A Tribute to Albert Schwartz,* edited by R. Powell and R. W. Henderson. *Contributions to Herpetology* 12. Ithaca, New York: Society for the Study of Amphibians and Reptiles. pp. 371–380.

Johnson, A. J., and J. F. X. Wellehan. 2005. Amphibian virology. *Veterinary Clinics of North America: Exotic Animal Practice* 8:53–65.

Johnson, C. N. 1998. Species extinction and the relationship between distribution and abundance. *Nature* 394:272–274.

Johnson, M. L., and R. Speare. 2003. Survival of *Batrachochytrium dendrobatidis* in water: Quarantine and disease control implications. *Emerging Infectious Diseases* 9:922–925.

Johnson, P. T. J., J. M. Chase, K. L. Dosch, et al. 2007. Aquatic eutrophication promotes pathogenic infection in amphibians. *Proceedings of the National Academy of Sciences (USA)* 104:15781–15786.

Johnson, P. T. J., K. B. Lunde, E. G. Richie, and A. E. Launer. 1999. The effect of trematode infection on amphibian limb development and survivorship. *Science* 284:802–804.

Johnson, P. T. J., K. B. Lunde, D. A. Zelmer, and J. K. Werners. 2003. Limb deformities as an emerging parasitic disease in amphibians: Evidence from museum specimens and resurvey data. *Conservation Biology* 17:1724–1737.

Joint Genome Institute (JGI), U.S. Department of Energy. 2008. Why sequence a frog-killing fungus? http://www.jgi.doe.gov/sequencing/why/3086.html.

Jones, M., A. Cockburn, R. Hamede, C. Hawkins, H. Hesterman, S. Lachish, D. Mann, H. McCallum, and D. Pemberton. 2008. Life-history change in disease-ravaged Tasmanian devil populations. *Proceedings of the National Academy of Sciences (USA)* 105:10023–10027.

Kagarise Sherman, C., and M. L. Morton. 1993. Population declines of Yosemite toads in the eastern Sierra Nevada of California. *Journal of Herpetology* 27:186–198.

Karr, J. R. 1996. Ecological integrity and ecological health are not the same. In *Engineering within Ecological Constraints,* edited by P. C. Schulze. Washington, D. C.: National Academy Press. pp. 97–109.

Kats, L. B., and R. P. Ferrer. 2003. Alien predators and amphibian declines: Review of two decades of science and the transition to conservation. *Diversity and Distributions* 9:99–110.

Kiesecker, J. M. 2002. Synergism between trematode infection and pesticide exposure: A link to amphibian limb deformities in nature? *Proceedings of the National Academy of Sciences (USA)* 99:9900–9904.

Kiesecker, J. M. 2003. Invasive species as a global problem: Toward understanding the worldwide decline of amphibians. In *Amphibian Conservation,* edited by R. D. Semlitsch. Washington, D. C.: Smithsonian Books. pp. 113–126.

Kiesecker, J. M., and A. R. Blaustein. 1995. Synergism between UV-B radiation and a pathogen magnifies amphibian embryo mortality in nature. *Proceedings of the National Academy of Science (USA)* 92:11049–11052.

Kiesecker, J. M., A. R. Blaustein, and L. K. Belden. 2001. Complex causes of amphibian population declines. *Nature* 410:681–684.

Kirshtein, J. D., C. W. Anderson, J. S. Wood, et al. 2007. Quantitative PCR detection of *Batrachochytrium dendrobatidis* DNA from sediments and water. *Diseases of Aquatic Organisms* 77:11–15.

Knapp, R. A. 2005. Effects of nonnative fish and habitat characteristics on lentic herpetofauna in Yosemite National Park, USA. *Biological Conservation* 121:265–279.

Knapp, R. A., Boiano, D. M., and Vredenberg, V. T. 2007. Removal of nonnative fish results in population expansion of a declining amphibian (mountain yellow-legged frog, *Rana muscosa*). *Biological Conservation* 135:11–20.

Knapp, R. A., P. S. Corn, and D. E. Schindler. 2001. The introduction of nonnative fish into wilderness lakes: Good intentions, conflicting mandates, and unintended consequences. *Ecosystems* 4:275–278.

Knapp, R. A., K. R. Matthews, and O. Sarnelle. 2001. Resistance and resilience of alpine lake fauna to fish introductions. *Ecological Monographs* 71: 401–421.

Kolpin, D. W., E. T. Furlong, M. T. Meyer, et al. 2002. Pharmaceuticals, hormones, and other organic wastewater contaminants in U. S. streams, 1999–2000: A national reconnaissance. *Environmental Science and Technology* 36:1202–1211.

Komak, S., and M. R. Crossland. 2000. An assessment of the introduced mosquitofish (*Gambusia affinis holbrooki*) as a predator of eggs, hatchlings and tadpoles of native and nonnative anurans. *Wildlife Research* 27:185–189.

Krajick, K. 2006. The lost world of the Kihansi toad. *Science* 311:1230–1232.

Kraus, F., and Campbell, E. 2002. Human-mediated escalation of a formerly eradicable problem: The invasion of Caribbean frogs in the Hawaiian Islands. *Biological Invasions* 4:327–332.

Kraus, F., E. W. Campbell, A. Allison, and T. Pratt. 1999. *Eleutherodactylus* frog introductions to Hawaii. *Herpetological Review* 30:21–25.

Kriger, K. M., and J-M. Hero. 2007. The chytrid fungus *Batrachochytrium dendrobatidis* is non-randomly distributed across amphibian breeding habitats. *Diversity and Distributions* 13:781–788.

Krimsky, S. 2000. *Hormonal Chaos.* Baltimore: Johns Hopkins University Press.

LaDeau, S. L., P. P. Marra, A. M. Kilpatrick. 2007. West Nile virus emergence and large-scale declines of North American birds. *Nature* 447:710–713.

La Marca, E., K. R. Lips, S. Lotters, et al. 2005. Catastrophic population declines and extinctions in neotropical harlequin frogs (Bufonidae: *Atelopus*). *Biotropica* 37:190–201.

Lannoo, M. J. 1998. Amphibian conservation and wetland management in the upper Midwest: A Catch-22 for the cricket frog? In *Status and Conservation of Midwestern Amphibians,* edited by M. J. Lannoo. Iowa City: University of Iowa Press. pp. 330–339.

Lannoo, M. J. 1998. *Status and Conservation of Midwestern Amphibians.* Iowa City: University of Iowa Press.

Lannoo, M. J., ed. 2005. *Amphibian Declines: The Conservation Status of United States Species.* Berkeley: University of California Press.

Lannoo, M. J. 2008. *Malformed Frogs: The Collapse of Aquatic Ecosystems.* Berkeley: University of California Press.

Lannoo, M. J., K. Lang, T. Waltz, and G. S. Phillips. 1994. An altered amphibian assemblage: Dickinson County, Iowa, 70 years after Frank Blanchard's survey. *American Midland Naturalist* 131:311–319.

Lauer, A., M. A. Simon, J. L. Banning, E. André, K. Duncan, and R. N. Harris. 2007. Common cutaneous bacteria from the eastern red-backed salamander can inhibit pathogenic fungi. *Copeia* 2007:630–640.

Laurance, W. F. 2008. Global warming and amphibian extinctions in eastern Australia. *Animal Ecology* 33:1–9.

Laurance, W. F., K. R. McDonald, and R. Speare. 1996. Catastrophic declines of Australian rain forest frogs: Support for the epidemic disease hypothesis. *Conservation Biology* 10:1–9.

Laurance, W. F., K. R. McDonald, and R. Speare. 1996. Epidemic disease and the catastrophic decline of Australian rain forest frogs. *Conservation Biology* 10:406–413.

Laurance, W. F., K. R. McDonald, and R. Speare. 1997. In defense of the epidemic disease hypothesis. *Conservation Biology* 11:1030–1034.

Lawton, J. H., and V. K. Brown. 1993. Redundancy in ecosystems. In *Biodiversity and Ecosystem Function,* edited by E-D. Schulze and H. A. Mooney. Berlin: Springer-Verlag. pp. 255–270.

Lee, S., K. Zippel, L. Ramos, and J. Searle. 2006. Captive-breeding programme for the Kihansi spray toad *Nectophrynoides asperginis* at the Wildlife Conservation Society, Bronx, New York. *International Zoo Yearbook* 40:241–253.

Lehtinen, R. M., S. M. Galatowitsch, and J. R. Tester. 1999. Consequences of habitat loss and fragmentation for wetland amphibian assemblages. *Wetlands* 19:1–12.

Leutwyler, K. 1999. Parasites or pollution? Biologists figure out what accounts for certain side-show frogs. *Scientific American.* http://www.sciam.com/article.cfm?id=parasites-or-pollution.

Levitus, S., J. I. Antonov, J. Wang, T. L. Delworth, K. W. Dixon, A. J. Broccoli. 2001. Anthropogenic warming of earth's climate system. *Science* 292:267–270.

Lewis, D. L., G. T. Baxter, K. M. Johnson, and M. D. Stone. 1985. Possible extinction of the Wyoming toad, *Bufo hemiophrys baxteri. Journal of Herpetology* 19:166–168.

Linder, G. L., S. K. Krest, and D. W. Sparling, eds. 2003. *Amphibian Decline: An Integrated Analysis of Multiple Stressor Effects.* Pensacola, Florida: Society of Environmental Toxicology and Chemistry (SETAC).

Lips, K. R. 1998. Decline of a tropical montane amphibian fauna. *Conservation Biology* 12:106–117.

Lips, K. R. 1999. Mass mortality and population declines of anurans at an upland site in western Panama. *Conservation Biology* 13:117–125.

Lips, K. R., and M. A. Donnelly. 2005. Lessons from the tropics. In *Amphibian Declines: The Conservation Status of United States Species,* edited by M. Lannoo. Berkeley: University of California Press. pp. 198–205.

Lips, K. R., F. Brem, R. Brenes, et al. 2006. Emerging infectious disease and the loss of biodiversity in a neotropical amphibian community. *Proceedings of the National Academy of Sciences (USA)* 103:3165–3170.

Lips, K. R., J. Diffendorfer, J. R. Mendelson III, and M. W. Sears. 2008. Riding the wave: Reconciling the roles of disease and climate change in amphibian declines. *PLoS Biology* 6(3):e72. doi:10.1371/journal.pbi0.0060072.

Lips, K. R., D. E. Green, and R. Papendick. 2003a. Chytridiomycosis in wild frogs from southern Costa Rica. *Journal of Herpetology* 37:215–218.

Lips, K. R., J. D. Reeve, and L. R. Witters. 2003b. Ecological traits predicting amphibian population declines in Central America. *Conservation Biology* 17:1078–1088.

Longcore, J. E., A. P. Pessier, and D. K. Nichols. 1999. *Batrachochytrium dendrobatidis,* gen. et sp. nov., a chytrid pathogenic to amphibians. *Mycologia* 91:219–227.

Loyola, R. D., C. G. Becker, U. Kubota, C. F. B. Haddad, C. R. Fonseca, and T. M. Lewinsohn. 2008. Hung out to dry: Choice of priority ecoregions for conserving threatened Neotropical anurans depends on life history traits. *PLoS ONE* 3:e2120.

Lyons, S. K., F. A. Smith, P. J. Wagner, E. P. White, and J. H. Brown. 2004. Was a 'hyperdisease' responsible for the late Pleistocene megafaunal extinction? *Ecology Letters* 7:859–868.

MacPhee, R. D. E., and P. A. Marx. 1997. The 40,000-year plague. In *Natural Change and Human Impact in Madagascar,* edited by S. M. Goodman and B. D. Patterson. Washington, D. C.: Smithsonian Institution Press. pp. 169–217.

MacPhee, R. D. E. 1998. Lightning strikes twice: Blitzkrieg, hyperdisease, and global explanations of the late quaternary catastrophic extinctions. Audio presentation and transcript. American Museum of Natural History. http://www.amnh.org/science/biodiversity/extinction/Day1/bytes/MacPheePres.html.

Marris, E. 2006. Should conservation biologists push policies? *Nature* 442:13.

Marris, E. 2008. Bagged and boxed: it's a frog's life. *Nature* 452:394–395.

Marsh, D. M. 2001. Fluctuations in amphibian populations: A meta-analysis. *Biological Conservation* 101:327–335.

Marsh, D. M., and N. G. Beckman. 2004. Effects of forest roads on the abundance and activity of terrestrial salamanders. *Ecological Applications* 14:1882–1891.

Martin, P. S., and R. G. Klein, eds. 1984. *Quaternary Extinctions: A Prehistoric Revolution.* Tucson: The University of Arizona Press.

Matson, P. A., W. J. Parton, A. G. Power, and M. J. Swift. 1997. Agricultural intensification and ecosystem properties. *Science* 277:504–509.

Mayden, R. L., ed. 1992. *Systematics, Historical Ecology, and North American Freshwater Fishes.* Palo Alto, California: Stanford University Press.

McCallum, M. L. 2007. Amphibian decline or extinction? Current declines dwarf background extinction rate. *Journal of Herpetology* 41:483–491.

McCallum, H., D. M. Tompkins, M. Jones, S. Lachish, S. Marvanek, B. Lazenby, G. Hocking, J. Wiersma, and C. E. Hawkins. 2007. Distribution and impacts of Tasmanian devil facial tumor disease. *EcoHealth* 4:318–325.

McCallum, H. 2008. Tasmanian devil facial tumor disease: lessons for conservation biology. *Trends in Ecology and Evolution* 23:631–637.

McCallum, H., and M. Jones. 2006. To lose both would look like carelessness: Tasmanian devil facial tumor disease. *PLoS Biology* 4(10):0001–0004.

McCoy, K. A., L. J. Bortnick, C. M. Campbell, H. J. Hamlin, L. J. Guillette, Jr., and C. M. St. Mary. 2008. Agriculture alters gonadal form and function in the toad *Bufo marinus. Environmental Health Perspectives* 116:1526–1532.

McDaniel, T. V., P. A. Martin, J. Struger, J. Sherry, C. H. Marvin, M. E. M. C. Master, S. Clarence, and G. Tetreault. 2008. Potential endocrine disruption of sexual development in free ranging male northern leopard frogs (*Rana pipiens*) and green frogs (*Rana clamitans*) from areas of intensive row crop agriculture. *Aquatic Toxicology* 88:230–242.

McDonald, K., and R. Alford. 1999. A review of declining frogs in northern Queensland. In *Declines and Disappearances of Australian Frogs,* edited by A. Campbell. Canberra: Environment Australia. pp. 14–22.

McGeoch, M. A. 1998. The selection, testing and application of terrestrial insects as bioindicators. *Biological Reviews (Cambridge Philosophical Society)* 73:181–201.

McKay, J. 2003. *An Evaluation of Captive Breeding and Sustainable Use of the Mexican Axolotl (*Ambystoma mexicanum*).* MSc Dissertation, University of Kent, Canterbury.

McMenamin, S. K., E. A. Hadley, and C. K. Wright. 2008. Climate change and wetland desiccation cause amphibian decline in Yellowstone National Park. *Proceedings of the National Academy of Sciences (USA)* 105:16988–16993.

Meegaskumbura, M., F. Bossuyt, R. Pethiyagoda, et al. 2002. Sri Lanka: An amphibian hot spot. *Science* 298:379.

Melvani, K. 1997. Frog tea? *Froglog* (23):2.

Mendelson, J. R., III, K. R. Lips, J. E. Diffendorfer, et al. 2006. Letters: Response to Pounds, J. A. et al. Responding to amphibian loss. *Science* 314:1541–1542.

Mendelson, J. R., III, K. R. Lips, R. W. Gagliardo, et al. (50 authors total) 2006. Confronting amphibian declines and extinctions. *Science* 313:48 + 13 pp. online supplemental material.

Meshaka, W. E., Jr. 2001. *The Cuban Treefrog in Florida: Life History of a Successful Colonizing Species.* Gainesville: University Press of Florida.

Meshaka, W. E., Jr. 2005. Exotic species. In *Amphibian Declines: The Conservation Status of United States Species,* edited by M. Lannoo. Berkeley: University of California Press. pp. 271–274.

Meyer, W. B., and B. L. Turner II. 1992. Human population growth and global land-use/cover change. *Annual Review of Ecology and Systematics.* 23:39–61.

Middleton, E. M., J. R. Herman, E. A. Celarier, J. W. Wilkinson, C. Carey, and R. J. Rusin. 2001. Evaluating ultraviolet radiation exposure with satellite data at sites of amphibian declines in Central and South America. *Conservation Biology* 15:914–929.

Minteer, B. A., and J. P. Collins. 2005. Ecological ethics: Building a new tool kit for ecologists and biodiversity managers. *Conservation Biology* 19:1803–1812.

Minteer, B. A., and J. P. Collins. 2005. Why we need an "ecological ethics." *Frontiers in Ecology and the Environment* 3:332–337.

Minteer, B. A., and J. P. Collins. 2008. From environmental to ecological ethics: Toward a practical ethics for ecologists and conservationists. *Science and Engineering Ethics* 14:483–501.

Mitchell, K. M., T. S. Churcher, T. W. J. Garner, and M. C. Fisher. 2008. Persistence of the emerging pathogen *Batrachochytrium dendrobatidis* outside the amphibian host greatly increases the probability of host extinction. *Proceedings of the Royal Society B* 275:329–334.

Moffat, A. S. 1998. Global nitrogen overload problem grows critical. *Science* 279:988–989.

Mongabay.com. 2008. Conservation. http://fish.mongabay.com/aquariast_role_conservation.htm.

Mooers, A. Ø. 2007. The diversity of biodiversity. *Nature* 445:717–718.

Moore, R. D., C. Gascon, and J. P. Collins. 2008. The Amphibian Specialist Group: Past present and future. In S. Stuart, M. Hoffmann, J. S. Chanson, N. A. Cox, R. J. Berridge, P. Ramani, and B. E. Young, editors. *Threatened Amphibians of the World*. Gland, Switzerland: IUCN Press.

Moran P. A. P. 1953. The statistical analysis of the Canadian lynx cycle: II. Synchronization and meteorology. *Australian Journal of Zoology* 1:291–298.

Morehouse, E. A., T. Y. James, A. R. D. Ganley, et al. 2003. Multilocus sequence typing suggests the chytrid pathogen of amphibians is a recently emerged clone. *Molecular Ecology* 12:395–403.

Morgan, J. A. T., V. T. Vredenberg, L. J. Rachowicz, et al. 2007. Population genetics of the frog-killing fungus *Batrachochytrium dendrobatidis*. *Proceedings of the National Academy of Sciences (USA)* 104:13845–13850.

Morse, S. S. 1995. Factors in the emergence of infectious diseases. *Emerging Infectious Diseases* 1:7–15.

Morrell, V. 2008. Into the wild: Reintroduced animals face daunting odds. *Science* 320:742–743.

Murphy, T. P. 2005. A Thousand Friends of Frogs: Its origins. In *Amphibian Declines: The Conservation Status of United States Species*, edited by M. Lannoo. Berkeley: University of California Press. pp. 341–342.

Murray, B. R., and G. C. Hose. 2005. Life-history and ecological correlates of decline and extinction in the endemic Australian frog fauna. *Austral Ecology* 30:564–571.

Muths, E., P. S. Corn, A. P. Pessier, and D. E. Green. 2003. Evidence for disease-related amphibian decline in Colorado. *Biological Conservation* 110:357–365.

Myers, N. 1988. Threatened biotas: "Hot-spots" in tropical forests. *Environmentalist* 8:187–208.

Myers, N. 1990. Mass extinctions: What can the past tell us about the present and the future? *Global and Planetary Change* 82:175–185.

Nace, G. W. 1968. The amphibian facility of the University of Michigan. *BioScience* 18:767–775.

National Amphibian Conservation Center (NACC). Web site. http://www.detroitzoo.org/Attractions/Amphibiville/NACC/.

National Heritage Trust (Australia) conference: Getting the jump! on amphibian disease. http://www.jcu.edu.au/school/phtm/PHTM/frogs/gjoad.htm.

New York Times. 2007. Andean wetlands and cities' water supplies threatened. July 21.

Nichols, D. K. 2003. Tracking down the killer chytrid of amphibians. *Herpetological Review* 34:101–104.

Nickerson, M. A., and J. Briggler. 2007. Harvesting as a factor in population decline of a long-lived salamander: The Ozark hellbender, *Cryptobranchus alleganiensis bishopi* Grobman. *Applied Herpetology* 4:207–216.

Nonindigineous Aquatic Species. 2008. Electronic database. U.S. Geological Survey. http://nas.er.usgs.gov/.

Noss, R. F. 1990. Indicators for monitoring biodiversity: A hierarchical approach. *Conservation Biology* 4:355–364.

Noss, R. F., and A. Y. Cooperrider. 1994. *Saving Nature's Legacy: Protecting and Restoring Biodiversity*. Washington, D. C.: Island Press.

Nussbaum, R. A., and M. E. Pfrender. 1998. Revision of the African caecilian genus *Schistometopum* Parker (Amphibia: Gymnophiona: Caeciliidae). *Miscellaneous Publications*. Museum of Zoology, University of Michigan, Ann Arbor 187:1–32.

Odum, E. P. 1992. Great ideas in ecology for the 1990s. *BioScience* 42:542–545.

Odum, R. A., and P. S. Corn. 2005. *Bufo baxteri* Porter, 1968. Wyoming toad. In *Amphibian Declines: The Conservation Status of United States Species,* edited by M. Lannoo. Berkeley: University of California Press. pp. 390–392.

Oliver, J. A. 1955. *The Natural History of North American Amphibians and Reptiles.* Princeton, New Jersey: D. Van Nostrand.

Oreskes, N. 2004. Beyond the ivory tower: The scientific consensus on climate change. *Science* 306:1686.

Oreskes, N. 2007. The long consensus on climate change. *The Washington Post,* February 1, 2007:A15.

Ouellet, M. 2000. Amphibian deformities: Current state of knowledge. In *Ecotoxicology of Amphibians and Reptiles,* edited by D. W. Sparling, D. W. Linder, and C. A. Bishop. Pensacola, Florida: Society of Environmental Toxicology and Chemistry (SETAC). pp. 617–661.

Ovaska, K. 1997. Vulnerability of amphibians in Canada to global warming and increased solar ultraviolet radiation. In *Amphibians in Decline: Canadian Studies of a Global Problem,* edited by D. M. Green. *Herpetological Conservation* 1. St. Louis: Society for the Study of Amphibians and Reptiles. pp. 206–225.

Owen, D., and D. Pemberton. 2005. *Tasmanian Devil: A Unique and Threatened Animal.* Crows Nest, Australia: Allen & Unwin.

Palen, W. J., D. E. Schindler, M. J. Adams, C. A. Pearl, R. B. Bury, S. A. Diamond. 2002. Optical characteristics of natural waters protect amphibians from UV-B radiation in the U. S. Pacific Northwest. *Ecology* 83:2951–2957.

Parmesan, C. 2006. Ecological and evolutionary responses to recent climate change. *Annual Review of Ecology and Systematics* 37:637–639.

Partners in Amphibian and Reptile Conservation (PARC). Web site. http://www.parcplace.org.

Patel, T. 1993. French may eat Indonesia out of frogs. *New Scientist* 138:7.

Pavajeau, L., K. C. Zippel, R. Gibson, and K. Johnson. 2008. Amphibian Ark and the 2008 Year of the Frog campaign. *International Zoo Yearbook* 42:24–29.

Payne, J. L., and S. Finnegan. 2007. The effect of geographic range on extinction risk during background and mass extinction. *Proceedings of the National Academy of Sciences (USA).* 104:10506–10511.

Pechmann, J. H. K. 2003. Natural population fluctuations and human influences. In *Amphibian Conservation,* edited by R. D. Semlitsch. Washington, D.C.: Smithsonian Books. pp. 85–93.

Pechmann, J. H. K., and H. M. Wilbur. 1994. Putting declining amphibian populations in perspective: Natural fluctuations and human impacts. *Herpetologica* 50:65–84.

Pechmann, J. H. K., D. E. Scott, R. D. Semlitsch, J. P. Caldwell, L. J. Vitt, and J. W. Gibbons. 1991. Declining amphibian populations: The problem of separating human impacts from natural fluctuations. *Science* 253:892–895.

Pennisi, E. 2002. 100 Frogs a-leaping for biodiversity. *Science* 298:339, 341.

Perrings, C. 1995. The economic value of biodiversity. In *Global Biodiversity Assessment,* edited by V. H. Heywood. Cambridge, England: Cambridge University Press. pp. 823–914.

Petranka, J. W. 1998. *Salamanders of the United States and Canada.* Washington, D. C.: Smithsonian Institution Press.

Pfennig, D. W., M. L. G. Loeb, and J. P. Collins. 1991. Pathogens as a factor limiting the spread of cannibalism in tiger salamanders. *Oecologia* 88:161–166.

Phillips, K. 1994. *Tracking the Vanishing Frogs: An Ecological Mystery.* New York: St. Martin's Press.

Picco, A. M., and J. P. Collins. 2008. Amphibian commerce as a likely source of pathogen pollution. *Conservation Biology.* 22:1582–1589.

Pimenta, B. V. S., C. F. B. Haddad, L. B. Nascimento, C. A. Goncalves Cruz, and J. P. Pombal Jr. 2005. Comment on "Status and Trends of Amphibian Declines and Extinctions Worldwide." *Science* 309:1999. Published online: http://www.sciencemag.org/cgi/content/full/309/5743/1999b.

Plowright, R. K., S. H. Sokolow, M. E. Gorman, P. Daszak, and J. E. Foley. 2008. Causal inference in disease ecology: investigating ecological drivers of disease emergence. *Frontiers in Ecology and the Environment* 6:420–429.

Poiani, K. A., B. D. Richter, M. G. Anderson, and H. G. Richter. 2000. Biodiversity conservation at multiple scales: Functional sites, landscapes, and networks. *BioScience* 50:133–146.

Pough, F. H., R. M. Andrews, J. E. Cadle, M. L. Crump, A. H. Savitzky, and K. D. Wells. 2004. *Herpetology,* 3rd edition. Upper Saddle River, New Jersey: Prentice-Hall.

Poulin, R. 2006. Global warming and temperature-mediated increases in cercarial emergence in trematode parasites. *Parasitology* 132:143–151.

Pounds, J. A., and M. L. Crump. 1994. Amphibian declines and climate disturbance: The case of the golden toad and the harlequin frog. *Conservation Biology* 8:72–85.

Pounds, J. A., and R. Puschendorf. 2004. Clouded futures. *Nature* 427:107–109.

Pounds, J. A., M . R. Bustamante, L. A. Coloma, et al. 2006. Widespread amphibian extinctions from epidemic disease driven by global warming. *Nature* 439:161–167.

Pounds, J. A., M . R. Bustamante, L. A. Coloma, et al. 2007. Global warming and amphibian losses: The proximate cause of frog declines? (Reply). *Nature* 447:E5–E6.

Pounds, J. A., A. C. Carnaval, R. Puschendorf, C. F. B. Haddad, and K. L. Masters. 2006. Letters: Responding to amphibian loss. *Science* 314:1541–1542.

Pounds, J. A., M. P. L. Fogden, and J. H. Campbell. 1999. Biological response to climate change on a tropical mountain. *Nature* 398:611–615.

Pounds, J. A., M. P. L. Fogden, J. M. Savage, and G. C. Gorman. 1997. Tests of null models for amphibian declines on a tropical mountain. *Conservation Biology* 11:1307–1322.

Primack, R. B. 2002. *Essentials of Conservation Biology,* 3rd edition. Sunderland, Massachusetts: Sinauer Associates.

Proyecto Coquí. 2008. http://coqui.uprrp.edu/eng/species.html.

Purvis, A., J. L. Gittleman, G. Cowlishaw, and G. M. Mace. 2000. Predicting extinction risk in declining species. *Proceedings of the Royal Society London B* 267:1947–1952.

Puschendorf, R. 2003. *Atelopus varius* (harlequin frog): Fungal infection. *Herpetological Review* 34:355.

Puschendorf, R., F. Bolaños, and G. Chaves. 2006. The amphibian chytrid fungus along an altitudinal transect before the first reported declines in Costa Rica. *Biological Conservation* 132:136–142.

Rabb, G. B. 2004. The evolution of zoos from menageries to centers of caring and conservation. *Curator* 47:237–246.

Rachowicz, L. J., J-M. Hero, R. A. Alford, et al. 2005. The novel and endemic pathogen hypotheses: Competing explanations for the origin of emerging diseases of wildlife. *Conservation Biology* 19:1441–1448.

Rachowicz, L. J., R. A. Knapp, J. A. T. Morgan, M. J. Stice, V. T. Vredenburg, J. M. Parker, and C. J. Briggs. 2006. Emerging infectious disease as a proximate cause of amphibian mass mortality. *Ecology* 87:1671–1683.

Rachowitz, L. J., and C. L. Briggs. 2007. Quantifing the disease transmission function: effects of density on *Batrachochytrium dendrobatidis* transmission in the mountain yellow-legged frog *Rana muscosa*. *Journal of Animal Ecology* 76:711–721.

Rahbek, C. 2007. Disease ecology: The silence of the robins. *Nature* 447:652–653.

Rahel, F. J. 2002. Homogenization of freshwater faunas. *Annual Review of Ecology and Systematics* 33:291–315.

Raloff, J. 1998. Common pesticide clobbers amphibians. *Science News* 154:150.

Raloff, J. 2003. Hawaii's hated frogs. *Science News* 163:11.

Ranvestel, A. W., K. R. Lips, C. M. Pringle, M. R. Whiles, and R. J. Bixby. 2004. Neotropical tadpoles influence stream benthos: evidence for the ecological consequences of decline in amphibian populations. *Freshwater Biology* 49:274–285.

Rastegar-Pouyani, N. 2003. Ecology and conservation of the genus *Neurergus* in the Zagros Mountains, western Iran. *Froglog* (56):1–2.

Raup, D. M. 1991. *Extinction: Bad Genes or Bad Luck.* New York: W. W. Norton.

Raven, P. H. 1990. The politics of preserving biodiversity. *BioScience* 40:769–774.

Raxworthy, C. J., R. G. Pearson, N. Rabibisoa, A. M. Rakotondrazafy, J-P Ramanamanjato, A. P. Raselimanana, S. Wu, R. A. Nussbaum, and D. A. Stone. 2008. Extinction vulnerability of tropical montane endemism from warming and upslope displacement: a preliminary appraisal for the highest massif in Madagascar. *Global Change Biology* 14:1703–1720.

Relyea, R., and J. Hoverman. 2006. Assessing the ecology in ecotoxicology: A review and synthesis in freshwater systems. *Ecology Letters* 9:1157–1171.

Relyea, R. A., N. M. Schoeppner, and J. T. Hoverman. 2005. Pesticides and amphibians: The importance of community context. *Ecological Applications* 15:1125–1134.

Relyea, R. 2009. A cocktail of contaminants: how mixtures of pesticides at low concentrations affect aquatic communities. *Oecologia* 159: 363–376.

Relyea, R. and N. Dieks. 2008. An unforeseen chain of events: Lethal effects of pesticides on frogs at sublethal concentrations. *Ecological Applications* 18:1728–1742.

Retallick, R. W. R., and V. Miera. 2007. Strain differences in the amphibian chytrid *Batrachochytrium dendrobatidis* and non-permanent, sub-lethal effects of infection. *Diseases of Aquatic Organisms* 75:201–207.

Retallick, R. W. R., H. McCallum, and R. Speare. 2004. Endemic infection of the amphibian chytrid fungus in a frog community post-decline. *PLoS Biology* 2(11):e351.

Revenga, C., I. Campbell, R. Abell, P. de Villiers, and M. Bryer. 2005. Prospects for monitoring. *Philosophical Transactions of the Royal Society B* 360:397–413.

Richards, S. J., K. R. McDonald, and R. A. Alford. 1993. Declines in populations of Australia's endemic tropical rainforest frogs. *Pacific Conservation Biology* 1:66–77.

Riley, S P. D., H. B. Shaffer, R. Voss, and B. M. Fitzpatrick. 2003. Hybridization between a rare, native tiger salamander (*Ambystoma californiense*) and its introduced congener. *Ecological Applications* 13:1263–1275.

Robinson, T., A. Hyatt, J. Pallister, N. Hamilton, and D. Halliday. 2005. The development of a cane toad biological control. Final research report for the period July 2004–30 June 2005. CSIRO, Australia.

Rodrigues, A. S. L., T. M. Brooks, and K. J. Gaston. 2005. Integrating phylogenetic diversity in the selection of priority areas for conservation: Does it make a difference? In *Phylogeny and Conservation,* edited by A. Purvis, J. L. Gittleman, and T. M. Brooks. Cambridge, England: Cambridge University Press. pp. 101–119.

Rivalan, P., V. Delmas, E. Angulo, et al. 2007. Can bans stimulate wildlife trade? *Nature* 447:529–530.

Roelants, K., D. J. Gower, M. Wilkinson, et al. 2007. Global patterns of diversification in the history of amphibians. *Proceedings of the National Academy of Sciences (USA)* 104:887–892.

Rohr, J. R., A. M. Schotthoefer, T. R. Raffel, H. J. Carrick, N. Halstead, J. T. Hoverman, C. M. Johnson, L. B. Johnson, C. Lieske, M. D. Piwoni, P. K. Schoff, V. R. Beasley. 2008. Agrochemicals increase trematode infections in a declining amphibian. *Nature* 455:1235–1239.

Rohr, J. R., T. R. Raffel, S. K. Sessions, and P. J. Hudson. 2008. Understanding the net effects of pesticides on amphibian trematode infections. *Ecological Applications* 18:1743–1753.

Roig-Franzia, M. 2006. Panama hotel is imperiled frogs' lifeboat. *Washington Post.* October 25, 2006.

Ron, S. R. 2005. Distribution of the amphibian pathogen *Batrachochytrium dendrobatidis* in the New World: Insights from niche models. *Biotropica* 37:209–221.

Rosenblum, E. B., J. E. Strajich, N. Maddox, and M. B. Eisen. 2008. Global gene expression profiles for life stages of the deadly amphibian pathogen *Batrachochytrium dendroatidis. Proceedings of the National Academy of Sciences (USA)* 105:17034–17039.

Rouse, J. D., C. A. Bishop, and J. Struger. 1999. Nitrogen pollution: An assessment of its threat to amphibian survival. *Environmental Health Perspectives* 107:799–803.

Rowe, C. L., and J. Freda. 2000. Effects of acidification on amphibians at multiple levels of biological organization. In *Ecotoxicology of Amphibians and Reptiles,* edited by D. W. Sparling, D. W. Linder, and C. A. Bishop. Pensacola, Florida: Society of Environmental Toxicology and Chemistry (SETAC). pp. 545–571.

Rowley, J. J. L., and R. A. Alford. 2007. Behaviour of Australian rainforest stream frogs may affect the transmission of chytridiomycosis. *Diseases of Aquatic Organisms* 77:1–9.

Russello, M. A., and G. Amato. 2007. On the horns of a dilemma: Molecular approaches refine *ex situ* conservation in crisis. *Molecular Ecology* 16:2405–2406.

Sakai, A. K., F. W. Allendorf, J. S. Holt, et al. 2001. The population biology of invasive species. *Annual Review of Ecology and Systematics* 32:305–332.

Salthe, S. N. 1969. Reproductive modes and the number and sizes of ova in the urodeles. *American Midland Naturalist* 81:467–490.

Salthe, S. N., and W. E. Duellman.1973. Quantitative constraints associated with reproductive mode in anurans. In *Evolutionary Biology of the Anurans,* edited by J. L. Vial. Columbia: University of Missouri Press. pp. 229–249.

Schipper, J., J. S. Chanson, F. Chiozza et al. (129 authors total). 2008. The status of the world's land and marine mammals: Diversity threat, and knowledge. *Science* 322:225–230 + 64 pp. online supplemental material.

Schlaepfer, M. A., C. Hoover, and C. K. Dodd, Jr. 2005. Challenges in evaluating the impact of the trade in amphibians and reptiles on wild populations. *BioScience* 55:256–264.

Schloegel L. M., J-M. Hero, L. Berger, R. Speare, K. McDonald, and P. Daszak. 2006. The decline of the sharp-snouted day frog (*Taudactylus acutirostris*): The first documented case of extinction by infection in a free-ranging wildlife species? *EcoHealth* 3:35–40.

Seimon, T. A., A. Seimon, P. Daszak, et al. 2007. Upward range extension of Andean anurans and chytridiomycosis to extreme elevations in response to tropical deglaciation. *Global Change Biology* 13:288–299.

Semlitsch, R. D., ed. 2003. *Amphibian Conservation.* Washington, D. C.: Smithsonian Books.

Semlitsch, R. D., D. E. Scott, J. H. K. Pechmann, and J. W. Gibbons. 1996. Structure and dynamics of an amphibian community: Evidence from a 16-year study of a natural pond. In *Long-term Studies of Vertebrate Communities,* edited by M. L. Cody and S. L. Smallwood. San Diego: Academic Press. pp. 217–248.

Sessions, S. K. 2003. What is causing deformed amphibians? In *Amphibian Conservation,* edited R. Semlitsch. Washington, D. C.: Smithsonian Books. pp. 168–186.

Sessions, S. K., and S. B. Ruth. 1990. Explanation for naturally occurring supernumerary limbs in amphibians. *Journal of Experimental Zoology* 254:38–47.

Sibley, D. A. 2000. *The Sibley Guide to Birds.* New York: Knopf.

Sigel, B. J., T. W. Sherry, and B. E. Young. 2006. Avian community response to lowland tropical rainforest isolation: 40 Years of change at La Selva Biological Station, Costa Rica. *Conservation Biology* 20:111–121.

Skelly, D. K. 2002. Experimental venue and estimation of interaction strength. *Ecology* 83:2097–2101.

Skelly, D. K., S. R. Bolden, L. K. Freidenburg, N. A. Friedenlands, and R. Levey. 2007. *Ribeiroia* infection is not responsible for Vermont amphibian deformities. *EcoHealth* 4:156–163.

Skelly, D. K., E. E. Werner, and S. A. Cortwright. 1999. Long-term distributional dynamics of a Michigan amphibian assemblage. *Ecology* 80:2326–2337.

Skerratt, L. F., L. Berger, R. Speare, et al. 2007. Spread of chytridiomycosis has caused the rapid global decline and extinction of frogs. *EcoHealth* 4:125–134.

Smith, H. M., and A. J. Kohler. 1978. A survey of herpetological introductions in the United States and Canada. *Transactions of the Kansas Academy of Science 1977* 80:1–24.

Smith, K. F., D. F. Sax, and K. D. Lafferty. 2006. Evidence for the role of infectious disease in species extinction and endangerment. *Conservation Biology* 20:1349–1357.

Smol, J. P., and M. S. V. Douglas. 2007. Crossing the final ecological threshold in high Arctic ponds. *Proceedings of the National Academy of Sciences (USA)* 104:12395–12397.

Snider, A. T., and E. Arbaugh. 2005. The National Amphibian Conservation Center. In *Amphibian Declines: The Conservation Status of United States Species,* edited by M. Lannoo. Berkeley: University of California Press. pp. 339–340.

Souder, W. 2000. *A Plague of Frogs: The Horrifying True Story.* New York: Hyperion.

Souder, W. 2005. Of men and deformed frogs: A journalist's lament. In *Amphibian Declines: The Conservation Status of United States Species,* edited by M. Lannoo. Berkeley: University of California Press. pp. 344–347.

Soulé, M. E. 1991. Conservation: Tactics for a constant crisis. *Science* 253:744–750.

Sparling, D. W. 2003. A review of the role of contaminants in amphibian declines. In *Handbook of Ecotoxicology,* edited by D. J., Hoffman, B. A. Rattner, G. A. Barton, Jr., and J. Cairns, Jr. Boca Raton, Florida: Lewis Publishers. pp. 1099–1128.

Sparling, D. W., and G. Fellers. 2007. Comparative toxicity of chlorpyrifos, diazinon, malathion and their oxon derivatives to larval *Rana boylii. Environmental Pollution* 147:535–539.

Sparling, D. W., Linder, D.W., and Bishop, C. A. 2000. *Ecotoxicology of Amphibians and Reptiles.* Pensacola, Florida: Society of Environmental Toxicology and Chemistry (SETAC).

Sparling, D. W., G. M. Fellers, and L. L. McConnell. 2001. Pesticides and amphibian population declines in California, USA. *Environmental Toxicology and Chemistry* 20:1591–1595.

Stallard, R. F. 2001. Possible environmental factors underlying amphibian decline in eastern Puerto Rico: Analysis of U. S. government data archives. *Conservation Biology* 15:943–953.

Stebbins, R. C. 1966. *A Field Guide to Western Reptiles and Amphibians.* Boston: Houghton Mifflin.

Stebbins, R. C., and N. W. Cohen. 1995. *A Natural History of Amphibians.* Princeton, New Jersey: Princeton University Press.

Stevens, G. C. 1989. The latitudinal gradient in geographical range: How so many species coexist in the tropics. *American Naturalist* 133:240–256.

Stokes, D. E. 1997. *Pasteur's Quadrant.* Washington, D. C.: Brookings Institution Press.

Storfer, A. 2003. Amphibian declines: Future directions. *Diversity and Distributions* 9:151–163.

Storfer A., M. E. Alfaro, B. J. Ridenhour, et al. 2007. Phylogenetic concordance analysis shows an emerging pathogen is novel and endemic. *Ecology Letters* 10:1075–1983.

Storfer, A. S., S. G. Mech, M. W. Reudink, R. E. Ziemba, J. Warren, and J. P. Collins. 2004. Evidence for introgression in the endangered Sonora tiger salamander, *Ambystoma tigrinum stebbinsi.*(Lowe). *Copeia* (4):783–796.

Stuart, S. N., J. S. Chanson, N. A. Cox, et al. 2004. Status and trends of amphibian declines and extinctions worldwide. *Science* 306:1783–1786.

Stuart, S. N., M. Hoffmann, J. S. Chanson, N. A. Cox, R. J. Berridge, P. Ramani, and B. E. Young, editors. 2008. *Threatened Amphibians of the World.* Barcelona, Spain: Lynx Editions; Gland, Switzerland: IUCN; and Arlington, VA.: Conservation International.

"Take action to aid amphibians." www.conservation.org/FMG/Articles/Pages/10160701.aspx. Go to link Take Action to Aid Amphibians.

Taylor, S. K., E. S. Williams, E. T. Thorne, K. W. Mills, D. I. Withers, and A. C. Pier. 1999. Causes of mortality of the Wyoming toad. *Journal of Wildlife Diseases* 35:49–57.

Teixeira, R. D., S. C. R. Periera Mello, and C. A. M. Lima dos Santos. 2001. *The World Market for Frog Legs.* Food and Agriculture Organization of the United Nations, GLOBEFISH Research Programme 82:1–44. Rome: FAO.

Thomas, J. A., M. G. Telfer, D. B. Roy, et al. 2004. Comparative losses of British butterflies, birds, and plants and the global extinction crisis. *Science* 303:1879–1881.

Thuiller, W. 2007. Climate change and the ecologist. *Nature* 448:550–552.

Tilman, D., K. G. Cassman, P. A. Matson, R. Naylor, and S. Polasky. 2002. Agricultural sustainability and intensive production practices. *Nature* 418:671–677.

Tilman, D., J. Fargione, B. Wolff, et al. 2001. Forecasting agriculturally driven global environmental change. *Science* 292:281–284.

Travis, J. 1994. Calibrating our expectations in studying amphibian populations. *Herpetologica* 50:104–108.

Trenerry, M. P., W. F. Laurance, and K. R. McDonald. 1994. Further evidence for the precipitous decline of endemic rainforest frogs in tropical Australia. *Pacific Conservation Biology* 1:150–153.

Turtle Survival Alliance: An IUCN Partnership Network for Sustainable Captive Management of Freshwater Turtles and Tortoises. 2008. Web site. http://www.turtlesurvival.org.

Twain, M. 1867. *The Celebrated Jumping Frog of Calaveras County and Other Sketches.* New York: C. H. Webb.

Tyler, M. J. 1991. Declining amphibian populations: A global phenomenon? An Australian perspective. *Alytes* 9:43–50.

Tyler, M. J. 1999. Distribution patterns of amphibians in the Australo-Papuan region. In *Patterns of Distribution of Amphibians: A Global Perspective,* edited by W. E. Duellman. Baltimore: John Hopkins University Press. pp. 541–563.

Tyler, M. J., R. Wassersug, and B. Smith. 2007. How frogs and humans interact: Influences beyond habitat destruction, epidemics and global warming. *Applied Herpetology* 4:1–18.

U.S. Endangered Species list. 2008. U.S. Fish & Wildlife Service, Threatened and Endangered Species System (TESS) database. http://ecos.fws.gov/tess_public/.

U.S. Geological Survey. 2008. Web site. http://www.usgs.gov/science/.

Vaira, M. 2003. Report of a breeding aggregation extirpation of an endemic marsupial frog, *Gastrotheca christiani,* in Argentina. *Froglog* (60):3.

Vences, M., J. Kosuch, M. O. Rödel, et al. 2004. Phylogeography of *Ptychadena mascareniensis* suggests transoceanic dispersal in a widespread African-Malagasy frog lineage. *Journal of Biogeography* 31:593–601.

Vertucci, F. A., and P. S. Corn. 1996. Evaluation of episodic acidification and amphibian declines in the Rocky Mountains. *Ecological Applications* 6:449–457.

Vieites, D. R., M. Min, and D. B. Wake. 2007. Rapid diversification and dispersal during periods of global warming by plethodontid salamanders. *Proceedings of the National Academy of Sciences (USA)* 104:19903–19907.

Vitousek, P. M. 1994. Beyond global warming: Ecology and global change. *Ecology* 75:1861–1876.

Vitousek, P., J. D. Aber, R. W. Howarth, et al. 1997. Human alterations of the global nitrogen cycle: Sources and consequences. *Ecological Applications* 7: 737–750.

Vitousek, P. M., H. A. Mooney, J. Lubchenco, and J. M. Melillo. 1997. Human domination of Earth's ecosystems. *Science* 277:494–499.

Vitt, L. J., J. P. Caldwell, H. M. Wilbur, and D. C. Smith. 1990. Amphibians as harbingers of decay. *BioScience* 40:418.

Voyles, J., L. Berger, S. Young, et al. 2007. Electrolyte depletion and osmotic imbalance in amphibians with chytridiomycosis. *Diseases of Aquatic Organisms* 77:113–118.

Vredenberg, V. T. 2004. Reversing introduced species effects: Experimental removal of introduced fish leads to rapid recovery of a declining frog. *Proceedings of the National Academy of Sciences (USA)* 101:7646–7650.

Vredenburg, V. T., R. Bingham, R. Knapp, J. A. T. Morgan, C. Moritz, and D. Wake. 2007. Concordant molecular and phenotypic data delineate new taxonomy and conservation priorities for the endangered mountain yellow-legged frog (Ranidae: *Rana muscosa*). *Journal of Zoology* 271:361–374.

Vucetich, J. A., D. W. Smith, and D. R. Stabler. 2005. Influence of harvest, climate and wolf predation on Yellowstone elk, 1961–2004. *Oikos* 111:259–270.

Wake, D. B. 2007. Climate change implicated in amphibian and lizard declines. *Proceedings of the National Academy of Sciences (USA)* 104: 8201.

Wake, D. B., and H. J. Morowitz. 1991. Declining amphibian populations: A global phenomenon? Findings and recommendations. Report to Board on Biology, National Research Council, Workshop on Declining Amphibian Populations, Irvine, California. Reprinted in *Alytes* 9:33–42.

Wake, D. B., and V. T. Vredenburg. 2008. Are we in the midst of the sixth mass extinction? A view from the world of amphibians. *Proceedings of the National Academy of Sciences* (USA) 105:11466–11473.

Wake, M. H. 1980. Reproduction, growth, and population structure of the Central American caecilian *Dermophis mexicanus*. *Herpetologica* 36:244–256.

Wake, M. H. 2008. "Eye of newt and toe of frog": Herpetology in 21st century science. *Herpetologica* 64:1–11.

Wake, M. H. 2008. Integrative biology: Science for the 21st century. *BioScience* 58:349–353.

Walker, B. H., D. Ludwig, C. S. Holling, and R. M. Peterman. 1969. Stability of semiarid savanna grazing systems. *Ecology* 69:473–498.

Walker, S. F., M. B. Salas, D. Jenkins, et al. 2007. Environmental detection of *Batrachochytrium dendrobatidis* in a temperate climate. *Diseases of Aquatic Organisms* 77:105–112.

Watanabe. 2007, January 12. Deadly frog fungus spreads to Japan." *Associated Press*. http://www.livescience.com/animals/070112_ap_japan_frogs.html.

Weir, L. A., and M. J. Mossman. 2005. North American Amphibian Monitoring Program (NAAMP). In *Amphibian Declines: The Conservation Status of United States Species*, edited by M. Lannoo. Berkeley: University of California Press. pp. 307–313.

Weldon, C., L. H. du Preez, A. D. Hyatt, R. Muller, and R. Speare. 2004. Origin of the amphibian chytrid fungus. *Emerging Infectious Diseases* 10:2100–2105.

Weller, W. F., and D. M. Green. 1997. Checklist and current status of Canadian amphibians. In *Amphibians in Decline: Canadian Studies of a Global Problem*, edited by D. M. Green. *Herpetological Conservation* 1:309–328. St. Louis: Society for the Study of Amphibians and Reptiles.

Wells, H. V. 1993. Frogs missing? *The Courier-News* (Clinton, Tennessee), April 7, 1993.

Werner, E. E., D. K. Skelly. R. A. Relyea, and K. L. Yurewicz. 2007. Amphibian species richness across environmental gradients. *Oikos* 116:1697–1712.

Werner, E. E., K. L. Yurewicz, D. K. Skelly, and R. A. Relyea. 2007. Turnover in an amphibian metacommunity: the role of local and regional factors. *Oikos* 116:1713–1725.

Weygoldt, P. 1989. Changes in the composition of mountain stream frog communities in the Atlantic Mountains of Brazil: Frogs as indicators of environmental deteriorations? *Studies on Neotropical Fauna and Environment* 243:249–255.

Whiles, M. R., K. R. Lips, C. M. Pringle, et al. 2006. The effects of amphibian population declines on the structure and function of neotropical stream ecosystems. *Frontiers in Ecology and the Environment* 4:27–34.

Whitfield, S. M., K. E. Bell, T. Philippi, et al. 2007. Amphibian and reptile declines over 35 years at La Selva. *Proceedings of the National Academy of Sciences (USA)* 104:8352–8356.

Williams, J. W., S. T. Jackson, and J. E. Kutzbach. 2007. Projected distributions of novel and disappearing climates by 2100 AD. *Proceedings of the National Academy of Sciences (USA)* 104:5738–5742.

Williams, S. E., and J-M. Hero. 1998. Rainforest frogs of the Australian wet tropics: Guild classification and the ecological similarity of declining species. *Proceedings of the Royal Society of London B* 265:597–602.

Wilson, A. B. 2005. Commercial trade. In *Amphibian Declines: The Conservation Status of United States Species,* edited by M. Lannoo. Berkeley: University of California Press. pp. 146–148.

Wilson, E. O. 1984. *Biophilia.* Cambridge, Massachusetts: Harvard University Press.

Wilson, E. O. 1992. *The Diversity of Life.* Cambridge, Massachusetts: The Belknap Press of Harvard University Press.

Wistar, O. 1902. *The Virginian: Horseman of the Plains,* 9th printing (1962). New York: Macmillan.

Woodhams, D. C., and R. A. Alford. 2005. Ecology of chytridiomycosis in rainforest stream frog assemblages of tropical Queensland. *Conservation Biology* 19:1449–1459.

Woodhams, D. C., R. A. Alford, and G. Marantelli. 2003. Emerging disease of amphibians cured by elevated body temperature. *Diseases of Aquatic Organisms* 55:65–67.

Woodhams, D. C., R. A. Alford, C. J. Briggs, M. Johnson, and L. A. Rollins-Smith. 2008. Life-history trade-offs influence disease in changing environments: strategies of an amphibian pathogen. *Ecology* 89:1627–1639.

Woodhams, D. C., K. Ardipradja, R. A. Alford, G. Marantelli, L. K. Reinert, and L. A. Rollins-Smith. 2007. Resistance to chytridiomycosis varies among amphibian species and is correlated with skin peptide defenses. *Animal Conservation* 10:409–417.

Woodhams, D. C., L. A. Rollins-Smith, C. Carey, L. Reinert, M. J. Tyler, and R. A. Alford. 2006. Population trends associated with skin peptide defenses against chytridiomycosis in Australian frogs. *Oecologia* 146:531–540.

Woolhouse, E. J., L. H. Taylor, and D. T. Haydon. 2001. Population biology of multihost pathogens. *Science* 292:1109–1112.

Wyman, R. L. 1990. What's happening to the amphibians? *Conservation Biology* 4:350–352.

Wynn, G. 2007. Climate change threatens Latam water supply—WBank. Reuters Foundation: AlertNet, July 20, 2007. Available at EcoEarth.Info News Archive, http://www.ecoearth.info/shared/reader/welcome.aspx?linkid=80438&keybold=Rainforest%20Birds%20Habitat.

Xie, F, M. W. N. Lau, S. Stuart, J. Chanson, N. Cox, and D. Fischman. 2007. Conservation needs of amphibians in China: A review. *Science in China Series C: Life Sciences* 50:265–276.

Young, B. E., K. R. Lips, J. K Reaser, et al. 2001. Population declines and priorities for amphibian conservation in Latin America. *Conservation Biology* 15:1213–1223.

Young S., L. Berger, and R. Speare. 2007. Amphibian chytridiomycosis: strategies for captive management and conservation. *International Zoological Yearbook* 41:1–11.

General Index

Note: page numbers followed by *f* and *t* indicate figures and tables, respectively.

Species Index

Taxonomic changes for species recommended by Frost et al. 2006 ("The Amphibian Tree of Life") are indicated in parentheses preceding the common names. Genera and family names indexed individually are in the "pre-Frost et al." sense. Note: page numbers followed by *f* and *t* indicate figures and tables, respectively.